Numerical Methods for Fractal-Fractional Differential Equations and Engineering

This book is about the simulation and modeling of novel chaotic systems within the frame of fractal-fractional operators. The methods used, their convergence, stability, and error analysis are given, and this is the first book to offer mathematical modeling and simulations of chaotic problems with a wide range of fractal-fractional operators, to find solutions.

Numerical Methods for Fractal-Fractional Differential Equations and Engineering: Simulations and Modeling provides details for stability, convergence, and analysis along with numerical methods and their solution procedures for fractal-fractional operators. The book offers applications to chaotic problems and simulations using multiple fractal-fractional operators and concentrates on models that display chaos. The book details how these systems can be predictable for a while and then can appear to become random.

Practitioners, engineers, researchers, and senior undergraduate and graduate students from mathematics and engineering disciplines will find this book of interest.

Mathematics and its Applications
Modelling, Engineering, and Social Sciences

Series Editor: Hemen Dutta, Department of Mathematics, Gauhati University

Tensor Calculus and Applications
Simplified Tools and Techniques
Bhaben Kalita

Discrete Mathematical Structures
A Succinct Foundation
Beri Venkatachalapathy Senthil Kumar and Hemen Dutta

Methods of Mathematical Modelling
Fractional Differential Equations
Edited by Harendra Singh, Devendra Kumar, and Dumitru Baleanu

Mathematical Methods in Engineering and Applied Sciences
Edited by Hemen Dutta

Sequence Spaces
Topics in Modern Summability Theory
Mohammad Mursaleen and Feyzi Başar

Fractional Calculus in Medical and Health Science
Devendra Kumar and Jagdev Singh

Topics in Contemporary Mathematical Analysis and Applications
Hemen Dutta

Sloshing in Upright Circular Containers
Theory, Analytical Solutions, and Applications
Alexander Timokha and Ihor Raynovskyy

Advanced Numerical Methods for Differential Equations
Applications in Science and Engineering
Edited by Harendra Singh, Jagdev Singh, S. D. Purohit, and Devendra Kumar

Concise Introduction to Logic and Set Theory
Edited by Iqbal H. Jebril, Hemen Dutta, and Ilwoo Cho

Integral Transforms and Engineering
Theory, Methods, and Applications
Abdon Atangana and Ali Akgül

Numerical Methods for Fractal-Fractional Differential Equations and Engineering
Simulations and Modeling
Muhammad Altaf Khan and Abdon Atangana

For more information about this series, please visit: https://www.routledge.com/Mathematics-and-its-Applications/book-series/MES

ISSN (online): 2689-0224

ISSN (print): 2689-0232

Numerical Methods for Fractal-Fractional Differential Equations and Engineering
Simulations and Modeling

Muhammad Altaf Khan
Abdon Atangana

CRC Press is an imprint of the
Taylor & Francis Group, an **informa** business

Designed cover image: Muhammad Altaf Khan and Abdon Atangana

First edition published 2023
by CRC Press
6000 Broken Sound Parkway NW, Suite 300, Boca Raton, FL 33487-2742

and by CRC Press
4 Park Square, Milton Park, Abingdon, Oxon, OX14 4RN

CRC Press is an imprint of Taylor & Francis Group, LLC

© 2023 Muhammad Altaf Khan and Abdon Atangana

Reasonable efforts have been made to publish reliable data and information, but the author and publisher cannot assume responsibility for the validity of all materials or the consequences of their use. The authors and publishers have attempted to trace the copyright holders of all material reproduced in this publication and apologize to copyright holders if permission to publish in this form has not been obtained. If any copyright material has not been acknowledged please write and let us know so we may rectify in any future reprint.

Except as permitted under U.S. Copyright Law, no part of this book may be reprinted, reproduced, transmitted, or utilized in any form by any electronic, mechanical, or other means, now known or hereafter invented, including photocopying, microfilming, and recording, or in any information storage or retrieval system, without written permission from the publishers.

For permission to photocopy or use material electronically from this work, access www.copyright.com or contact the Copyright Clearance Center, Inc. (CCC), 222 Rosewood Drive, Danvers, MA 01923, 978-750-8400. For works that are not available on CCC please contact mpkbookspermissions@tandf.co.uk

Trademark notice: Product or corporate names may be trademarks or registered trademarks and are used only for identification and explanation without intent to infringe.

ISBN: 978-1-032-41522-2 (hbk)
ISBN: 978-1-032-41689-2 (pbk)
ISBN: 978-1-003-35925-8 (ebk)

DOI: 10.1201/9781003359258

Typeset in Nimbus Roman
by KnowledgeWorks Global Ltd.

Publisher's note: This book has been prepared from camera-ready copy provided by the authors.

Dedication

The true knowledge is from God who created nature. We will thank God the king of kings for giving us the strength, good health, and wisdom to write this book. May all the glory be to Him forever.

Contents

Preface xv

Acknowledgement xvii

Contributors xix

Chapter 1 Basic Principle of Nonlocalities 1

    1.1 Introduction 1
    1.2 Chaotic dynamics 2
    1.3 Strange attractors 3
    1.4 Some important concepts 4
    1.5 Some important concepts of numerical approximation 9
        1.5.1 Interpolation 9
        1.5.2 Linear interpolation 9
        1.5.3 Lagrange interpolation 9
        1.5.4 Middle point method 10
    1.6 Basic Reproduction number 10
    1.7 Stable 10
        1.7.1 Unstable 11
        1.7.2 Asymptotically stable 11

Chapter 2 Basic of Fractional Operators 13

    2.1 Introduction 13
    2.2 Some properties of the fractional operators 16
    2.3 Fundamental theorem of fractional calculus 17
    2.4 Fractal-Fractional operators 25

Chapter 3 Definitions of Fractal-Fractional Operators with Numerical Approximations 31

    3.1 Introduction 31
    3.2 Numerical schemes for fractal-fractional derivative 34
        3.2.1 Numerical scheme for Caputo fractal-fractional model 34
        3.2.2 Numerical scheme for Caputo-Fabrizio fractal-fractional operator 36
        3.2.3 Numerical scheme for Atangana-Baleanu fractal-fractional operator 38

	3.3	Numerical solution of fractional differential equations (FDEs) 41
		3.3.1 Numerical schemes for Atangana-Baleanu FDEs.... 41
Chapter 4		Error Analysis 43
	4.1	Introduction 43
	4.2	Error analysis for fractal-fractional RL Cauchy problems ... 43
	4.3	Error analysis for fractal-fractional CF cauchy problem 45
	4.4	Error analysis for fractal-fractional cauchy problem with Mittag-Leffler Kernel 47
Chapter 5		Existence and Uniqueness of Fractal Fractional Differential Equations 51
	5.1	Introduction 51
	5.2	Existence and uniqueness for power law case 52
	5.3	Existence and uniqueness for Mittag-Leffler case 54
	5.4	Existence and uniqueness for exponential case 56
	5.5	Existence and uniqueness for the case with Delta-Dirac Kernel 57
Chapter 6		A Numerical Solution of Fractal-Fractional ODE with Linear Interpolation 61
	6.1	Introduction 61
	6.2	Case with the Delta-Dirac Kernel 62
		6.2.1 Examples of fractal differential equations 64
	6.3	The case of power law kernel 64
	6.4	Case with exponential decay kernel 68
		6.4.1 Examples of fractal-fractional with exponential decay function 69
	6.5	Case with generalised Mittag-Leffler Kernel 70
Chapter 7		Numerical Scheme of Fractal-Fractional ODE with Middle Point Interpolation 73
	7.1	Introduction 73
	7.2	Numerical scheme for Delta-Dirac case 73
	7.3	Numerical scheme for exponential case 75
	7.4	Numerical scheme for power law case 76
	7.5	Numerical scheme for the Mittag-Leffler case 77
Chapter 8		Fractal-Fractional Euler Method 81
	8.1	Introduction 81
	8.2	Euler method with Dirac-Delta 81
	8.3	Fractal-fractional Euler method with the exponential kernel 84

Contents   ix

8.4	Fractal-fractional Euler method for power law kernel 85
8.5	Fractal-fractional Euler method with the generalised Mittag-Leffler 88

Chapter 9   Application of Fractal-Fractional Operators to a Chaotic Model .. 91

- 9.1 Introduction 91
- 9.2 Model 92
  - 9.2.1 Fixed points 92
- 9.3 Existence and uniqueness 93
- 9.4 Stability of the used numerical scheme 95
- 9.5 Case for power law 96
- 9.6 Numerical schemes and its simulations 100
  - 9.6.1 Numerical procedure in the sense of fractal-fractional-Caputo operator 100
  - 9.6.2 Numerical procedure for fractal-fractional Caputo-Fabrizio operator 102
  - 9.6.3 Numerical procedure for fractal-fractional Atangana-Baleanu operator 105
- 9.7 Numerical results 107
- 9.8 Conclusion 107

Chapter 10   Fractal-Fractional Modified Chua Chaotic Attractor 117

- 10.1 Introduction 117
- 10.2 Model framework 117
- 10.3 Existence and uniqueness conditions 118
- 10.4 Consistency of the scheme 121
  - 10.4.1 For the case of power law 123
- 10.5 Numerical procedure for the chaotic model 125
  - 10.5.1 Numerical procedure in the sense of fractal-fractional-Caputo operator 125
  - 10.5.2 Numerical procedure for fractal-fractional Caputo-Fabrizio operator 127
  - 10.5.3 Numerical procedure for fractal-fractional Atangana-Baleanu operator 130
- 10.6 Numerical results 132
- 10.7 Conclusion 136

Chapter 11   Application of Fractal-Fractional Operators to Study a New Chaotic Model 137

- 11.1 Introduction 137
- 11.2 Model framework 138
- 11.3 Existence and Uniqueness 138
  - 11.3.1 Equilibrium points and its analysis 142

	11.4	Numerical procedure for the chaotic model 142
		11.4.1 Numerical procedure in the sense of fractal-fractional-Caputo operator 142
		11.4.2 Numerical procedure for fractal-fractional Caputo-Fabrizio operator 145
		11.4.3 Numerical procedure for fractal-fractional Atangana-Baleanu operator 147
	11.5	Numerical results 149
	11.6	Conclusion 153
Chapter 12	Fractal-Fractional Operators and Their Application to a Chaotic System with Sinusoidal Component 155
	12.1	Introduction 155
	12.2	Model descriptions 156
	12.3	Existence and Uniqueness 156
	12.4	Equilibrium points 159
	12.5	Numerical procedure for the chaotic model 159
		12.5.1 Numerical procedure in the sense of fractal-fractional-Caputo operator 159
		12.5.2 Numerical procedure for fractal-fractional Caputo-Fabrizio operator 162
		12.5.3 Numerical procedure for fractal-fractional Atangana-Baleanu operator 164
	12.6	Numerical results 166
	12.7	Conclusion 166
Chapter 13	Application of Fractal-Fractional Operators to Four-Scroll Chaotic System 171
	13.1	Introduction 171
	13.2	Model descriptions 172
	13.3	Existence and uniqueness 172
	13.4	Equilibrium points 175
	13.5	Numerical procedure for the chaotic model 176
		13.5.1 Numerical scheme for power law kernel using linear interpolation 176
		13.5.2 Numerical scheme for exponential decay kernel using linear interpolations 180
		13.5.3 Numerical scheme for generalised Mittag-Leffler Kernel using linear interpolations .. 183
	13.6	Numerical results 186
	13.7	Conclusion 188

Contents   xi

Chapter 14   Application of Fractal-Fractional Operators to a Novel Chaotic Model 191

                 14.1 Introduction 191
                 14.2 Model descriptions 192
                 14.3 Existence and uniqueness 193
                          14.3.1 Equilibrium points and their analysis 195
                 14.4 Numerical schemes based on linear interpolations 195
                 14.5 Numerical scheme for power law kernel 195
                          14.5.1 Numerical scheme for exponential decay kernel using linear interpolations 201
                          14.5.2 Numerical scheme for generalised Mittag-Leffler Kernel using linear interpolations .. 203
                 14.6 Conclusion 209

Chapter 15   A 4D Chaotic System under Fractal-Fractional Operators 211

                 15.1 Introduction 211
                 15.2 Model details 212
                 15.3 Existence and uniqueness 212
                 15.4 Schemes based on linear interpolations 215
                          15.4.1 Numerical scheme for power law kernel using linear interpolations 215
                          15.4.2 Numerical scheme for exponential decay kernel using linear interpolations 220
                          15.4.3 Numerical scheme for generalised Mittag-Leffler Kernel using linear interpolations .. 224
                 15.5 Conclusion 233

Chapter 16   Self-Excited and Hidden Attractors through Fractal-Fractional Operators 237

                 16.1 Introduction 237
                 16.2 Chaotic model and its dynamical behaviour 238
                 16.3 Existence and uniqueness 238
                 16.4 Equilibrium points analysis 239
                 16.5 Numerical procedure for the chaotic model 240
                 16.6 Numerical scheme for power law kernel 240
                          16.6.1 Numerical scheme for exponential decay kernel using linear interpolations 245
                          16.6.2 Numerical scheme for generalised Mittag-Leffler Kernel using linear interpolations .. 249
                 16.7 Conclusion 252

Chapter 17 Dynamical Analysis of a Chaotic Model in Fractal-Fractional Operators 255

17.1 Introduction 255
17.2 Model descriptions 256
17.3 Existence and uniqueness 256
    17.3.1 Model analysis 258
17.4 Numerical schemes based on middle-point interpolations . 258
    17.4.1 Numerical scheme for power law case 258
    17.4.2 Numerical scheme based on middle-point interpolation for exponential case 261
    17.4.3 Numerical scheme for the Mittag-Leffler case 265
17.5 Conclusion 270

Chapter 18 A Chaotic Cancer Model in Fractal-Fractional Operators 273

18.1 Introduction 273
18.2 Model framework 274
18.3 Existence and uniqueness 275
    18.3.1 Equilibrium points 276
18.4 Numerical procedure for the chaotic model 277
    18.4.1 Numerical scheme for power law case 277
    18.4.2 Numerical scheme for exponential case 280
    18.4.3 Numerical scheme for the Mittag-Leffler case 283
18.5 Conclusion 286

Chapter 19 A Multiple Chaotic Attractor Model under Fractal-Fractional Operators 289

19.1 Introduction 289
19.2 Model descriptions 290
19.3 Existence and uniqueness 290
    19.3.1 Equilibria and their stability 292
19.4 Numerical procedure for the chaotic model 293
    19.4.1 Numerical scheme for power law case 293
    19.4.2 Numerical scheme for exponential case 295
    19.4.3 Numerical scheme for the Mittag-Leffler case 299
19.5 Conclusion 302

Chapter 20 The Dynamics of Multiple Chaotic Attractor with Fractal-Fractional Operators 305

20.1 Introduction 305
20.2 Model descriptions 306
20.3 Existence and uniqueness of the model 306
20.4 Numerical procedure for the chaotic model 309
    20.4.1 Numerical scheme for power law case 309

	20.4.2	Numerical scheme for exponential case 311
	20.4.3	Numerical scheme for the Mittag-Leffler case 315
20.5	Conclusion 319

Chapter 21 Dynamics of 3D Chaotic Systems with Fractal-Fractional Operators 321

- 21.1 Introduction 321
- 21.2 Model descriptions and their analysis 322
- 21.3 Existence and uniqueness 322
  - 21.3.1 Equilibrium points and their analysis 323
- 21.4 Numerical procedure for the chaotic model using Euler-based method 324
  - 21.4.1 Euler-based numerical scheme for FF-Caputo operator 324
  - 21.4.2 Euler-based numerical scheme for FF-CF operator 325
  - 21.4.3 Euler-based numerical scheme for FF Atangana-Baleanu operator 329
- 21.5 Conclusion 331

Chapter 22 The Hidden Attractors Model with Fractal-Fractional Operators 335

- 22.1 Introduction 335
- 22.2 Model and its analysis 336
- 22.3 Existence and uniqueness 336
  - 22.3.1 Equilibrium points and their analysis 337
- 22.4 Numerical procedure for the chaotic model 338
  - 22.4.1 Numerical scheme with Euler for FF-Caputo operator 338
  - 22.4.2 Numerical scheme with Euler FF Caputo-Fabrizio operator 339
  - 22.4.3 Numerical scheme with Euler FF Atangana-Baleanu 343
- 22.5 Conclusion 346

Chapter 23 An SIR Epidemic Model with Fractal-Fractional Derivative 349

- 23.1 Introduction 349
- 23.2 Model formulation 349
- 23.3 Positivity of the model 350
- 23.4 Existence and uniqueness 352
  - 23.4.1 Equilibrium points and their analysis 354
  - 23.4.2 Global stability 355
- 23.5 Numerical results and the schemes 356
  - 23.5.1 Euler scheme with power law case 356

	23.5.2 Euler scheme with exponential kernel	357
	23.5.3 Euler scheme with Mittag-Leffler Kernel	361
	23.6 Conclusion	365
Chapter 24	Application of Fractal-Fractional Operators to COVID-19 Infection	367
	24.1 Introduction	367
	24.2 Mathematical model	368
	24.2.1 Fractal-fractional order COVID-19 model	369
	24.3 Existence and uniqueness	370
	24.4 Equilibrium points and their analysis	374
	24.5 Data fitting, numerical schemes, and their graphical results	376
	24.5.1 Numerical scheme for COVID-19 infection model with power law	376
	24.5.2 Numerical scheme for COVID-19 infection model with the exponential kernel	380
	24.5.3 Numerical scheme for COVID model with Mittag-Leffler Kernel	383
	24.6 Conclusion	389
References		393
Index		411

Preface

One of the key mathematical methods for simulating real-world issues is differentiation. The rate of change technique, which allows for the estimation of a function's variation in two points, was used to launch the idea. While a zero rate of change implies that the function is constant and a positive rate of change suggests that the process is increasing, a negative rate of change indicates that the portion is decreasing. In many domains, including classical mechanics, where heterogeneity has not been taken into account, the notion has been employed with tremendous effectiveness. To account for the influence of heterogeneity in mathematical models, differential operators with integrals were proposed. Numerous sectors of engineering, science, and technology have used this novel idea. In a recently developed idea, the differential operators are given two orders: fractional-order and fractal dimension. A new class of differential and integral equations being led by the new differential and integral operators has the potential to describe actual complicated issues. This monograph will concentrate on models that exhibit chaos in particular. It is important to note that deterministic systems are included in a chaotic system, and their behaviour may be anticipated using specific mathematical models. However, these systems appear to turn random at times after being predictable for a while. Several processes found in numerous natural systems show chaotic characteristics. These traits, for instance, are present in many engineering and technological systems, such as irregular heartbeats, climate, weather, fluid flow, and electronics. The underlying principles of this theory include complex systems, the edge of chaos theory, and self-similar techniques. Numerous scientific fields, including sociology, computer science, economics, ecology, pandemic crisis management, environmental science, meteorology, anthropology, engineering, and many more, have found a use for this idea. Nonlinear differential equations are frequently utilised as chaotic behaviour mathematical models. As a result, analytical techniques are not a good choice for finding the answers. As a result, researchers use numerical techniques to solve these models. Numerical methods have been proposed in recent years to resolve differential equations with fractal-fractional derivatives, and they have been effectively used in a number of challenging issues. This book focuses on numerical methods for resolving these novel equations and their associated applications in epidemiological modelling and chaos theory. The first eight chapters of the book's 24 chapters discuss theories and numerical methods for fractal-fractional ordinary differential equations. Applications of these novel differential operators to chaos and epidemiology were covered in the last chapters.

Acknowledgement

The fear of the Lord is the foundation of all wisdom and the knowledge of the holy one understands. We thank God for his mercy, protection, heath, and wisdom. Behind great men stand great women, we would like to thank God for giving us understanding, wise, and caring wives as they stood by us to help us finish this book. We want to thank the department of research of the University of the Free State for their financial support.

Contributors

Muhammad Altaf Khan
Faculty of Natural and Agricultural Sciences
University of the Free State, South Africa

Abdon Atangana
Faculty of Natural and Agricultural Sciences
University of the Free State, South Africa

1 Basic Principle of Nonlocalities

1.1 INTRODUCTION

The study of chaotic states in dynamic systems, where the state variables are extremely sensitive to the beginning circumstances, is the subject of the mathematical field known as chaos theory. It is believed that the chaos theory is an interdisciplinary theory that addresses the unpredictability associated with large chaotic systems, self-organisation, their ongoing feedback loops, self-similarity, recurrence, and fractals. A key illustration of chaos theory is the butterfly effect, which demonstrates how slight changes in the state variable of nonlinear deterministic systems may result in significant changes in the state later. It implies that there is a delicate reliance on the variables' initial values or initial conditions.

Long-term forecasts of the behaviour of dynamical systems are typically not attainable because of the tiny changes in the initial values of the state variables, such as those caused by rounding mistakes in numerical computations that may lead to differing outcomes. It can occur even in deterministic systems, which means that the given beginning circumstances completely determine the future outcomes, leading to a unique evolution without the inclusion of random factors. Simply put, the fact that these systems are deterministic makes them unpredictable, and this behaviour is known as deterministic chaos or just chaos. Edward Lorenz was the author of this summary.

Chaos: When the present determines the future, but the approximate present does not approximately determine the future [39, 58, 100, 271]. A chaotic system is a deterministic system that is difficult to predict.

Many natural systems, such as fluid movement, irregular heartbeats, weather, and climate, exhibit chaotic behaviour. Additionally, it happens on its own in certain artificially constructed systems, such as the stock market and traffic. Analytical methods like recurrence plots and Poincar maps can be used to analyse a chaotic mathematical model of this behaviour. Meteorology, anthropology, sociology, environmental science, computer science, engineering, economics, ecology, and pandemic crisis management are just a few of the fields where chaos theory has applications. Complex dynamical systems, edge of chaos theory, and self-assembly processes were all developed from this idea. Numerous natural systems, including heartbeat abnormalities, fluid movement, climate, and weather, exhibit chaotic behaviour. Systems with artificial components, such as the stock market and traffic, can also exhibit chaotic behaviour. Such systems' chaotic behaviour can be efficiently studied using a mathematical model or analytical methods like Poincare maps and recurrence plots. Henri Poincar was a pioneer in the chaos theory movement. He discovered that there can be nonperiodic orbits that are neither rising continuously nor moving towards a fixed

DOI: 10.1201/9781003359258-1

point in the 1880s while researching the three-body issue. "Hadamard's billiards," Jacques Hadamard's important investigation of the chaotic motion of a free particle gliding without resistance over a surface of continuous negative curvature, was published in 1898. Hadamard was able to demonstrate the instability of all particle trajectories by demonstrating how they all diverge exponentially from one another with a positive Lyapunov exponent. Ergodic theory gave rise to chaos theory. George David Birkhoff, Andrey Nikolaevich Kolmogorov, Mary Lucy Cartwright and John Edensor Littlewood, and Stephen Smale later conducted work on the subject of nonlinear differential equations. These investigations, with the exception of Smale, were all directly motivated by physics: in the case of Birkhoff, the three-body issue; in the case of Kolmogorov, turbulence; and in the case of Cartwright and Littlewood, radio engineering. Experimentalists have seen turbulence in fluid motion and nonperiodic oscillation in radio circuits without having a theory to explain what they were witnessing, despite the fact that chaotic planetary motion had not been seen. Despite early discoveries in the first decade of the twentieth century, it wasn't until the middle of the century that several scientists realised that linear theory, the dominant system theory at the time, could not possibly account for the observed behaviour of some tests, such as the logistic map. Chaos theorists viewed what had previously been assigned to measure imprecision and simple noise as a whole component of the system under study [88, 89, 150, 170, 205].

1.2 CHAOTIC DYNAMICS

Chaos, as used in everyday language, is a condition of disorganisation. But chaos theory gives a more specific definition of this phrase. There is no agreed formal definition of chaos, but one that is often used comes from Robert L. Devaney. He states that for a dynamical system to be classified as chaotic, it must have certain characteristics below [81]:

- Sensitivity to the initial conditions implying each point in a chaotic system is arbitrarily closely approximated by other points that have radically different future routes or trajectories when a system is sensitive to beginning circumstances. The present trajectory may therefore be arbitrarily slightly altered or perturbed, which might result in drastically different future behaviour. As a result of the system's sensitivity to beginning circumstances, if we begin with little knowledge about the system (as is typically the case in practise), the system would eventually stop being predictable. This is especially common when it comes to the weather, which is often only foreseeable about a week in advance. This does not imply that one cannot predict events that will take place in the far future, merely that the system is subject to some limitations.
- Topologically mixing: When a system exhibits topological mixing, each given area or open set in its phase space ultimately overlaps with every other region. The mixing of coloured fluids or dyes is an example of a chaotic system, and this mathematical idea of mixing is consistent with common

Basic Principle of Nonlocalities

sense. Topological mixing is frequently left out of common descriptions of chaos, which solely consider chaos as sensitivity to beginning circumstances. Sensitive reliance on the starting conditions alone, however, does not result in pandemonium. Take the straightforward dynamical system, for instance, that results from constantly doubling a starting value. Since every close-by pair of points ultimately grow far apart, this system displays sensitive dependency on beginning circumstances everywhere. However, there is no topological mixing in this illustration, and as a result, there is no chaos.

- Having dense periodic orbits: Every point in the space is arbitrarily near to being approached by periodic orbits if a chaotic system has dense periodic orbits.
- Strange attractor: Strange attractors, which result from chaotic systems, differ from fixed-point attractors and limit cycles in their level of intricacy and complexity. Strange attractors can be found in certain discrete systems as well as continuous dynamical systems.

The latter two definitions of the four aforementioned qualities really suggest sensitivity to original circumstances. The "sensitivity to beginning circumstances" is the most important and useful of these traits; hence, it is not necessary to mention it in the definition. If attention is limited to intervals, the second characteristic implies the other two. As an alternative, the general weaker type definition of chaos took into account the first and just two of the aforementioned qualities [65, 165, 253].

1.3 STRANGE ATTRACTORS

The logistic one-dimensional map may be described by the following in a dynamical system such as

$$x \to 4x(1-x), \tag{1.1}$$

is chaotic everywhere, yet in many instances, it can be seen that the chaotic behaviour only exists in a portion of the phase space. The chaotic behaviour of an attractor is of interest because it results in a large number of beginning circumstances that lead to orbits and coverage of the chaotic region [226].

The chaotic attractor may be imagined as simply beginning at a location in the attractor's basin of attraction and sketching its subsequent orbit. The best chaotic complex system with a pattern and the butterfly effect is the Lorenz attractor. The weird attractors with tremendous complexity are the term used to describe the attractor that emerges in a chaotic system. Strange attractors appear in dynamical systems, whether they are continuous or discrete systems. We include the Lorenz system as an example of a continuous dynamical system, and the Henon map as an example of a discrete dynamical system.

1.4 SOME IMPORTANT CONCEPTS

SYSTEM

A set of things/objects that is governed through some specific rules is known as system.

PHYSICAL SYSTEM

A physical system is a system that appears in nature. The solar system's planets, a fluid in a heated jar, the weather, and a colony of fleas kept in a container under the ecologist's watch are examples. The molecules of liquid and gas serve as symbols for fluid and weather.

MATHEMATICAL SYSTEM

A mathematical system is one that may be expressed by equations with one or more variables. For this, take a look at the following instance,

$$P(t+dt) = A*P(t)*(P^* - P(t)) \qquad (1.2)$$

where $P(t)$ represents the species' current population at any given moment. The time period is represented by t, dt, whereas A and $P*$ take values related to the problem's interest. The equation (1.2) is usually referred to as mathematical models in mathematical systems that was created to roughly represent a physical or practical issue.

PHASE SPACE

Phase space refers to all possible combinations of circumstances in a system. The phase space of the system is defined by the locations of each ball on the pool table.

COMPUTER SIMULATION

It is a computer-based mathematical modelling approach that aims to forecast the behaviour or result of a physical or real-world system. By comparing the findings to the events in real life that the mathematical models are trying to predict, it is possible to assess the accuracy of certain of these models. Computer simulations have developed into an effective tool for the mathematical modelling of a variety of natural systems in physics, astronomy, climatology, chemistry, biology, and manufacturing, as well as human systems in economics, psychology, social science, health care, and engineering. Running the system's model is how one simulates a system. It may be utilised to investigate and get fresh perspectives on cutting-edge technology as well as to gauge the functioning of systems that are too complicated for analytical solutions.

Basic Principle of Nonlocalities

DYNAMICS

Dynamics is the investigation of forces and how they impact motion through time. It belongs to the field of classical mechanics. Three laws of motion created by Isaac Newton are the cornerstones of dynamics. Kinematics and kinetics are sometimes thought of as the components of dynamics. Some people do not use the name "kinetics," believing dynamics and kinematics to be distinct disciplines. According to the second perspective, kinematics deals with defining how a system is right now, whereas dynamics deals with how and why the system evolves over time.

PARAMETERS

Parameters usually refers to any quality that may be used to characterise or categorise a certain system. In other words, a parameter is a component of a system that is crucial or useful for identifying the system or assessing its functionality, status, and state.

A DYNAMICAL SYSTEM

A dynamical system is one in which the time dependence of a point in the surrounding space is described by a function. By allowing for several choices in terms of space and how time is measured, the most generic definition combines a number of mathematical ideas, including ergodic theory and ordinary differential equations. Without remembering its physical origin, time can be measured as an integer, a real or complex number, or a more generic algebraic object. Similarly, space can be a manifold or just a set without the necessity for a smooth space-time structure to be constructed on it.

A COMPLEX SYSTEM

Due to dependencies, rivalries, relationships, or other forms of interactions between its elements or between a specific system and its surroundings, complex systems have behaviour that is inherently challenging to describe. These interactions give birth to certain characteristics that distinguish complex systems, including nonlinearity, emergence, spontaneous order, adaptability, and feedback loops, among others. These systems are found in a wide range of domains; therefore, the similarities between them have given rise to a separate area of study. It is sometimes helpful to depict such a system as a network, where the nodes stand in for the various parts and the links for their interactions.

NONLINEAR

The adjective nonlinear can have several meanings depending on the context of the subject, but in this case, it refers to a system that must be used to identify a response that is not linearly connected to changes in input. If we use a linear system as an example, the output would also double if the input did. If a guy puts forth twice the effort, he will receive twice the reward. Nonlinear systems are challenging to solve, yet linear systems are simple to handle and have their precise solution in hand. For instance, if a nonlinear system's solution were discovered, it would no longer be chaotic but rather predictable.

LINEARISATION OF A SYSTEM

Finding the linear approximation of a function at a certain location is the mathematical concept of linearisation. The first-order Taylor expansion around the point of interest represents the linear approximation of a function. In the study of discrete dynamical systems or systems of nonlinear differential equations, linearisation is a technique for determining the local stability of an equilibrium point. In disciplines including engineering, physics, economics, and ecology, this approach is employed.

CHAOS

Chaos is a nonlinear complicated dynamical system's erratic behaviour. The unpredictability, seeming randomness, but not arbitraryness, sensitivity to the starting values of the variables, and the description of the presence of a weird attractor, which is frequently a fractal, are all characteristics of its behaviour over long time scales. It should be highlighted that whereas chaotic behaviour has an odd attractor that controls its structure, random activity is purely random.

RANDOM

Completely unpredictable, arbitrary, and highly varying.

SENSITIVITY TO INITIAL CONDITIONS

A small change at a particular time leads to large differences at a later time.

THE BUTTERFLY EFFECT

It is a crucial phrase that has to do with sensitivity to early circumstances. It has to do with how unpredictable weather may be. For example, some people think that if a butterfly flaps its wings just slightly in one place that minor disturbance might eventually cause a significant shift in the weather all over the planet.

UNSTABLE

Simply means not stable.

STABLE

A system is known to be stable or unstable if a tiny adjustment is made to it, and it goes back to its initial condition. A jar's flat bottom serves as an illustration of stability: The jar begins to tilt slightly, and after a while it rocks back to rest in a vertical posture. While a little alteration is made to an unstable system, it causes distinct behaviour. For example, when balancing a pen with a pencil tip, the pencil tip falls and cannot return to its initial position. A minor adjustment might result in drastically different behaviour in an unstable system. However, it should be emphasised that a system may occasionally be stable in certain directions but unstable in others.

LYAPUNOV EXPONENTS

It displays the speed at which neighbouring trajectories diverge. Such movements diverge exponentially quickly, with the Lyapunov exponent serving as the exponential's coefficient. To put it another way, a system with at least one big Lyapunov exponent is extremely sensitive to starting circumstances, whereas a system with tiny Lyapunov exponents initially deviates from the norm more gradually. Every independent direction has a Lyapunov exponent.

AN ATTRACTOR

An attractor is a location in phase space that the dynamical motion often approaches. The six pockets or holes in the pool table can be compared to the billiard balls in the previous example as an attractor.

A STRANGE ATTRACTOR

A strange attractor is the attractor for chaotic behaviour. It is often a fractal.

A FRACTAL

A group of points with a fractional dimension is referred to as a fractal. Another straightforward definition is "an item that possesses self-similarity at any scale." For instance, a surface with numerous holes may have a dimension that is less than two, while a curve with a very asymmetric form may have a dimension that is really more than one. At tiny sizes, fractals are self-similarity. Alternatively, we may claim that the objects exhibit self-similarity at all scales. Fractal behaviour may be produced in a variety of methods, but utilising a complex number, one might produce a two-dimensional fractal shape using the well-known Julia set. Indeed, the following polynomial yields this collection.

$$Z_{n+1} = Z_n^2 + C. \tag{1.3}$$

From the above, the following iterative scheme can be obtained,

$$\begin{cases} x_{n+1} = x_n^2 + c_x - y_n^2, \\ y_{n+1} = 2x_n y_n + c_y. \end{cases} \tag{1.4}$$

Using the fact that
$$Z_{n+1} = x_{n+1} + iy_{n+1}. \tag{1.5}$$
Another set of complex number was suggested by Harmitton called the quaternion, where the complex number is defined like
$$Z = a + ib + jc + kd, \tag{1.6}$$
where I, j, and k are fictitious numbers. However, given that our environment is three dimensional, this set was not realistic. As a result, other than its cross section, the Julia set of such a set is unimaginable in our reality. Atangana and Toufik had just proposed a new complex number called Trinition, which is defined as
$$Z = a + ib + jc. \tag{1.7}$$
The associative Julia set is given as [21]
$$\begin{cases} x_{n+1} = x_n^2 - y_n^2 + z_n^2 + c_x \\ y_{n+1} = 2x_n y_n + c_y \\ z_{n+1} = 2x_n z_n + c_z \end{cases} \tag{1.8}$$

PERIODIC BEHAVIOUR

Periodic behaviour is non-chaotic since the system returns to its original state after a certain time, at which point it repeats what it previously did.

PERIOD

The minimum time it takes for the motion to repeat.

NON-PERIODIC MOTION

Non-periodic motion is motion that does not repeat.

BIFURCATION

A bifurcation is described as the abrupt shift that takes place in a system when a parameter is significantly altered. It can also refer to a particular instance where a system is asymptotically periodic and the periodic behaviour doubles. A system could oscillate between two states, for instance. The system then oscillates between four states, two of which are similar to one of the initial states, and two of which are similar to the other original state. This happens when a parameter is raised.

PERIOD-DOUBLING

The term "period-doubling path to chaos" describes a situation in which, when a parameter is adjusted to a limiting value, a sequence of bifurcations take place at an ever-increasing pace. Chaos develops when the value reaches its limit.

UNIVERSALITY

The concept of universality in chaos holds that the same kind of disorder may appear in quite diverse systems. The group of systems that travel down the period-doubling path to chaos serves as an illustration.

THE POINCARE MAP

The motion of a body in three dimensions is projected onto a plane to create the Poincare map. It reduces the problem's dimension by one, which makes it easier to analyse a trajectory.

1.5 SOME IMPORTANT CONCEPTS OF NUMERICAL APPROXIMATION

1.5.1 INTERPOLATION

Interpolation is a sort of estimating used in the mathematical discipline of numerical analysis. It is a technique for creating new data points depending on the range of a discrete set of known data points. In engineering and research, it is common to have several data points that, by sampling or experimentation, indicate the values of a function for a small set of independent variable values. Interpolation, or estimating the value of that function for an intermediate value of the independent variable, is frequently necessary.

1.5.2 LINEAR INTERPOLATION

When a function f has two known values at other places, linear interpolation is frequently employed to estimate that value. This approximation's inaccuracy is described as

$$\Pi_T(t) = f(t) - P(t),$$

where in general P is the polynomial expressed as

$$P(t) = f(t_j) + \frac{f(t_{j+1}) - f(t_j)}{t_{j+1} - t_j}(t - t_j).$$

The error is bounded by

$$\Pi_T(t) \leq \frac{(t_{j+1} - t_j)^2}{8} \max_{t_j \leq t \leq t_{j+1}} |f''(t)|.$$

1.5.3 LAGRANGE INTERPOLATION

The given values are combined linearly to generate the interpolating polynomial's Lagrange form. A linear combination of the provided values using previously known coefficients is frequently an effective and practical polynomial interpolation method. Formula for Lagrange interpolation, for

$$P(t) = \frac{t-t_{j-1}}{t_{j+1}-t_j} f(t_j) - \frac{t-t_j}{t_{j+1}-t_j} f(t_{j-1}).$$

Here,

$$f(t) - P(t) = \frac{f^{(2)}}{2!}(\xi) \Pi_{i=0}^{1}(t-t_j) = \Pi_T(t).$$

If $t_j = b + ih$, where $h = t_{j+1} - t_j$, then

$$|\Pi_T(t)| \leq \frac{h^2}{8} \max_{\xi \in [t_j, t_{j+1}]} |f^{(2)}(\xi)|.$$

1.5.4 MIDDLE POINT METHOD

The midpoint of a line segment is known as the midpoint in geometry. It is the centroid of the segment and of the ends, and it is equally distant from both of them. It cuts the section in half.

Foe middle point, we have

$$\left(\frac{t_j + t_{j+1}}{2}, \frac{y_j + y_{j+1}}{2}\right).$$

1.6 BASIC REPRODUCTION NUMBER

The basic reproduction number is defined as the average number of infection introduced into a population purely susceptible, produced a number of secondary infection. Commonly, the basic reproduction number is denoted by $\mathscr{R}_0$. For disease elimination, it is required that its value should be less than 1 while for disease to be exists in the population, their value should greater than 1.

1.7 STABLE

A steady state x^* is called stable, if the solution starts nearby stays nearby. More formally, if an equilibrium point, x^*, is stable for every $\varepsilon > 0$, there exists a $\delta > 0$ such that the solution to the initial data x^0 with

$$|x^0 - x^*| < \delta,$$

satisfy

$$|x(t) - x^*| < \varepsilon,$$

for all $t > 0$.

1.7.1 UNSTABLE

A steady state which is not stable is called unstable.

1.7.2 ASYMPTOTICALLY STABLE

A steady state, say x^*, is called asymptotically stable, if it is stable and all the solutions near x^* converge x^*. More formally, an equilibrium point, say x^* is asymptotically stable if x^* is stable there exists a $\delta > 0$, such that all solutions with initial data, say x^0, with

$$|x^0 - x^*| < \delta,$$

and

$$\lim_{t \to \infty} |x - x^*| = 0.$$

2 Basic of Fractional Operators

2.1 INTRODUCTION

The fractal derivative, also known as the Hausdorff derivative, is a non-Newtonian modification of the derivative that deals with the measurement of fractals as they are specified in fractal geometry. It is used in applied mathematics and mathematical analysis. For the study of anomalous diffusion, where conventional methods fall short of taking into account the media's fractal character, fractal derivatives were developed. The scale of a fractal measure t is determined by t. In contrast to the similarly applied fractional derivative, such a derivative is local. Such processes are present in many real-world phenomena, including turbulence, porous media, aquifers, and other phenomena that frequently have fractal characteristics. Traditional diffusion or dispersion rules, such as Fick's laws of diffusion, Darcy's law, and Fourier's law, which are all based on random walks in empty space and produce roughly the same outcome, do not apply to fractal media. In order to handle this, ideas like distance and velocity must be redefined for fractal media. In particular, scales for space and time must be altered in accordance with the equation (x^a, t^b). Fractal spacetime (x^a, t^b) allows for the redefinition of fundamental physical concepts like velocity. On the other hand, a fractional derivative in applied mathematics and mathematical analysis is a derivative of any arbitrary order, real or complex. It first appears in a letter that Leibniz wrote to Guillaume de l'Hpital in 1695. In a letter to one of the Bernoulli brothers written around the same time, Leibniz discussed the relationship between the binomial theorem and the Leibniz rule for the fractional derivative of a product of two functions. The concept of fractional-order differentiation and integration, their mutually inverse relationship, the realisation that fractional-order differentiation and integration can be thought of as the same generalised operation, and even the unified notation for differentiation and integration of fractional order can all be found in one of Abel's early papers, which introduced fractional calculus. Separately, Liouville established the subject's foundations in a paper dated 1832. Around 1890, the self-taught Oliver Heaviside developed the practical application of fractional differential operators in the analysis of electrical transmission lines. Over the course of the 19th and 20th centuries, the theory and applications of fractional calculus significantly developed, and various authors contributed definitions of fractional derivatives and integrals. Nonlocal operators' geometrical significance has typically been interpreted incorrectly. This might be in part because even our understanding of nonlocal operators' properties is still incomplete. It is important to keep in mind that the Stieltjes-Riemann integral was created by Stieltjes by extending the Riemann integral in classical integral calculus. This integral has

DOI: 10.1201/9781003359258-2

recently been used to a few practical problems, including statistics and the expectation formula. Additionally, this integral was used to calculate the curved surface, which the Riemann integral was unable to do. The associate derivative of this operator has not been proposed, even though it theoretically and practically created major new study opportunities. The first effort was made using the fractal derivative. Under the assumption that the function is differentiable, the basic theorem of calculus was used to produce an integral that is a subclass of the Stieltjes-Riemann integral. By utilising the fact that an integral operator is differentiable, Atangana enlarged the fractional derivatives to a fractal-fractional derivative, which is more suited to recreate complexity. The reason for this is straightforward: when the fractal order or dimension is 1, we recover all fractional derivatives, the fractal derivative when the fractional order is 1, and the classical derivative when both are 1. We'll use a simple decay problem which Chen has already addressed to demonstrate the impact of fractal derivatives. This chapter's goal is to define the terms for the corresponding fractal-fractional operators as well as the fractional and fractal operators themselves. There are significant disputes concerning current nonlocal operator tendencies.

Definition 1 *The derivative of a function f is defined by*

$$D^1 f(x) = \lim_{h \to 0} \frac{f(x+h) - f(x)}{h}. \tag{2.1}$$

Definition 2 *Unlike classical Newtonian derivatives, a fractional derivative can be defined through a convolution. Consider a function f that is defined on some open interval $(0,b)$, then the fractional derivative of the given function f can be defined as follows:*

$$D_0^\alpha f(y) = \frac{1}{\Gamma(\alpha)} \int_0^y (y-t)^{\alpha-1} f(t) dt. \tag{2.2}$$

The Riemann–Liouville (RL) derivative based on the Cauchy formula that used for the calculation of iterated integrals, which can be defined as [59, 155, 202]:

$$D_0^\alpha f(t) = \frac{1}{\Gamma(m-\alpha)} \frac{d^m}{dt^m} \int_0^t \frac{f(y)dy}{(t-y)^{\alpha+1-m}}, \; t > 0. \tag{2.3}$$

Another derivative that is proposed by M. Caputo in 1967, which is known as Caputo derivative and illustrated in [22, 41, 114], is given by

$$_a^C D_t^\alpha f(t) = \frac{1}{\Gamma(m-\alpha)} \int_a^t \frac{f^m(y)dy}{(t-y)^{\alpha+1-m}}. \tag{2.4}$$

Definition 3 A function $f(t) \in H^1(q_1, q_2)$ and $0 < \alpha < 1$. Then, it follows from [42] the definition of Caputo-Fabrizio (CF) operator is

$$^{CF}D_t^\alpha f(t) = \frac{M(\alpha)}{1-\alpha} \int_a^t f'(y) \exp\left(\frac{-\alpha(t-y)}{1-\alpha}\right) dy, \tag{2.5}$$

Basic of Fractional Operators

where $M(0) = M(1) = 1$ and $M(\alpha)$ denote the normalisation function. If $f(t)$ does not belong to $H^1(q_1, q_2)$, then another shape of the derivative is

$$^{CF}D_t^\alpha f(t) = \frac{\alpha M(\alpha)}{1-\alpha} \int_a^t (f(t) - f(y)) \exp\left(\frac{-\alpha(t-y)}{1-\alpha}\right) dy. \tag{2.6}$$

Remark 1 Choosing $\beta = \frac{1-\alpha}{\alpha} \in (0, \infty)$, then $\alpha = \frac{1}{1+\beta}$, and so the equation (2.6) becomes

$$^{CF}D_t^\alpha f(t) = \frac{M(\beta)}{\beta} \int_a^t (v'(y)) \exp\left(\frac{-(t-y)}{\beta}\right) dy, \tag{2.7}$$

where $M(0) = M(\infty) = 1$,

Remark 2 We present the following property:

$$\lim_{\beta \to 0} \frac{1}{\beta} \exp\left(\frac{-(t-y)}{\beta}\right) = \delta(y-t), \tag{2.8}$$

where $\delta(y-t)$ represents the Hamilton function

Definition 4 *[151]* Later on Losada and Nieto modified this derivative and presented the following:

$$^{CF}D_t^\alpha f(t) = \frac{(2-\alpha)M(\alpha)}{2(1-\alpha)} \int_a^t f'(y) \exp\left(\frac{-\alpha(t-y)}{1-\alpha}\right) dy, \tag{2.9}$$

and their integral is given by

$$^{CF}I_t^\alpha \left(f(t)\right) = \frac{2(1-\alpha)}{(2-\alpha)M(\alpha)} f(t) + \frac{2\alpha}{(2-\alpha)M(\alpha)} \int_a^t f(y) dy, \ t \geq 0. \tag{2.10}$$

Definition 5 *A function $f \in (a,b)$ with $b > a$, $0 \leq \alpha \leq 1$, and not necessarily differentiable, then the Atangana-Baleanu derivative in RL sense can be defined as*

$$^{ABR}_aD_t^\alpha f(t) = \frac{B(\alpha)}{1-\alpha} \frac{d}{dt} \int_0^t f(y) E_\alpha\left[-\alpha \frac{(t-y)^\alpha}{1-\alpha}\right] dy, \tag{2.11}$$

where $B(\alpha)$ possesses the same properties as described for Caputo and CF operator.

Definition 6 *The fractional integral or the RL integral with fractional order α is defined through the following equation*

$$I_t^\alpha f(t) = \frac{1}{\Gamma(\alpha)} \int_0^t (t-\tau)^{\alpha-1} f(\tau) d\tau. \tag{2.12}$$

Definition 7 *A function $f \in (a,b)$ with $b > a$, $0 \leq \alpha \leq 1$, then the Atangana-Balenau derivative in Caputo sense can be defined as*

$$^{ABC}_aD_t^\alpha f(t) = \frac{B(\alpha)}{1-\alpha} \int_0^t f'(y) E_\alpha\left[-\alpha \frac{(t-y)^\alpha}{1-\alpha}\right] dy, \tag{2.13}$$

where $B(\alpha)$ possesses the same properties as described for Caputo and CF operator.

Definition 8 *The fractional integral for the Atangana-Baleanu derivative is given by*

$$I_t^\alpha f(t) = \frac{1-\alpha}{B(\alpha)} f(t) + \frac{\alpha}{B(\alpha)\Gamma(\alpha)} \int_0^t f(y)(t-y)^{\alpha-1} dy. \qquad (2.14)$$

2.2 SOME PROPERTIES OF THE FRACTIONAL OPERATORS

We include some of the defined differential and integral operators' key characteristics. The Laplace transform of the relevant fractional differential operators is first shown. The Laplace transform of the Caputo derivative for general case is

$$\begin{aligned}
L_t\left({}_0^C D_t^\alpha f(t)\right) &= \int_0^\infty e^{-st} {}_0^C D_t^\alpha f(t) dt, \\
&= \int_0^\infty e^{-st} \left(\frac{d^n}{dt^n} f(t) \frac{t^{n-\alpha-1}}{\Gamma(n-\alpha)}\right) dt, \\
&= s^\alpha L_t(f(t)) - \sum_{k=0}^{n-1} s^{\alpha-k-1} f^{(k)}(0). \qquad (2.15)
\end{aligned}$$

The Laplace transform of the CF derivative is given as

$$L_t\left({}_0^{CF} D_t^\alpha f(t)\right) = \frac{M(\alpha)}{1-\alpha} \frac{sL(f(t)) - f(0)}{s + \frac{\alpha}{1-\alpha}}. \qquad (2.16)$$

The Laplace transform of the Atangana-Baleanu fractional derivative is given as

$$L_t\left({}_0^{ABC} D_t^\alpha f(t)\right) = \frac{AB(\alpha)}{1-\alpha}\left(\frac{s^\alpha L_t(f(t)) - s^{\alpha-1} f(0)}{s^\alpha + \frac{\alpha}{1-\alpha}}\right). \qquad (2.17)$$

The Summudu transform of the Caputo derivative is given as

$$S_t\left({}_0^C D_t^\alpha f(t)\right) = \frac{S_t(f(t)) - f(0)}{u^\alpha}. \qquad (2.18)$$

The Summudu transform of the CF derivative is given as

$$S_t\left({}_0^{CF} D_t^\alpha f(t)\right) = \frac{S_t(f(t)) - f(0)}{\alpha u + 1 - \alpha} M(\alpha). \qquad (2.19)$$

The Summudu transform of the Atangana-Baleanu derivative is given as

$$S_t\left({}_0^{ABC} D_t^\alpha f(t)\right) = \frac{S_t(f(t)) - f(0)}{1 - \alpha + \alpha s^\alpha} AB(\alpha). \qquad (2.20)$$

2.3 FUNDAMENTAL THEOREM OF FRACTIONAL CALCULUS

The fundamental theory of calculus, which acts as a link between differential and integral calculus, is one of the most significant theorems in calculus. Many times, this theorem has been applied to get precise answers to specific equations or even to change differential equations into integral equations. This idea is also crucial in fractional calculus since it allows for the conversion indicated above. The RL type and the Caputo time are two forms of fractional differential operators that are studied in the area of fractional calculus. But after carefully researching the literature, we discovered that these two derivatives are related to a single integral operator. Nevertheless, the fractional integral was found to meet the basic. However, the RL derivative was used to derive the fractional integral in order to prove the calculus foundational theorem. Atangana recently argued that two fractional calculus should be considered, the first being the Caputo type and the second should be the RL type, both of which were derived differently. This is because it is not always possible for this theorem to be satisfied between the Caputo type of derivative and the Riemann-Liouville fractional integral. The outcomes of Atangana are presented in this section [13].

$$\begin{aligned}{}_0^C D_t^\alpha \Theta(t) &= \frac{1}{\Gamma(1-\alpha)} \int_0^t \frac{d}{d\tau}\Theta(\tau)(t-\tau)^{-\alpha} d\tau, \\ &= \frac{d}{dt}\Theta(t) * \frac{t^{-\alpha}}{\Gamma(1-\alpha)}, t > 0, 0 < \alpha \leqslant 1, \end{aligned} \quad (2.21)$$

while

$$\begin{aligned}{}_0^{RL} D_t^\alpha \Theta(t) &= \frac{1}{\Gamma(1-\alpha)} \frac{d}{dt} \int_0^t \Theta(\rho)(t-\rho)^{-\alpha} d\rho, \\ &= \frac{d}{dt}\left(\Theta(t) * \frac{t^{-\alpha}}{\Gamma(1-\alpha)}\right), t > 0, 0 < \alpha \leqslant 1. \end{aligned} \quad (2.22)$$

It should be noticed that the space is different in the scenarios mentioned above. The function Θ is differentiable in the case of the Caputo derivative but not necessarily in the RL case. The function must only be continuous, which is our need. Replacing allow us

$$H(t) = \int_0^t \Theta(\rho) \frac{(t-\rho)^{-\alpha}}{\Gamma(1-\alpha)} d\rho, \quad (2.23)$$

then,

$${}_0^{RL} D_t^\alpha \Theta(t) = \frac{d}{dt} H(t). \quad (2.24)$$

This describes that ${}_0^{RL}D_t^\alpha$ is a derivative in the classical way. This means

$$0^{RL} D_t^\alpha \Theta(t_0) = \lim_{t \to t_0} \frac{H(t) - H(t_0)}{t - t_0}. \quad (2.25)$$

This confers a local character on the fractional derivative RL. For the Caputo type, we put

$$K(t,\rho) = \frac{d}{d\rho}\Theta(\rho)(t-\rho)^{-\alpha}, \qquad (2.26)$$

so, we get

$$^{C}D_t^{\alpha}\Theta(t) = \int_0^t K(t,\rho)d\rho. \qquad (2.27)$$

So, the Caputo derivative is an integral that follows from its definition. It is well known that

$$\int_0^t \Theta'(\rho)d\rho = \Theta(t) - \Theta(0), \qquad (2.28)$$

but

$$\frac{d}{dt}\int_0^t \Theta(\rho)d\rho = \Theta(t). \qquad (2.29)$$

This implies that while the original condition is retained with the integral, it is lost with the derivative. It demonstrates that the integration has memory, but the derivative does not and can only recall the beginning condition. Additionally, the operator integral is well-posed but the derivative is not, which has ramifications for inverse issues. Let Q and w, for instance, represent an unknown function and the inverse problem's data, respectively. Therefore, one may formulate the inverse issue as follows in terms of operator equations:

$$\Xi Q = w, \qquad (2.30)$$

where Ξ denotes an operator that is defined in the space Ξ to the data W. The following classification can be made in calculus:

well-posed problems

$$w(y) = w(0) + \int_0^y Q(\varpi)d\varpi, \qquad (2.31)$$

ill-posed problems

$$w'(y) = Q(y). \qquad (2.32)$$

Recalling once again that the RL is a derivative and the Caputo derivative is an integral. It is well known that the RL is a derivative of a convolution of a function and a kernel, whereas the Caputo type is the convolution of a first derivative and a kernel. The Stieltjes integral, which arises in differential equation models of high order elasticity and viscoelasticity, is where the notion of convolution originates. The time shift rho is a summation of elementary relation processes brought on by unit step and evenly spaced time working, in contrast to the convolution itself. We may differentiate between the RL family of fractional derivatives and the Caputo family of fractional derivatives using the principles we have already established [13]. First off, Caputo type acts differently on any function than RL does since Caputo type has integral features, whereas RL has derivative properties. While there are two derivatives, the Caputo derivative and the RL derivative every publication in the literature

Basic of Fractional Operators

refers to the RL fractional integral. The fractional integral for the Caputo derivative has never before been defined. Perhaps it causes confusion for the scientists that study this subject. We'll try to solve it and separate the two calculus as they don't share the same objectives. The power-law, exponential decay law, and Mittag-Leffler function cases will be discussed in the following.

Starting for the simple case when $0 < \alpha \leq 1$, we have the Laplace transform of these operators in the sense of Caputo as

$$\begin{aligned} L\left({}_0^C D_t^\alpha \Theta(t)\right) &= \Gamma^\alpha \widetilde{\Theta}(\Gamma) - \Gamma^{\alpha-1}\Theta(0), \\ L\left({}_0^{CF} D_t^\alpha \Theta(t)\right) &= \frac{\Gamma \widetilde{\Theta}(\Gamma) - \Theta(0)}{\Gamma + \frac{\alpha}{1-\alpha}} \frac{M(\alpha)}{1-\alpha}, \\ L\left({}_0^{ABC} D_t^\alpha \Theta(t)\right) &= \frac{\Gamma^\alpha \widetilde{\Theta}(\Gamma) - \Gamma^{\alpha-1}\Theta(0)}{\Gamma^\alpha + \frac{\alpha}{1-\alpha}} \frac{AB(\alpha)}{1-\alpha}. \end{aligned} \quad (2.33)$$

The Caputo derivative in the Laplace space clearly reveals information about the function, the starting condition, and the process kernel. More specifically, the kernel's contribution or memory impact is on the starting circumstance. The Caputo-type being a convolution, the convolution theorem helped to explain how successfully these conclusions were reached. We pose the question, "Does this operator have its own anti-inverse?" Consider the equation below

$$L\left({}_0^C D_t^\alpha \Theta(t)\right) = L(v(t)), \quad (2.34)$$

where $v(t)$ is known and is having the laplace transform. Since $\Theta(0)$ is known, we write

$$\Gamma^\alpha \widetilde{\Theta}(\Gamma) = L(v(t)) + \Gamma^{\alpha-1}\Theta(0). \quad (2.35)$$

$$\Theta(t) = \frac{1}{\Gamma(\alpha)} \int_0^t v(\rho)(t-\rho)^{\alpha-1} d\rho + \Theta(0).$$

With the CF

$$\begin{aligned} L\left({}_0^{CF} D_t^\alpha \Theta(t)\right) &= L(v(t)), \\ \frac{M(\alpha)}{1-\alpha} \frac{\Gamma \widetilde{\Theta}(\Gamma) - \Theta(0)}{\Gamma + \frac{\alpha}{1-\alpha}} &= L(v(t)) = \widetilde{v}(\Gamma) \\ \Gamma \widetilde{\Theta}(\Gamma) - \Theta(0) &= \frac{1-\alpha}{M(\alpha)}\left(\Gamma + \frac{\alpha}{1-\alpha}\right)\widetilde{v}(\Gamma) \\ \widetilde{\Theta}(\Gamma) &= \frac{\Theta(0)}{\Gamma} + \frac{1}{\Gamma}\frac{1-\alpha}{M(\alpha)}\left(\Gamma + \frac{\alpha}{1-\alpha}\right)\widetilde{v}(\Gamma) \\ \Theta(t) &= \Theta(0) + \frac{1-\alpha}{M(\alpha)}v(t) + \frac{\alpha}{M(\alpha)}\int_0^t v(\rho)d\rho. \end{aligned} \quad (2.36)$$

We obtained the same for the ABC-derivative

$$\Theta(t) = \Theta(0) + \frac{1-\alpha}{AB(\alpha)} v(t) + \frac{\alpha}{AB(\alpha)\Gamma(\alpha)} \int_0^t v(\rho)(t-\rho)^{\alpha-1} d\rho, t > 0, 0 < \alpha \leq 1 \quad (2.37)$$

$v(0) = 0$. It is therefore mathematically convenient by concluding that the corresponding Caputo integral should be in the case of power-law

$$_0^C I_t^\alpha \Theta(t) = \phi(t) + \frac{1}{\Gamma(\alpha)} \int_0^t \Theta(\rho)(t-\rho)^{\alpha-1} d\rho, \quad (2.38)$$

with $t > 0, 0 < \alpha \leq 1$.
For the case of exponential decay

$$_0^{CF} I_t^\alpha \Theta(t) = \phi(t) + \frac{1-\alpha}{M(\alpha)} \Theta(t) + \frac{\alpha}{M(\alpha)} \int_0^t \Theta(\rho) d\rho, t > 0, 0 < \alpha \leq 1. \quad (2.39)$$

where $\phi(0) + \frac{1-\alpha}{M(\alpha)} \Theta(0) = 0$. While for the case of Atangana-Baleanu

$$_0^{ABC} I_t^\alpha \Theta(t) = \phi(t) + \frac{1-\alpha}{AB(\alpha)} \Theta(t) + \frac{\alpha}{AB(\alpha)\Gamma(\alpha)} \int_0^t \Theta(\rho)(t-\rho)^{\alpha-1} d\rho, \quad (2.40)$$

$t > 0$, $0 < \alpha \leq 1$, and $\phi(0) + \frac{1-\alpha}{AB(\alpha)} \Theta(0) = 0$. The function $\phi(t)$ selection should be in the form to satisfy the fundamental theorem of calculus such that

$$\begin{aligned} _0^C D_t^\alpha \left[_0^{RL} I_t^\alpha \Theta(t) \right] &= {}_0^C D_t^\alpha \phi(t) + {}_0^C D_t^\alpha \left[_0^{RL} I_t^\alpha \Theta(t) \right], \\ &= {}_0^C D_t^\alpha \phi(t) + \Theta(t) - \frac{t^{-\alpha}}{\Gamma(1-\alpha)^0} {}_0^{RL} I_t^\alpha \Theta(t) \bigg|_{t=0^+}. \end{aligned} \quad (2.41)$$

To satisfy the concepts of mathematics, it is only fair by choosing that

$$_0^C D_t^\alpha \phi(t) = \frac{t^{-\alpha}}{\Gamma(1-\alpha)} \left(_0^{RL} I_t^\alpha \Theta \right)(0^+), \quad (2.42)$$

such that

$$_0^C D_t^\alpha \left[_0^C I_t^\alpha \Theta(t) \right] = \Theta(t), \quad (2.43)$$

Reversely, we have

$$\begin{aligned} _0^C I_t^\alpha \left[_0^C D_t^\alpha \Theta(t) \right] &= \phi(t) + {}_0^{RL} I_t^\alpha \left[_0^C D_t^\alpha \Theta(t) \right], \\ &= \phi(t) + \Theta(t) - \Theta(0). \end{aligned} \quad (2.44)$$

In this case the only reasonability by concluding it is $\phi(t) = \Theta(0)$. Therefore, by taking into consideration, the Cauchy problem with Caputo derivative

$$_0^C D_t^\alpha \eta(t) = \Omega(t, \eta(t)), \quad \eta(0) = \eta_0, \quad (2.45)$$

then we use the Caputo integral instead of RL integral,

$$\eta(t) = \phi(t) + {}_0^{RL}I_t^\alpha \Omega(t, \eta(t)) \tag{2.46}$$

On $\eta(t)$ applying the Caputo derivative, we write

$$\begin{aligned}
{}_0^C D_t^\alpha \eta(t) &= {}_0^C D_t^\alpha \phi(t) + {}_0^C D_t^\alpha \left[{}_0^{RL}I_t^\alpha \Omega(t, \eta(t)) \right], \\
&= \Omega(t, \eta(t)) + {}_0^C D_t^\alpha \phi(t) - \frac{t^{-\alpha}}{\Gamma(1-\alpha)} \left[{}^{RL}I_t^\alpha \Omega(t, \eta(t)) \right] \Big|_{t=0^+}
\end{aligned} \tag{2.47}$$

Thus,

$$ {}_0^C D_t^\alpha \phi(t) = \frac{t^{-\alpha}}{\Gamma(1-\alpha)} \left[{}^{RL}I_t^\alpha \Omega(t, \eta(t)) \right] \Big|_{t=0^+} \tag{2.48}$$

therefore,

$$ {}_0^C D_t^\alpha \eta(t) = \Omega(t, \eta(t)). \tag{2.49}$$

The same for the CF case

$$\begin{aligned}
{}_0^{CF} D_t^\alpha \left[{}_0^{CF} I_t^a \Theta(t) \right] &= {}_0^{CF} D_t^\alpha \phi(t) + {}_0^{CF} D_t^\alpha \left[\frac{1-\alpha}{M(\alpha)} \Theta(t) + \frac{\alpha}{M(\alpha)} \int_0^t f(\rho) d\rho \right], \\
&= \Theta(t) + {}_0^{CF} D_t^\alpha \phi(t) - \exp\left(-\frac{\alpha}{1-\alpha} t\right) \Theta(0).
\end{aligned} \tag{2.50}$$

For fairness, we choose

$$ {}_0^{CF} D_t^\alpha \phi(t) = \exp\left(-\frac{\alpha}{1-\alpha} t\right) \Theta(0). \tag{2.51}$$

Thus,

$$ {}_0^{CF} D_t^\alpha \left[{}_0^{CF} I_t^a \Theta(t) \right] = \Theta(t). \tag{2.52}$$

Reversely,

$$\begin{aligned}
{}_0^{CF} I_t^\alpha \left[{}_0^{CF} D_t^\alpha \Theta(t) \right] &= \phi(t) + {}_0^{CFR} I_t^\alpha \left[{}_0^{CF} D_t^a \Theta(t) \right], \\
&= \phi(t) + \Theta(t) - \Theta(0).
\end{aligned} \tag{2.53}$$

It is only fair by selecting $\phi(t) = \Theta(0)$, and we obtain $\Theta(t)$. Thus, by considering the Cauchy problem with the CF derivative [13]

$$ {}_0^{CF} D_t^\alpha \Theta(t) = w(t, \Theta(t)) \tag{2.54}$$

Using $^{CF}_0 I^\alpha_t$, we have

$$\Theta(t) = \phi(t) + \frac{1-\alpha}{M(\alpha)} w(t, \Theta(t)) + \frac{\alpha}{M(\alpha)} \int_0^t w(\rho, f(\rho)) d\rho \qquad (2.55)$$

$$\begin{aligned}
^{CF}_0 D^\alpha_t \Theta(t) &= {}^{CF}_0 D^\alpha_t \phi(t) + {}^{CF}_0 D^\alpha_t \left[{}^{CFR}_0 I^\alpha_t w(t, \Theta(t)) \right] \\
&= {}^{CF}_0 D^\alpha_t \phi(t) + w(t, \Theta(t)) - \exp\left(-\frac{\alpha}{1-\alpha} t\right) w(0, \Theta(0)), \\
&= w(t, \Theta(t)). \qquad (2.56)
\end{aligned}$$

Since

$$^{CF}_0 D^a_t \phi(t) = \exp\left(-\frac{\alpha}{1-\alpha} t\right) w(0, \Theta(0)).$$

With ABC, we have the following:

$$\begin{aligned}
^{ABC}_0 D^\alpha_t \left[{}^{ABC}_0 I^\alpha_t \Theta(t) \right] &= {}^{ABC}_0 D^a_t \phi(t) + {}^{ABC}_0 D^a_t \left({}^{AB}_0 I^\alpha_t \Theta(t) \right) \\
&= {}^{ABC}_0 D^\alpha_t \phi(t) + \Theta(t) - E_\alpha\left(-\frac{\alpha}{1-\alpha} t^\alpha\right) \Theta(0). \quad (2.57)
\end{aligned}$$

We give the following in order to satisfy the fundamental theorem of calculus,

$$^{ABC}_0 D^\alpha_t \phi(t) = E_a\left(-\frac{\alpha}{1-\alpha} t^\alpha\right) \Theta(0). \qquad (2.58)$$

Reversely,

$$\begin{aligned}
^{ABC}_0 I^\alpha_t \left[{}^{ABC}_0 D^\alpha_t \Theta(t) \right] &= \phi(t) + {}^{AB}_0 I^\alpha_t \left[{}^{AB}_0 D^a_t \Theta(t) \right], \\
&= \phi(t) + \Theta(t) - \Theta(0) \qquad (2.59)
\end{aligned}$$

Also, we have $\phi(t) = \Theta(0)$. Using the similar fashion, we consider

$$\begin{aligned}
^{ABC}_0 D^a_t x(t) &= \Theta(t, x(t)) \\
x(t) &= \phi(t) + {}^{AB}_0 I^\alpha_t \Theta(t, x(t)) \\
^{AB}_0 D^\alpha_t x(t) &= {}^{ABC}_0 D^\alpha_t \phi(t) + {}^{ABC}_0 D^\alpha_t \left[{}^{AB}_0 I^\alpha_t \Theta(t, x(t)) \right]. \qquad (2.60)
\end{aligned}$$

It is important to note that the AB integral is not equivalent to the ABC derivative since both are conceptualised differently from a theoretical to a practical standpoint [13]. Anybody can confirm this by

$$\begin{aligned}
^{ABR}_0 D^\alpha_t \left[{}^{AB}_0 I^\alpha_t \Theta(t) \right] &= \Theta(t), \\
^{AB}_0 I^\alpha_t \left[{}^{ABR}_0 D^\alpha_t \Theta(t) \right] &= \Theta(t). \qquad (2.61)
\end{aligned}$$

Basic of Fractional Operators

It plainly proves the calculus foundational theorem. Applying the AB-integral to the ABC derivative or the CFR-integral to the CF derivative and expecting them to hold the calculus fundamental theorem is not accurate scientifically. The Atangana-Baleanu integral in the RL sense and the relationship between it and the Caputo sense of the RL is

$$^{ABC}_{0}D^{\alpha}_{t}\Theta(t) = {}^{ABR}_{0}D^{\alpha}_{t}\Theta(t) - \frac{AB(\alpha)}{1-\alpha}\Theta(0)E_{\alpha}\left(-\frac{\alpha}{1-\alpha}t^{a}\right) \quad (2.62)$$

Replacing $\Theta(t)$ by

$$\frac{1-\alpha}{AB(\alpha)}\Theta(t,y(t)) + \frac{a}{AB(\alpha)\Gamma(\alpha)}\int_{0}^{t}\Theta(\rho,y(\rho))(t-\rho)^{a-1}d\rho.$$

Thabet [96] proved the same results even for the case of discrete. It can be said that the Caputo derivative is more stronger than that of RL type.

Remark 3. By addressing the requirement for an operator that is anti-derivative of the Caputo derivative, we are correcting the error of the past. The two derivatives are distinct from one another, although the RL integral is the sole one used in the literature to complete the analysis. The calculus for the Caputo integral and its derivative should be distinct. There is no justification for anticipating the basic theorem of calculus to be met by applying the Caputo derivative to the RL integral. Additionally, it is incorrect technically to understand the RL integral on the Caputo derivative. Using the classical differential operator known as the Caputo derivative as an example [13]

$$^{ABC}_{0}D^{\alpha}_{t}\eta(t) = \Theta(t,\eta(t)), \quad \eta(0) = \eta_{0} \quad (2.63)$$

On applying mistakenly the RL integral, we have

$$\eta(t) = \eta(0) + {}^{RL}_{0}I^{\alpha}_{t}\Theta(t,\eta(t)) \quad (2.64)$$

Then by applying again the Caputo, we get

$$^{C}_{0}D^{\alpha}_{t}\eta(t) = {}^{C}_{0}D^{\alpha}_{t}\left[{}^{RL}_{0}I^{\alpha}_{t}\Theta(t,\eta(t))\right],$$

$$\eta(t) = \Theta(t,\eta(t)) - \frac{t^{-\alpha}}{\Gamma(1-\alpha)^{0}}{}^{RL}I^{\alpha}_{t}\Theta(t,\eta(t))\bigg|_{t=0}. \quad (2.65)$$

The term ${}^{RL}_{0}I^{\alpha}_{t}\Theta(t,\eta(t))$ does not vanish when $t = 0$, if the function Θ is not continuous; thus imagine that if the function Θ has a singularity, the function will still exist. This limits the use of space to continuous space exclusively; what about non-continuous space? The Caputo integral has been used to resolve this issue [13].

- In the case of power-law if the function is continuous, then
  $${}^{C}_{0}D^{\alpha}_{t}\phi(t) = 0.$$

- If the kernel used is $\frac{M(\alpha)}{1-\alpha} \exp\left(-\frac{\alpha}{1-\alpha}t\right)$, then
$${}_0^{CF}D_t^\alpha \phi(t) = \Theta(0)\exp\left(-\frac{\alpha}{1-\alpha}t\right)$$

- In case of Atangana-Baleanu,
$${}_0^{ABC}D_t^\alpha \phi(t) = \Theta(0)E_\alpha\left(-\frac{\alpha}{1-\alpha}t^\alpha\right).$$

- $\phi(t)$ can be obtained to have the explicit Caputo integral.

We provide the existence and unique results of the Cauchy problem with Caputo derivative

$$\substack{C\\0}D_t^\alpha x(t) = \Theta(t,x(t)),\ x(0) = x_0. \tag{2.66}$$

We have by using the Caputo integral,

$$x(t) = \phi(t) + \frac{1}{\Gamma(\alpha)}\int_0^t \Theta(\rho,x(\rho))(t-\rho)^{\alpha-1}d\rho. \tag{2.67}$$

Example 1. We can identify a function, let's call $\phi(t)$, that is connected to the CF situation. To move on, it's important to first remember that the $\phi(t)$ specifies a memory function similar to the CF. Angel et al. [234] have demonstrated that the kernel used in the CF derivative has a cross-over influence on both the probability distribution and the mean square displacement [13]. As a result, the function associated with memory must be specified in accordance with the implication that there is a fixed period $t_1 = a$ such that, until that time, CF portrays ordinary diffusion and, after that time, restricted diffusion. $\phi(t)$ indicate the piecewise function, which has a value of zero at the start and a value of zero after the crossover since the prior history has been recorded. In order to define the function denoted by [13]

$$\begin{cases} \phi'(t) = \frac{(1-\alpha)\Theta(0)}{aM(\alpha)\Gamma(2-\alpha)\Gamma(\alpha)} \exp\left[-\frac{\alpha}{1-\alpha}t\right] t^{1-a}(a-t)^{\alpha-1}, a<t \\ 0, a \geqslant t. \end{cases} \tag{2.68}$$

$$\begin{aligned}
{}_0^{CF}D_t^\alpha \phi(t) &= \frac{M(\alpha)}{(1-\alpha)}\int_0^t \phi'(\rho)\exp\left[-\frac{\alpha}{1-\alpha}(t-\rho)\right]d\rho, \\
&= \frac{M(\alpha)}{(1-\alpha)}\int_0^a \frac{(1-\alpha)\Theta(0)}{aM(\alpha)\Gamma(2-\alpha)\Gamma(\alpha)}\exp\left[-\frac{\alpha}{1-\alpha}\rho\right] \\
&\quad \times \rho^{1-a}(a-\rho)^{\alpha-1}\exp\left[-\frac{\alpha}{1-\alpha}(t-\rho)\right]d\rho, \\
&= \frac{\Theta(0)}{a\Gamma(2-\alpha)\Gamma(\alpha)}\exp\left[-\frac{\alpha}{1-\alpha}t\right]\int_0^a \rho^{1-\alpha}(a-\rho)^{\alpha-1}d\rho, \\
&= \frac{\Theta(0)}{a\Gamma(2-\alpha)\Gamma(\alpha)}\exp\left[-\frac{\alpha}{1-\alpha}t\right]a\int_0^1 \rho^{1-\alpha}(1-\rho)^{\alpha-1}d\rho, \\
&= \Theta(0)\exp\left[-\frac{\alpha}{1-\alpha}t\right] \tag{2.69}
\end{aligned}$$

Basic of Fractional Operators

which satisfies the equation. We have for the case of classical Caputo

$$ {}^C_0 D_t^\alpha \phi(t) = \frac{t^{-\alpha}}{\Gamma(1-\alpha)} \left[{}^{RL}_0 I_t^\alpha \Theta(t) \right] \Big|_{t=0^+} \tag{2.70}$$

If f is not continuous at $t = 0$, then we assume that

$$ \lim_{t \to 0^+} \left[{}^{RL}_0 I_t^\alpha \Theta(t) \right] = c, \tag{2.71}$$

then

$$ {}^C_0 D_t^\alpha \phi(t) = \frac{t^{-a}}{\Gamma(1-\alpha)} c. \tag{2.72}$$

Thus,

$$ \phi(t) = c. \tag{2.73}$$

This also refers to the outcomes that Angel et al. came to [234]. The waiting time distribution, mean square displacement, and probability distribution all show that the power-law kernel does not display any process with a cross-over effect, but provides long-tailed behaviours.

2.4 FRACTAL-FRACTIONAL OPERATORS

Definition 9 *[10] Consider $f(t)$ which is continuous and differentiable over the interval (a,b) with q order, then one can define the fractal-fractional derivative of $f(t)$ with fractional order p in the sense of RL having the kernel of the power-law is given by [10]:*

$$ {}^{FFP} D_{0,t}^{p,q} \left(f(t) \right) = \frac{1}{\Gamma(m-p)} \frac{d}{dt^q} \int_0^t (t-s)^{m-p-1} f(s) ds, \tag{2.74}$$

with $m-1 < p, q \leq m \in \mathbb{N}$, and $\frac{df(s)}{ds^q} = \lim_{t \to s} \frac{f(t)-f(s)}{t^q - s^q}$.

Definition 10 *Let a function $f(t)$ that is continuous and fractal differentiable on some interval, say, (a,b) with fractal order q, then we can define the fractal-fractional derivative of $f(t)$ with fractional order p in RL sense with exponential decay kernel shown by:*

$$ {}^{FFE} D_{0,t}^{p,q} \left(f(t) \right) = \frac{M(p)}{1-p} \frac{d}{dt^q} \int_0^t \exp\left(-\frac{p}{1-p}(t-s) \right) f(s) ds, \tag{2.75}$$

with $p > 0$, $q \leq m \in \mathbb{N}$ and $M(0) = M(1) = 1$.

Definition 11 *A function $f(t)$ that posseses the properties of continuity and fractal differentiability on some interval (a,b) with fractal order q, and we can define the fractal-fractional derivative of $f(t)$ with fractional order p in RL sense with generalised Mittag-Leffler kernel as shown below:*

$$^{FFM}D_{0,t}^{p,q}\left(f(t)\right) = \frac{AB(p)}{1-p}\frac{d}{dt^q}\int_0^t E_p\left(-\frac{p}{1-p}(t-s)^p\right)f(s)ds, \quad (2.76)$$

with p, q, and $AB(p) = 1 - p + \frac{p}{\Gamma(p)}$.

Definition 12 *If a continuous function $f(t)$ is defined on some interval (a,b), then one can define the fractal-fractional integral of $f(t)$ with fractional order p with power-law kernel as*

$$^{FFP}J_{0,t}^{p}\left(f(t)\right) = \frac{q}{\Gamma(p)}\int_0^t (t-s)^{p-1}s^{q-1}f(s)ds. \quad (2.77)$$

Definition 13 *If a continuous function $f(t)$ is on some interval (a,b) then one can define the fractal-fractional integral of $f(t)$ with fractional order p with exponential decay kernel given as*

$$^{FFE}J_{0,t}^{p}\left(f(t)\right) = \frac{pq}{M(p)}\int_0^t s^{p-1}f(s)ds + \frac{q(1-p)t^{q-1}f(t)}{M(p)}. \quad (2.78)$$

Definition 14 *Consider that $f(t)$ is continuous on an open interval (a,b), then the fractal-fractional integral of $f(t)$ with order p having generalised Mittag-Leffler type kernel is defined as follows:*

$$^{FFM}J_{0,t}^{p,q}\left(f(t)\right) = \frac{pq}{AB(p)}\int_0^t s^{q-1}(t-s)^{p-1}f(s)ds + \frac{q(1-p)t^{q-1}f(t)}{AB(p)}. \quad (2.79)$$

Theorem 1 *Consider $\alpha \in (0,1)$, then there exists a function g well-defined that equalises the Caputo and the Atangana-Baleanu derivatives with fractional order.*

Proof 1 *For the solution of this problem, we consider the fractional equation as below:*

$$^{ABC}_0D_y^{\alpha}[g(y)] = {}^C_0D_y^{\alpha}[g(y)]. \quad (2.80)$$

Applying the laplace transform on both sides of the above equation, we have

$$\frac{M(\alpha)}{1-\alpha}\frac{q^\alpha G(q) - q^{\alpha-1}g(0)}{q^\alpha + \frac{\alpha}{1-\alpha}} = q^\alpha G(q) - g(0),$$

$$G(q)\left[\frac{M(\alpha)}{1-\alpha}\frac{q^\alpha}{q^\alpha + \frac{\alpha}{1-\alpha}} - q^\alpha\right] = \left[\frac{M(\alpha)}{1-\alpha}\frac{q^{\alpha-1}}{q^\alpha + \frac{\alpha}{1-\alpha}} - 1\right]g(0),$$

Basic of Fractional Operators

$$G(q) = \frac{\left[\frac{M(\alpha)}{1-\alpha}\frac{q^{\alpha-1}}{q^{\alpha}+\frac{\alpha}{1-\alpha}} - 1\right]g(0)}{\left[\frac{M(\alpha)}{1-\alpha}\frac{q^{\alpha}}{q^{\alpha}+\frac{\alpha}{1-\alpha}} - q^{\alpha}\right]}$$

$$= \frac{\left[\frac{M(\alpha)}{1-\alpha} - 1\right]q^{\alpha} - \frac{\alpha}{1-\alpha}}{\left[\frac{M(\alpha)}{1-\alpha} - \frac{\alpha}{1-\alpha}\right]q^{\alpha} - q^{2\alpha}},$$

$$= \frac{g(0)\left[\frac{M(\alpha)}{1-\alpha} - 1\right]}{\left[\frac{M(\alpha)}{1-\alpha} - \frac{\alpha}{1-\alpha}\right] - q^{\alpha}}$$

$$- \frac{\frac{\alpha}{1-\alpha}g(0)}{\left[\frac{M(\alpha)}{1-\alpha} - \frac{\alpha}{1-\alpha}\right] - q^{\alpha}}\frac{1}{q^{\alpha}}. \quad (2.81)$$

Now applying the inverse laplace transform on both sides of the above equation, we obtain

$$g(t) = g(0)\left\{1 - \frac{M(\alpha)}{1-\alpha}\right\}y^{\alpha-1}E_{\alpha,\alpha}\left\{\left\{\frac{M(\alpha)}{1-\alpha} - \frac{\alpha}{1-\alpha}\right\}y^{\alpha}\right\}$$

$$+ \frac{1}{\Gamma(\alpha)}\int_0^t g(0)x^{\alpha-1}(t-x)^{\alpha-1}\frac{\alpha}{\alpha-1}g(0)$$

$$E_{\alpha,\alpha}\left\{\left\{\frac{M(\alpha)}{1-\alpha} - \frac{\alpha}{1-\alpha}\right\}x^{\alpha}\right\}dy. \quad (2.82)$$

We present here the effect of fractal derivative on a given function, the effect of fractional derivative on the same function, and finally the effect of fractal fractional derivative. We consider the following fractal differential equations

$$\begin{cases} \frac{dy(t)}{dt^{\beta}} = e^{t^{\beta}}, \\ y(0) = y_0. \end{cases} \quad (2.83)$$

We assume that $y(t)$ is differentiable, therefore

$$\begin{cases} \frac{dy(t)}{dt} = \beta t^{\beta-1}e^{t^{\beta}}, \\ y(0) = y_0. \end{cases} \quad (2.84)$$

then we can write

$$y(t) = y(0) + \int_0^t \beta \tau^{\beta-1}e^{\tau^{\beta}}d\tau, \ y(0) = y_0, \quad (2.85)$$

and

$$y(t) = y(0) + e^{t^{\beta}} - 1, \ y(0) = y_0 = 1. \quad (2.86)$$

then $y(t) = e^{t^{\beta}}$, if $\beta = 1$,

$$\frac{dy(t)}{dt} = e^t, \ y(0) = y_0. \quad (2.87)$$

and
$$y(t) = e^t, \ y(0) = 1. \tag{2.88}$$

For power-law function
$$\frac{dy(t)}{dt^\beta} = t^\alpha, \ y(0) = y_0, \tag{2.89}$$

then
$$\frac{dy(t)}{dt} = \beta t^{\beta+\alpha-1}, \ y(0) = y_0. \tag{2.90}$$

$$y(t) = y(0) + \frac{\beta t^{\alpha+\beta}}{\beta+\alpha}, \ y(0) = 0. \tag{2.91}$$

So,
$$y(t) = \frac{\beta t^{\alpha+\beta}}{\beta+\alpha}. \tag{2.92}$$

if $y(0) = 1$, then
$$y(t) = 1 + \frac{\beta t^{\alpha+\beta}}{\beta+\alpha}.$$

With fractional derivative, we have
$$\begin{cases} {}^C D_t^\beta y(t) = t^\alpha \\ y(0) = y_0, \end{cases} \tag{2.93}$$

then, we write
$$y(t) = y(0) + \frac{1}{\Gamma(\beta)} \int_0^t \tau^\alpha (t-\tau)^{\beta-1} d\tau, \ y(0) = y_0, \tag{2.94}$$

then, we get
$$y(t) = y(0) + \frac{t^{\alpha+\beta}}{\Gamma(\beta)} B(\alpha+1,\beta), \ y(0) = y_0. \tag{2.95}$$

We get using the relation of Beta and Gamma function,
$$y(t) = y(0) + \frac{t^{\alpha+\beta} \Gamma(\alpha+1)}{\Gamma(\alpha+\beta+1)}, \ y(0) = y_0. \tag{2.96}$$

If $y(0) = 0$, then
$$y(t) = \frac{t^{\alpha+\beta} \Gamma(\alpha+1)}{\Gamma(\alpha+\beta+1)} \tag{2.97}$$

Basic of Fractional Operators

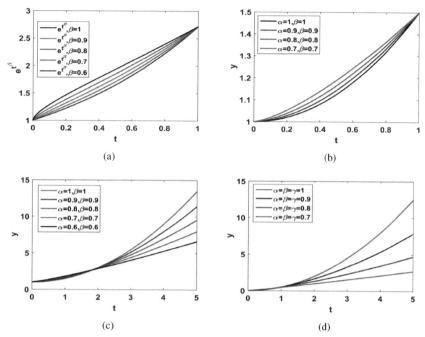

Figure 2.1 The solution behaviour of the functions: (a) $y = e^{t^{\beta}}$, (b) $y = 1 + \frac{\beta t^{\alpha+\beta}}{\beta+\alpha}$, (c) $y = 1 + \frac{t^{\alpha+\beta}\Gamma(\alpha+1)}{\Gamma(\alpha+\beta+1)}$, (d) $y = \frac{\beta t^{\beta+\gamma+\alpha-1}}{\Gamma(\alpha)} \frac{\Gamma(\alpha+\gamma)}{\Gamma(\alpha+\beta+\gamma)}$.

If $y(0) = 1$, then

$$y(t) = 1 + \frac{t^{\alpha+\beta}\Gamma(\alpha+1)}{\Gamma(\alpha+\beta+1)}. \tag{2.98}$$

For fractal-fractional derivative, we have

$$^{FFP}D_t^{\alpha,\beta}y(t) = t^{\gamma}, \ y(0) = y_0. \tag{2.99}$$

then, we write

$$^{RL}D_t^{\alpha}y(t) = \beta t^{\beta+\gamma-1}, \ y(0) = y_0, \tag{2.100}$$

so,

$$y(t) = \frac{\beta}{\Gamma(\alpha)}\int_0^t \tau^{\beta+\gamma-1}(t-\tau)^{\alpha-1}d\tau, \ y(0) = y_0. \tag{2.101}$$

also, we write

$$y(t) = \frac{\beta}{\Gamma(\alpha)}t^{\beta+\gamma+\alpha-1}B(\beta+\gamma,\alpha), \ y(0) = y_0. \tag{2.102}$$

Finally, we have

$$y(t) = \frac{\beta}{\Gamma(\alpha)}t^{\beta+\gamma+\alpha-1}\frac{\Gamma(\alpha+\gamma)}{\Gamma(\beta+\gamma+\alpha)}, \ y(0) = y_0. \tag{2.103}$$

The graphical representation of the above function discussed is given in Figure 2.1.

3 Definitions of Fractal-Fractional Operators with Numerical Approximations

3.1 INTRODUCTION

To accomplish two distinct goals, two classes of non-conventional differential and integrals were established. On the one hand, the idea of fractional calculus was developed to account for memory effects seen in a variety of real-world issues, namely, the power process, fading memories, and cross-over behaviours. The idea of fractal calculus, on the other hand, is essentially a differential operator with a power law setting. In order to solve these physical issues that reveal self-similar patterns, new differential and integral operators were devised. These new operators were inspired by the formulation of fractal representation. However, these two ideas have successfully served the purpose for which they were initially introduced.

Researchers soon discovered that more advanced physical issues with both behaviours exist in nature. Therefore, neither the fractional nor the fractal concepts were enough to deal with such issues. This unsolved issue was never addressed before to 2016. Atangana [10] introduced the idea of fractal-fractional calculus around the middle of 2016. The novel differential operators can be viewed as a convolution of the fractal derivative with the generalised Mittag-Leffler function, the exponential decay law, or the power law. The novel idea has been used in numerous scientific fields. Using the novel idea, researchers were able to capture more intricate chaotic attractors in chaotic issues. This chapter contains some fresh, comprehensive material on fractal-fractional. The fundamentals of fractal-fractional operators were covered in this chapter. We offer the fundamental concepts of fractal-fractional operators in the sense of the derivatives of Caputo, Caputo-Fabrizio, and Atangana-Baleanu. Additionally, a numerical solution is provided to resolve the suggested fractal-fractional problems.

Definition 15 *Let f be a continuous function, let $\alpha > 0$, the fractal derivative of f with order α is given as*

$$_{0}^{F}D_{t}^{\alpha}f(t) = \lim_{t_1 \to t} \frac{f(t_1) - f(t)}{t_1^{\alpha} - t^{\alpha}}. \tag{3.1}$$

Definition 16 *Let f be a continuous function, let $\alpha, \beta > 0$, the fractal derivative of f with orders α and β is given as*

$$ {}_0^F D_t^{\alpha,\beta} f(t) = \lim_{t_1 \to t} \frac{f^\beta(t_1) - f^\beta(t)}{t_1^\alpha - t^\alpha}. \tag{3.2}$$

Remark 1 *If the function f is differentiable, then fractal derivative can be represented as*

$$ {}_0^F D_t^\alpha f(t) = \frac{t^{1-\alpha}}{\alpha} f'(t). \tag{3.3}$$

It is very easy to see that if the fractal dimension order is 1, we have

$$ {}_0^F D_t^\alpha f(t) = f'(t). \tag{3.4}$$

Remark 2 *If the function f is differentiable, then the associative integral can be given as*

$$ {}_0^F J_t^\alpha f(t) = \alpha \int_0^t \tau^{\alpha-1} f(\tau) d\tau. \tag{3.5}$$

Therefore, when $\alpha \to 1$, we state

$$ {}_0^F J_t^1 f(t) = \int_0^t f(\tau) d\tau. \tag{3.6}$$

Definition 17 *Let f be a continuous function, let $\alpha, \beta > 0$, be two positive real numbers such that $0 < \alpha \leq 1$ and $\beta > 0$, then a fractal-fractional differential operator with power law kernel is given as*

$$ {}_0^{FFP} D_{t,R}^{\alpha,\beta} f(t) = \frac{1}{\Gamma(1-\alpha)} \frac{d}{dt^\beta} \int_0^t f(\tau)(t-\tau)^{-\alpha} d\tau \tag{3.7}$$

while

$$ {}_0^{FFP} D_{t,C}^{\alpha,\beta} f(t) = \frac{1}{\Gamma(1-\alpha)} \int_0^t \frac{d}{d\tau^\beta} f(\tau)(t-\tau)^{-\alpha} d\tau. \tag{3.8}$$

Remark 3 *Since the integral is differentiable, equation (3.7) can be represented as*

$$ {}_0^{FFP} D_t^{\alpha,\beta} f(t) = \frac{1}{\Gamma(1-\alpha)} \frac{d}{dt} \int_0^t f(\tau)(t-\tau)^{-\alpha} d\tau \frac{1}{\beta t^{\beta-1}}. \tag{3.9}$$

and equation (3.7) can be reformulated as

$$ {}_0^{FFP} D_t^{\alpha,\beta} f(t) = \frac{1}{\Gamma(1-\alpha)} \int_0^t \frac{d}{dt} \frac{1}{\beta \tau^{\beta-1}} f(\tau)(t-\tau)^{-\alpha} d\tau. \tag{3.10}$$

Remark 4 When $\beta \to 1$, we have

$$_{0}^{FFP}D_{t,R}^{\alpha,1}f(t) = \frac{1}{\Gamma(1-\alpha)}\frac{d}{dt}\int_{0}^{t}f(\tau)(t-\tau)^{-\alpha}d\tau. \tag{3.11}$$

when $\alpha \to 1$, we have

$$_{0}^{FFP}D_{t,R}^{1,\beta}f(t) = f'(t)\frac{t^{1-\beta}}{\beta}. \tag{3.12}$$

Also,

$$_{0}^{FFP}D_{t,C}^{\alpha,1}f(t) = \frac{1}{\Gamma(1-\alpha)}\int_{0}^{t}\frac{d}{dt}f(\tau)(t-\tau)^{-\alpha}d\tau. \tag{3.13}$$

and

$$_{0}^{FFP}D_{t,C}^{1,\beta}f(t) = f'(t)\frac{t^{1-\beta}}{\beta}. \tag{3.14}$$

Definition 18 Let f be a continuous function, let $0 < \alpha \leq 1$ and $\beta > 0$, then the fractal-fractional derivative of f with fractional order α and fractal order dimension β with exponential decay kernel is given as

$$_{0}^{FFP}D_{t,R}^{\alpha,\beta}f(t) = \frac{M(\alpha)}{1-\alpha}\frac{d}{dt^{\beta}}\int_{0}^{t}f(\tau)\exp\left[\frac{-\alpha}{1-\alpha}(t-\tau)\right]d\tau \tag{3.15}$$

and

$$_{0}^{FFP}D_{t,C}^{\alpha,\beta}f(t) = \frac{M(\alpha)}{1-\alpha}\int_{0}^{t}\frac{d}{dt^{\beta}}f(\tau)\exp\left[\frac{-\alpha}{1-\alpha}(t-\tau)\right]d\tau \tag{3.16}$$

Remark 5 Since the integral is differentiable equation (3.15) can be reformulated as

$$_{0}^{FFP}D_{t,R}^{\alpha,\beta}f(t) = \frac{M(\alpha)}{1-\alpha}\frac{d}{dt}\int_{0}^{t}f(\tau)\exp\left[\frac{-\alpha}{1-\alpha}(t-\tau)\right]d\tau\frac{1}{\beta t^{\beta-1}} \tag{3.17}$$

and equation (3.16) can be reformulated as

$$_{0}^{FFP}D_{t,C}^{\alpha,\beta}f(t) = \frac{M(\alpha)}{1-\alpha}\int_{0}^{t}\frac{d}{dt}f(\tau)\exp\left[\frac{-\alpha}{1-\alpha}(t-\tau)\right]d\tau \tag{3.18}$$

when $\alpha = 1$, then

$$_{0}^{FFR}D_{t,R}^{1,\beta}f(t) = {}_{0}^{FFC}D_{t,C}^{1,\beta}f(t) = \frac{t^{1-\beta}}{\beta}f'(t) \tag{3.19}$$

However, when $\beta = 1$, we have

$$_{0}^{FFP}D_{t,R}^{\alpha,1}f(t) = \frac{M(\alpha)}{1-\alpha}\frac{d}{dt}\int_{0}^{t}f(\tau)\exp\left[\frac{-\alpha}{1-\alpha}(t-\tau)\right]d\tau\frac{1}{\beta t^{\beta-1}} \tag{3.20}$$

and

$$_{0}^{FFP}D_{t,C}^{\alpha,1}f(t) = \frac{M(\alpha)}{1-\alpha}\int_{0}^{t}\frac{d}{d\tau}f(\tau)\exp\left[\frac{-\alpha}{1-\alpha}(t-\tau)\right]d\tau. \tag{3.21}$$

3.2 NUMERICAL SCHEMES FOR FRACTAL-FRACTIONAL DERIVATIVE

We briefly describe the numerical methods used to compute the fractal-fractional differential operators. There are three examples given: the generalised Mittag-Leffler function, the fractal-fractional with power law kernel, and the fractal-fractional with exponential decay function. Numerical analysis will benefit from these findings.

3.2.1 NUMERICAL SCHEME FOR CAPUTO FRACTAL-FRACTIONAL MODEL

Here, we provide a broad description of a numerical method for the Caputo fractal-fractional derivative. Since the fractional integral is differentiable and the Voltera integral is used, the Caputo specified model can be represented in the sense of Riemann-Liouville (RL), as demonstrated by:

$$^{FFP}D_{0,t}^{\rho,\kappa}f(t) = \frac{1}{\Gamma(1-\rho)}\frac{d}{dt}\int_0^t (t-\kappa)^{-\rho}f(\kappa)d\kappa \frac{1}{\kappa t^{\kappa-1}}, \qquad (3.22)$$

then, the following is obtained,

$$^{RL}D_{0,t}^{\rho}\left(x_1(t)\right) = \kappa t^{\kappa-1}f_1(x_1,x_2,x_3),$$

$$^{RL}D_{0,t}^{\rho}\left(x_2(t)\right) = \kappa t^{\kappa-1}f_2(x_1,x_2,x_3),$$

$$^{RL}D_{0,t}^{\rho}\left(x_3(t)\right) = \kappa t^{\kappa-1}f_3(x_1,x_2,x_3). \qquad (3.23)$$

The following outcome is obtained by switching out the RL to Caputo derivative and then applying the RL fractional integral on both sides:

$$x_1(t) = \frac{\kappa}{\Gamma(\rho)}\int_0^t \lambda^{\kappa-1}(t-\lambda)^{\rho-1}f_1(x_1,x_2,x_3,\lambda)d\lambda,$$

$$x_2(t) = \frac{\kappa}{\Gamma(\rho)}\int_0^t \lambda^{\kappa-1}(t-\lambda)^{\rho-1}f_2(x_1,x_2,x_2,\lambda)d\lambda,$$

$$x_3(t) = \frac{\kappa}{\Gamma(\rho)}\int_0^t \lambda^{\kappa-1}(t-\lambda)^{\rho-1}f_3(x_1,x_2,x_2,\lambda)d\lambda, \qquad (3.24)$$

where the functions f_1, f_2, and f_3 represent generally a system with three variables x_1, x_2, and x_3. Further, we replace t by t_{n+1} and have the following,

$$x_1^{n+1} = \frac{\kappa}{\Gamma(\rho)}\int_0^t \lambda^{\kappa-1}(t_{n+1}-\lambda)^{\rho-1}f_1(x_1,x_2,x_3,\lambda)d\lambda,$$

$$x_2^{n+1} = \frac{\kappa}{\Gamma(\rho)}\int_0^t \lambda^{\kappa-1}(t_{n+1}-\lambda)^{\rho-1}f_2(x_1,x_2,x_3,\lambda)d\lambda,$$

$$x_3^{n+1} = \frac{\kappa}{\Gamma(\rho)}\int_0^t \lambda^{\kappa-1}(t_{n+1}-\lambda)^{\rho-1}f_3(x_1,x_2,x_3,\lambda)d\lambda. \qquad (3.25)$$

Further simplifying the expressions in (3.25) lead to the following:

$$x_1^{n+1} = \frac{\kappa}{\Gamma(\rho)}\sum_{j=0}^{n}\int_{t_j}^{t_{j+1}} \lambda^{\kappa-1}(t_{n+1}-\lambda)^{\rho-1}f_1(x_1,x_2,x_3,\lambda)d\lambda,$$

Definitions of Fractal-Fractional Operators with Numerical Approximations

$$x_2^{n+1} = \frac{\kappa}{\Gamma(\rho)} \sum_{j=0}^{n} \int_{t_j}^{t_{j+1}} \lambda^{\kappa-1}(t_{n+1}-\lambda)^{\rho-1} f_2(x_1,x_2,x_3,\lambda)d\lambda,$$

$$x_3^{n+1} = \frac{\kappa}{\Gamma(\rho)} \sum_{j=0}^{n} \int_{t_j}^{t_{j+1}} \lambda^{\kappa-1}(t_{n+1}-\lambda)^{\rho-1} f_2(x_1,x_2,x_3,\lambda)d\lambda, \quad (3.26)$$

The functions $\lambda^{\kappa-1} f_i(x_1,x_2,x_3,\lambda)$ for $i = 1,2,3$ in equation (3.26) are approximated in $[t_j,t_{j+1}]$ using the Lagrangian piece-wise interpolation, given by:

$$P_j(\lambda) = \frac{\lambda-t_{j-1}}{t_j-t_{j-1}} t_j^{\kappa-1} f_1(x_1^j,x_2^j,x_3^j,t_j) - \frac{\lambda-t_j}{t_j-t_{j-1}} t_{j-1}^{\kappa-1} f_1(x_1^{j-1},x_2^{j-1},x_3^{j-1},t_{j-1}),$$

$$Q_j(\lambda) = \frac{\lambda-t_{j-1}}{t_j-t_{j-1}} t_j^{\kappa-1} f_2(x_1^j,x_2^j,x_3^j,t_j) - \frac{\lambda-t_j}{t_j-t_{j-1}} t_{j-1}^{\kappa-1} f_2(x_1^{j-1},x_2^{j-1},x_3^{j-1},t_{j-1})$$

$$R_j(\lambda) = \frac{\lambda-t_{j-1}}{t_j-t_{j-1}} t_j^{\kappa-1} f_3(x_1^j,x_2^j,x_3^j,t_j) -$$
$$\frac{\lambda-t_j}{t_j-t_{j-1}} t_{j-1}^{\kappa-1} f_3(x_1^{j-1},x_2^{j-1},x_3^{j-1},t_{j-1}). \quad (3.27)$$

with the above, we have the following,

$$x_1^{n+1} = \frac{\kappa}{\Gamma(\rho)} \sum_{j=0}^{n} \int_{t_j}^{t_{j+1}} \lambda^{\kappa-1}(t_{n+1}-\lambda)^{\rho-1} P_j(\lambda) d\lambda,$$

$$x_2^{n+1} = \frac{\kappa}{\Gamma(\rho)} \sum_{j=0}^{n} \int_{t_j}^{t_{j+1}} \lambda^{\kappa-1}(t_{n+1}-\lambda)^{\rho-1} Q_j(\lambda) d\lambda$$

$$x_3^{n+1} = \frac{\kappa}{\Gamma(\rho)} \sum_{j=0}^{n} \int_{t_j}^{t_{j+1}} \lambda^{\kappa-1}(t_{n+1}-\lambda)^{\rho-1} R_j(\lambda) d\lambda. \quad (3.28)$$

The results given in (3.28) has the final numerical solution, given by,

$$x_1^{n+1} = +\frac{\kappa h^\rho}{\Gamma(\rho+2)} \sum_{j=1}^{n} \Big[t_j^{\kappa-1} f_1(x_1^j,x_2^j,x_3^j,t_j)$$
$$\times \Big((n+1-j)^\rho(n-j+2+\rho) - (n-j)^\rho(n-j+2+2\rho) \Big)$$
$$- t_{j-1}^{\kappa-1} f_1(x_1^{j-1},x_2^{j-1},x_3^{j-1},t_{j-1})$$
$$\times \Big((n-j+1)^{\rho+1} - (n-j)^\rho(n-j+1+\rho) \Big) \Big],$$

$$x_2^{n+1} = +\frac{\kappa h^\rho}{\Gamma(\rho+2)} \sum_{j=1}^{n} \Big[t_j^{\kappa-1} f_2(x_1^j,x_2^j,x_3^j,t_j)$$
$$\times \Big((n+1-j)^\rho(n-j+2+\rho) - (n-j)^\rho(n-j+2+2\rho) \Big)$$
$$- t_{j-1}^{\kappa-1} f_2(x_1^{j-1},x_2^{j-1},x_3^{j-1},t_{j-1})$$

$$\times\left((n-j+1)^{\rho+1}-(n-j)^{\rho}(n-j+1+\rho)\right)\Big]$$

$$x_3^{n+1} = +\frac{\kappa h^{\rho}}{\Gamma(\rho+2)}\sum_{j=1}^{n}\left[t_j^{\kappa-1}f_3(x_1^j,x_2^j,x_3^j,t_j)\right.$$

$$\times\left((n+1-j)^{\rho}(n-j+2+\rho)-(n-j)^{\rho}(n-j+2+2\rho)\right)$$

$$-t_{j-1}^{\kappa-1}f_3(x_1^{j-1},x_2^{j-1},x_3^{j-1},t_{j-1})$$

$$\left.\times\left((n-j+1)^{\rho+1}-(n-j)^{\rho}(n-j+1+\rho)\right)\right]. \quad (3.29)$$

The scheme presented in (3.29) represents the numerical scheme for the fractal-fractional model in Caputo operator sense with the fractional order κ and the fractal ρ. Next, we describe a procedure to handle the numerical solution of fractal-fractional system composed in Caputo-Fabrizio derivative.

3.2.2 NUMERICAL SCHEME FOR CAPUTO-FABRIZIO FRACTAL-FRACTIONAL OPERATOR

In this subsection, we give a brief details of the scheme for the fractal-fractional Caputo-Fabrizio operator. A numerical approach for a model in three variables, say x_1, x_2, and x_3 based on the Adams-Bashforth method is shown below in the form:

$$^{CF}D_{0,t}^{\rho}\left(x_1(t)\right) = \kappa t^{\kappa-1}f_1(x_1,x_2,x_3,t),$$

$$^{CF}D_{0,t}^{\rho}\left(x_2(t)\right) = \kappa t^{\kappa-1}f_2(x_1,x_2,x_3,t)$$

$$^{CF}D_{0,t}^{\rho}\left(x_3(t)\right) = \kappa t^{\kappa-1}f_3(x_1,x_2,x_3,t). \quad (3.30)$$

On equation (3.30) by applying the CF integral leads to the expression below:

$$x_1(t) = \frac{\kappa t^{\kappa-1}(1-\rho)}{M(\rho)}f_1(x_1,x_2,x_3,t)+\frac{\rho\kappa}{M(\rho)}\int_0^t\lambda^{\kappa-1}f_1(x_1,x_2,x_3,\lambda)d\lambda,$$

$$x_2(t) = \frac{\kappa t^{\kappa-1}(1-\rho)}{M(\rho)}f_2(x_1,x_2,x_3,t)+\frac{\rho\kappa}{M(\rho)}\int_0^t\lambda^{\kappa-1}f_2(x_1,x_2,x_3,\lambda)d\lambda,$$

$$x_3(t) = \frac{\kappa t^{\kappa-1}(1-\rho)}{M(\rho)}f_3(x_1,x_2,x_3,t)$$

$$+\frac{\rho\kappa}{M(\rho)}\int_0^t\lambda^{\kappa-1}f_3(x_1,x_2,x_3,\lambda)d\lambda, \quad (3.31)$$

where f_1, f_2, and f_3 are generally functions associated to mathematical model of fractional differential equations.

When $t = t_{n+1}$, we obtain,

$$x_1^{n+1}(t) = x_1^n + \frac{\kappa t^{\kappa-1}(1-\rho)}{M(\rho)}f_1(x_1^n,x_2^n,x_3^n,t_n)$$

Definitions of Fractal-Fractional Operators with Numerical Approximations 37

$$+ \frac{\rho \kappa}{M(\rho)} \int_0^{t_{n+1}} \lambda^{\kappa-1} f_1(x_1, x_2, x_3, \lambda) d\lambda,$$

$$x_2^{n+1}(t) = x_2^n + \frac{\kappa t_n^{\kappa-1}(1-\rho)}{M(\rho)} f_2(x_1^n, x_2^n, x_3^n, t_n)$$

$$+ \frac{\rho \kappa}{M(\rho)} \int_0^{t_{n+1}} \lambda^{\kappa-1} f_2(x_1, x_2, x_3, \lambda) d\lambda,$$

$$x_3^{n+1}(t) = x_3^n + \frac{\kappa t_n^{\kappa-1}(1-\rho)}{M(\rho)} f_3(x_1^n, x_2^n, x_3^n, t_n)$$

$$+ \frac{\rho \kappa}{M(\rho)} \int_0^{t_{n+1}} \lambda^{\kappa-1} f_3(x_1, x_2, x_3, \lambda) d\lambda. \quad (3.32)$$

Furthermore, we have

$$x_1^{n+1}(t) = x_1^n + \frac{\kappa t_n^{\kappa-1}(1-\rho)}{M(\rho)} f_1(x_1^n, x_2^n, x_3^n, t_n) - \frac{\kappa t_{n-1}^{\kappa-1}(1-\rho)}{M(\rho)}$$

$$\times f_1(x_1^{n-1}, x_2^{n-1}, x_3^{n-1}, t_{n-1})$$

$$+ \frac{\rho \kappa}{M(\rho)} \int_{t_n}^{t_{n+1}} \lambda^{\kappa-1} f_1(x_1, x_2, x_3, \lambda) d\lambda,$$

$$x_2^{n+1}(t) = x_2^n + \frac{\kappa t_n^{\kappa-1}(1-\rho)}{M(\rho)}$$

$$\times f_2(x_1^n, x_2^n, x_3^n, t_n) - \frac{\kappa t_{n-1}^{\kappa-1}(1-\rho)}{M(\rho)} f_2(x_1^{n-1}, x_2^{n-1}, x_3^{n-1}, t_{n-1})$$

$$+ \frac{\rho \kappa}{M(\rho)} \int_{t_n}^{t_{n+1}} \lambda^{\kappa-1} f_2(x_1, x_2, x_3, \lambda) d\lambda,$$

$$x_3^{n+1}(t) = x_3^n + \frac{\kappa t_n^{\kappa-1}(1-\rho)}{M(\rho)}$$

$$\times f_3(x_1^n, x_2^n, x_3^n, t_n) - \frac{\kappa t_{n-1}^{\kappa-1}(1-\rho)}{M(\rho)} f_3(x_1^{n-1}, x_2^{n-1}, x_3^{n-1}, t_{n-1})$$

$$+ \frac{\rho \kappa}{M(\rho)} \int_{t_n}^{t_{n+1}} \lambda^{\kappa-1} f_3(x_1, x_2, x_3, \lambda) d\lambda. \quad (3.33)$$

Simplifying the integrals in the above equations leads to the following system,

$$x_1^{n+1} = x_1^n + \frac{\kappa t_n^{\kappa-1}(1-\rho)}{M(\rho)} f_1(x_1^n, x_2^n, x_3^n, t_n) - \frac{\kappa t_{n-1}^{\kappa-1}(1-\rho)}{M(\rho)}$$

$$\times f_1(x_1^{n-1}, x_2^{n-1}, x_3^{n-1}, t_{n-1})$$

$$+ \frac{\rho \kappa}{M(\rho)} \left[\frac{3h}{2} t_n^{\kappa-1} f_1(x_1^n, x_2^n, x_3^n, t_n) - \frac{h}{2} t_{n-1}^{\kappa-1} f_1(x_1^{n-1}, x_2^{n-1}, x_3^{n-1}, t_{n-1}) \right],$$

$$x_2^{n+1} = x_2^n + \frac{\kappa t_n^{\kappa-1}(1-\rho)}{M(\rho)}$$
$$\times f_2(x_1^n, x_2^n, x_3^n, t_n) - \frac{\kappa t_{n-1}^{\kappa-1}(1-\rho)}{M(\rho)} f_2(x_1^{n-1}, x_2^{n-1}, x_3^{n-1}, t_{n-1})$$
$$+ \frac{\rho \kappa}{M(\rho)} \left[\frac{3h}{2} t_n^{\kappa-1} f_2(x_1^n, x_2^n, x_3^n, t_n) - \frac{h}{2} t_{n-1}^{\kappa-1} f_2(x_1^{n-1}, x_2^{n-1}, x_3^{n-1}, t_{n-1}) \right],$$

$$x_3^{n+1} = x_3^n + \frac{\kappa t_n^{\kappa-1}(1-\rho)}{M(\rho)}$$
$$\times f_3(x_1^n, x_2^n, x_3^n, t_n) - \frac{\kappa t_{n-1}^{\kappa-1}(1-\rho)}{M(\rho)} f_3(x_1^{n-1}, x_2^{n-1}, x_3^{n-1}, t_{n-1})$$
$$+ \frac{\rho \kappa}{M(\rho)} \left[\frac{3h}{2} t_n^{\kappa-1} f_3(x_1^n, x_2^n, x_3^n, t_n) \right.$$
$$\left. - \frac{h}{2} t_{n-1}^{\kappa-1} f_3(x_1^{n-1}, x_2^{n-1}, x_3^{n-1}, t_{n-1}) \right]. \quad (3.34)$$

Simplifications of (3.34) further, we have

$$x_1^{n+1} = x_1^n + \kappa t_n^{\kappa-1} \left[\frac{1-\rho}{M(\rho)} + \frac{3\rho h}{2M(\rho)} \right] f_1(x_1^n, x_2^n, x_3^n, t_n)$$
$$- \kappa t_{n-1}^{\kappa-1} \left[\frac{1-\rho}{M(\rho)} + \frac{\rho h}{2M(\rho)} \right] f_1(x_1^{n-1}, x_2^{n-1}, x_3^{n-1}, t_{n-1}),$$
$$x_2^{n+1} = x_2^n + \kappa t_n^{\kappa-1} \left[\frac{1-\rho}{M(\rho)} + \frac{3\rho h}{2M(\rho)} \right] f_2(x_1^n, x_2^n, x_3^n, t_n)$$
$$- \kappa t_{n-1}^{\kappa-1} \left[\frac{1-\rho}{M(\rho)} + \frac{\rho h}{2M(\rho)} \right] f_2(x_1^{n-1}, x_2^{n-1}, x_3^{n-1}, t_{n-1}),$$
$$x_3^{n+1} = x_3^n + \kappa t_n^{\kappa-1} \left[\frac{1-\rho}{M(\rho)} + \frac{3\rho h}{2M(\rho)} \right] f_3(x_1^n, x_2^n, x_3^n, t_n)$$
$$- \kappa t_{n-1}^{\kappa-1} \left[\frac{1-\rho}{M(\rho)} + \frac{\rho h}{2M(\rho)} \right] f_3(x_1^{n-1}, x_2^{n-1}, x_3^{n-1}, t_{n-1}). \quad (3.35)$$

Next, we provide a brief details of fractal-fractional numerical scheme for the Atangana-Baleanu operator.

3.2.3 NUMERICAL SCHEME FOR ATANGANA-BALEANU FRACTAL-FRACTIONAL OPERATOR

We formulate the following generic system of fractal-fractional differential equations in order to offer a numerical technique for the resolution of a fractal-fractional model in the Atangana-Baleanu operator:

$$^{FF}D_{0,t}^{\rho,\kappa}\left(x_1(t)\right) = f_1(x_1, x_2, x_3),$$
$$^{FF}D_{0,t}^{\rho,\kappa}\left(x_2(t)\right) = f_2(x_1, x_2, x_3),$$

Definitions of Fractal-Fractional Operators with Numerical Approximations

$$^{FF}D_{0,t}^{\rho,\kappa}\left(x_3(t)\right) = f_3(x_1,x_2,x_3). \tag{3.36}$$

We write the system (3.36), in the form below,

$$^{ABR}D_{0,t}^{\rho}\left(x_1(t)\right) = \kappa t^{\kappa-1} f_1(x_1,x_2,x_3,t),$$

$$^{ABR}D_{0,t}^{\rho}\left(x_1(t)\right) = \kappa t^{\kappa-1} f_2(x_1,x_2,x_3,t),$$

$$^{ABR}D_{0,t}^{\rho}\left(x_1(t)\right) = \kappa t^{\kappa-1} f_3(x_1,x_2,x_3,t). \tag{3.37}$$

We obtain by using the Atangana-Baleanu integral,

$$x_1(t) = \frac{\kappa t^{\kappa 1}(1-\rho)}{AB(\rho)} f_1(x_1,x_2,x_3,t)$$
$$+ \frac{\rho \kappa}{AB(\rho)\Gamma(\rho)} \int_0^t \lambda^{\kappa-1}(t-\lambda)^{\rho-1} f_1(x_1,x_2,x_3,\lambda) d\lambda,$$

$$x_2(t) = \frac{\kappa t^{\kappa 1}(1-\rho)}{AB(\rho)} f_2(x_1,x_2,x_3,t)$$
$$+ \frac{\rho \kappa}{AB(\rho)\Gamma(\rho)} \int_0^t \lambda^{\kappa-1}(t-\lambda)^{\rho-1} f_2(x_1,x_2,x_3,\lambda) d\lambda,$$

$$x_3(t) = \frac{\kappa t^{\kappa 1}(1-\rho)}{AB(\rho)} f_3(x_1,x_2,x_3,t)$$
$$+ \frac{\rho \kappa}{AB(\rho)\Gamma(\rho)} \int_0^t \lambda^{\kappa-1}(t-\lambda)^{\rho-1} f_3(x_1,x_2,x_3,\lambda) d\lambda. \tag{3.38}$$

When $t = t_{n+1}$, we get

$$x_1^{n+1} = \frac{\kappa t_n^{\kappa-1}(1-\rho)}{AB(\rho)} f_1(x_1^n,x_2^n,x_3^n,t_n)$$
$$+ \frac{\rho \kappa}{AB(\rho)\Gamma(\rho)} \int_0^{t_{n+1}} \lambda^{\kappa-1}(t_{n+1}-\lambda)^{\rho-1} f_1(x_1,x_2,x_3,\lambda) d\lambda,$$

$$x_2^{n+1} = \frac{\kappa t_n^{\kappa-1}(1-\rho)}{AB(\rho)} f_2(x_1^n,x_2^n,x_3^n,t_n)$$
$$+ \frac{\rho \kappa}{AB(\rho)\Gamma(\rho)} \int_0^{t_{n+1}} \lambda^{\kappa-1}(t_{n+1}-\lambda)^{\rho-1} f_2(x_1,x_2,x_3,\lambda) d\lambda,$$

$$x_3^{n+1} = \frac{\kappa t_n^{\kappa-1}(1-\rho)}{AB(\rho)} f_3(x_1^n,x_2^n,x_3^n,t_n)$$
$$+ \frac{\rho \kappa}{AB(\rho)\Gamma(\rho)} \int_0^{t_{n+1}} \lambda^{\kappa-1}(t_{n+1}-\lambda)^{\rho-1} f_3(x_1,x_2,x_3,\lambda) d\lambda. \tag{3.39}$$

Simplifying further the equations (3.39) and obtain the following:

$$x_1^{n+1} = \frac{\kappa t_n^{\kappa-1}(1-\rho)}{AB(\rho)} f_1(x_1^n,x_2^n,x_3^n,t_n)$$

$$+\frac{\rho\kappa}{AB(\rho)\Gamma(\rho)}\sum_{j=0}^{n}\int_{t_j}^{t_{j+1}}\lambda^{\kappa-1}(t_{n+1}-\lambda)^{\rho-1}f_1(x_1,x_2,x_3,\lambda)d\lambda,$$

$$x_2^{n+1}=\frac{\kappa t_n^{\kappa-1}(1-\rho)}{AB(\rho)}f_2(x_1^n,x_2^n,x_3^n,t_n)$$

$$+\frac{\rho\kappa}{AB(\rho)\Gamma(\rho)}\sum_{j=0}^{n}\int_{t_j}^{t_{j+1}}\lambda^{\kappa-1}(t_{n+1}-\lambda)^{\rho-1}f_2(x_1,x_2,x_3,\lambda)d\lambda,$$

$$x_3^{n+1}=\frac{\kappa t_n^{\kappa-1}(1-\rho)}{AB(\rho)}f_3(x_1^n,x_2^n,x_3^n,t_n)$$

$$+\frac{\rho\kappa}{AB(\rho)\Gamma(\rho)}\sum_{j=0}^{n}\int_{t_j}^{t_{j+1}}\lambda^{\kappa-1}(t_{n+1}-\lambda)^{\rho-1}f_3(x_1,x_2,x_3,\lambda)d\lambda.$$

(3.40)

Simplifications of the integrals leads finally to below numerical scheme:

$$x_1^{n+1}=\frac{\kappa t_n^{\kappa-1}(1-\rho)}{AB(\rho)}f_1(x_1^n,x_2^n,x_3^n,t_n)$$

$$+\frac{\kappa(\Delta t)^\rho}{AB(\rho)\Gamma(\rho+2)}\sum_{j=1}^{n}\left[t_j^{\kappa-1}f_1(x_1^j,x_2^j,x_3^j,t_j)\right.$$

$$\times\left((n+1-j)^\rho(n-j+2+\rho)-(n-j)^\rho(n-j+2+2\rho)\right)$$

$$-t_{j-1}^{\kappa-1}f_1(x_1^{j-1},x_2^{j-1},x_3^{j-1},t_{j-1})$$

$$\left.\times\left((n-j+1)^{\rho+1}-(n-j)^\rho(n-j+1+\rho)\right)\right],$$

$$x_2^{n+1}=\frac{\kappa t_n^{\kappa-1}(1-\rho)}{AB(\rho)}f_2(x_1^n,x_2^n,x_3^n,t_n)$$

$$+\frac{\kappa(\Delta t)^\rho}{AB(\rho)\Gamma(\rho+2)}\sum_{j=1}^{n}\left[t_j^{\kappa-1}f_2(x_1^j,x_2^j,x_3^j,t_j)\right.$$

$$\times\left((n+1-j)^\rho(n-j+2+\rho)-(n-j)^\rho(n-j+2+2\rho)\right)$$

$$-t_{j-1}^{\kappa-1}f_2(x_1^{j-1},x_2^{j-1},x_3^{j-1},t_{j-1})$$

$$\left.\times\left((n-j+1)^{\rho+1}-(n-j)^\rho(n-j+1+\rho)\right)\right],$$

$$x_3^{n+1}=\frac{\kappa t_n^{\kappa-1}(1-\rho)}{AB(\rho)}f_3(x_1^n,x_2^n,x_3^n,t_n)$$

$$+\frac{\kappa(\Delta t)^\rho}{AB(\rho)\Gamma(\rho+2)}\sum_{j=1}^{n}\left[t_j^{\kappa-1}f_3(x_1^j,x_2^j,x_3^j,t_j)\right.$$

$$\times\left((n+1-j)^\rho(n-j+2+\rho)-(n-j)^\rho(n-j+2+2\rho)\right)$$

$$-t_{j-1}^{\kappa-1} f_3(x_1^{j-1}, x_2^{j-1}, x_3^{j-1}, t_{j-1})$$
$$\times \left((n-j+1)^{\rho+1} - (n-j)^{\rho}(n-j+1+\rho) \right) \bigg]. \tag{3.41}$$

3.3 NUMERICAL SOLUTION OF FRACTIONAL DIFFERENTIAL EQUATIONS (FDES)

3.3.1 NUMERICAL SCHEMES FOR ATANGANA-BALEANU FDES

In general, we provide a numerical method for the Atangana-Baleanu derivative solution of a fractional differential equation. We look at the differential equations for fractions provided by

$$^{ABC}_{0}D^{\rho}x(t) = g(t, x(t)), \quad x(0) = x_0. \tag{3.42}$$

The equation (3.42) can be written in the form below by using the fundamental theorem of fractional calculus:

$$\begin{aligned} x(t) - x(0) &= \frac{(1-\rho)}{ABC(\rho)} g(t, x(t)) \\ &\quad + \frac{\rho}{AB(\rho)\Gamma(\rho)} \int_0^t g(\xi, x(\xi))(t-\xi)^{\rho-1} d\xi. \end{aligned} \tag{3.43}$$

Using $t = t_{n+1}$, $n = 0, 1, 2, \ldots,$ the following is shown

$$\begin{aligned} x(t_{n+1}) - x_0 &= \frac{1-\rho}{ABC(\rho)} g(t_n, x(t_n)) + \\ &\quad \frac{\rho}{ABC(\rho)\Gamma(\rho)} \int_0^{t_{n+1}} g(\xi, x(\xi))(t_{n+1}-\xi)^{\rho-1} d\xi, \\ &= \frac{1-\rho}{ABC(\rho)} g(t_n, x(t_n)) + \\ &\quad \frac{\rho}{ABC(\rho)\Gamma(\rho)} \sum_{k=0}^{n} \int_{t_k}^{t_{k+1}} g(\xi, x(\xi))(t_{n+1}-\xi)^{\rho-1} d\xi. \end{aligned} \tag{3.44}$$

The function $g(\xi, x(\xi))$ is approximated in the interval $[t_k, t_{k+1}]$ using the interpolation polynomial

$$\begin{aligned} P_k(\xi) &= \frac{\xi - t_{k-1}}{t_k - t_{k-1}} g(t_k, x(t_k)) - \frac{\xi - t_{k-1}}{t_k - t_{k-1}} g(t_{k-1}, x(t_{k-1})) \\ &= \frac{g(t_k, x(t_k))}{h}(\xi - t_{k-1}) - \frac{g(t_{k-1}, x(t_{k-1}))}{h}(\xi - t_k) \\ &\cong \frac{g(t_k, x_k)}{h}(\xi - t_{k-1}) - \frac{g(t_{k-1}, x_{k-1})}{h}(\xi - t_k), \end{aligned} \tag{3.45}$$

giving the following,

$$x_{n+1} = x_0 + \frac{1-\rho}{AB(\rho)} g(t_n, x(t_n)) +$$
$$\frac{\rho}{AB(\rho) \times \Gamma(\rho)} \sum_{k=0}^{n} \left(\frac{g(t_k, x_k)}{h} \int_{t_k}^{t_{k+1}} (\xi - t_{k-1})(t_{n+1} - \xi)^{\rho-1} d\xi \right.$$
$$\left. - \frac{g(t_{k-1}, x_{k-1})}{h} \int_{t_k}^{t_{k+1}} (\xi - t_k)(t_{n+1} - \xi)^{\rho-1} d\xi \right). \tag{3.46}$$

Now,

$$A_{\rho,1} = \int_{t_k}^{t_{k+1}} (\xi - t_{k-1})(t_{n+1} - \xi)^{\rho-1} d\xi, \tag{3.47}$$

and

$$A_{\rho,2} = \int_{t_k}^{t_{k+1}} (\xi - t_k)(t_{n+1} - \xi)^{\rho-1} d\xi. \tag{3.48}$$

Solution of these integrals is given by

$$A_{\rho,1} = h^{\rho+1} \frac{(n+1-k)^\rho (n-k+2+\rho) - (n-k)^\rho (n-k+2+2\rho)}{\rho(\rho+1)}, \tag{3.49}$$

$$A_{\rho,2} = h^{\rho+1} \frac{(n+1-k)^{\rho+1} - (n-k)^\rho (n-k+1+\rho)}{\rho(\rho+1)}. \tag{3.50}$$

The final approximate solution is given by,

$$x_{n+1} = x_0 + \frac{1-\rho}{ABC(\rho)} g(t_n, x(t_n)) + \frac{\rho}{ABC(\rho) \times \Gamma(\rho)} \sum_{k=1}^{n}$$
$$\left(\frac{h^\rho g(t_k, x_k)}{\Gamma(\rho+2)} ((n+1-k)^\rho (n-k+2+\rho) - (n-k)^\rho (n-k+2+2\rho)) \right.$$
$$\left. - \frac{h^\rho g(t_{k-1}, x_{k-1})}{\Gamma(\rho+2)} ((n+1-k)^{\rho+1} - (n-k)^\rho (n-k+1+\rho)) \right). \tag{3.51}$$

4 Error Analysis

4.1 INTRODUCTION

For various nonlinear Cauchy problems, we have presented a number of numerical algorithms, including differential operators ranging from power laws to exponential decay laws to extended Mittag-Leffler kernels and their expansions. We will provide a thorough explanation of the reasoning that supports the analysis of error in this part. The numerical error analysis of the numerical methodology for fractal-fractional issues will be the first step in the analysis. We want to be clear that no consistency or convergence analysis is presented in this section. In particular, when the fractal size is 1, the polynomial interpolations utilised to construct these schemes naturally lead to a stable and convergent scheme.

4.2 ERROR ANALYSIS FOR FRACTAL-FRACTIONAL RL CAUCHY PROBLEMS

In this section, we consider a general Cauchy problem where the derivative is that type of Caputo type.

$$_0^{FFP}D_t^{\alpha,\beta}y(t) = f(t,y(t)), \tag{4.1}$$

using the integral associated with this derivative yields

$$y(t) = \frac{\beta}{\Gamma(\alpha)} \int_0^t \tau^{\beta-1}(t-\tau)^{\alpha-1} f(\tau,y(\tau))d\tau. \tag{4.2}$$

Then at $t_{n+1} = (n+1)\Delta t$, we have

$$\begin{aligned} y(t_{n+1}) &= \frac{\beta}{\Gamma(\alpha)} \int_0^{t_{n+1}} \tau^{\beta-1}(t_{n+1}-\tau)^{\alpha-1} f(\tau,y(\tau))d\tau, \\ &= \frac{\beta}{\Gamma(\alpha)} \sum_{j=0}^n \int_{t_j}^{t_{j+1}} \tau^{\beta-1}(t_{n+1}-\tau)^{\alpha-1} f(\tau,y(\tau))d\tau, \end{aligned} \tag{4.3}$$

within the interval $[t_j, t_{j+1}]$, we approximate the function $F(\tau) = \tau^{\beta-1}f(t,y(t))$ using the Lagrange polynomial interpolation

$$F(\tau) = P_j(\tau) = \frac{\tau - t_{j-1}}{\Delta t} F(y^j, t_j) - \frac{\tau - t_j}{\Delta t} F(y^{j-1}, t_{j-1}). \tag{4.4}$$

By considering the error, we get

$$y(t_{n+1}) = \frac{\beta}{\Gamma(\alpha)} \sum_{j=0}^n \int_{t_j}^{t_{j+1}} (P_j(\tau) + R_j(\tau))(t_{n+1}-\tau)^{\alpha-1} \tau^{\beta-1} d\tau,$$

DOI: 10.1201/9781003359258-4

$$= \frac{\beta}{\Gamma(\alpha)} \sum_{j=0}^{n} \int_{t_j}^{t_{j+1}} P_j(\tau)(t_{n+1}-\tau)^{\alpha-1}\tau^{\beta-1}d\tau +$$
$$+ \frac{\beta}{\Gamma(\alpha)} \sum_{j=0}^{n} \int_{t_j}^{t_{j+1}} R_j(\tau)(t_{n+1}-\tau)^{\alpha-1}\tau^{\beta-1}d\tau. \quad (4.5)$$

$$R(\alpha,n,\beta) = \frac{\beta}{\Gamma(\alpha)} \sum_{j=0}^{n} \int_{t_j}^{t_{j+1}} R_j(\tau)(t_{n+1}-\tau)^{\alpha-1}d\tau,$$
$$= \frac{\beta}{\Gamma(\alpha)} \sum_{j=0}^{n} \int_{t_j}^{t_{j+1}} \left(\frac{(\tau-t_j)(\tau-t_{j-1})}{2!}\right) \frac{\partial^2}{\partial \tau^2}\left[F(\tau,y(\tau))\big|_{\tau=\gamma_\tau}\right]$$
$$\times (t_{n+1}-\tau)^{\alpha-1}d\tau,$$
$$= \frac{\beta}{\Gamma(\alpha)} \sum_{j=0}^{n} \int_{t_j}^{t_{j+1}} \left(\frac{(\tau-t_j)(\tau-t_{j-1})}{2!}\right) \frac{\partial^2}{\partial \tau^2}\left[\tau^{\beta-1}f(\tau,y(\tau))\big|_{\tau=\gamma_\tau}\right]$$
$$\times (t_{n+1}-\tau)^{\alpha-1}d\tau. \quad (4.6)$$

Now, we shall point out that the formula $(\tau-t_{j-1})(t_{n+1}-\tau)^{\alpha-1}$ is positive within the interval $[t_j, t_{j+1}]$, thus one can find $\gamma_j \in [t_j, t_{j+1}]$ verifying

$$R(\alpha,n,\beta) = \frac{\beta\alpha}{\Gamma(\alpha)} \sum_{j=0}^{n} \frac{\partial^2}{\partial \tau^2}\left[\tau^{\beta-1}f(\tau,y(\tau))\big|_{\tau=\gamma_j}\right]\frac{(\gamma_j-\tau_j)}{2}$$
$$\times \int_{t_j}^{t_{j+1}} (\tau-t_{j-1})(t_{n+1}-\tau)^{\alpha-1}d\tau. \quad (4.7)$$

$$R(\alpha,n,\beta) = \frac{\beta\alpha}{\Gamma(\alpha)} \sum_{j=0}^{n} \frac{\partial^2}{\partial \tau^2}\left[\tau^{\beta-1}f(\tau,y(\tau))\big|_{\tau=\gamma_j}\right]\frac{(\gamma_j-\tau_j)}{2}$$
$$\times \left\{\frac{(n+1-j)^\alpha(n-j+\alpha+2)-(n-j)^\alpha(n-j+2+2\alpha)}{\alpha(\alpha+1)}\right\},$$
$$= \frac{\beta\alpha}{\Gamma(\alpha+2)} \sum_{j=0}^{n} \frac{\partial^2}{\partial \tau^2}\left[\tau^{\beta-1}f(\tau,y(\tau))\big|_{\tau=\gamma_j}\right]\frac{(\gamma_j-\tau_j)}{2}$$
$$\times \left\{(n+1-j)^\alpha(n-j+\alpha+2)\right.$$
$$\left.-(n-j)^\alpha(n-j+2+2\alpha)\right\}, \quad (4.8)$$

where

$$\int_{t_j}^{t_{j+1}} (\tau-t_{j-1})(t_{n+1}-\tau)^{\alpha-1}d\tau = \left\{\frac{A_{1,n,j}-A_{2,n,j}}{\alpha(\alpha+1)}\right\}.$$

where $A_{1,n,j} = (n+1-j)^\alpha(n-j+\alpha+2)$, and $A_{2,n,j} = (n-j)^\alpha(n-j+2+2\alpha)$.

Error Analysis

Therefore, by applying the absolute value on both sides yields,

$$\|R(\alpha,n,\beta)\| \leq \frac{\beta\alpha(\Delta t)^{\alpha+2}}{2\Gamma(\alpha+2)} \max_{\tau \in [0, t_{n+1}]} \left\| \frac{\partial^2}{\partial \tau^2}\left[\tau^{\beta-1}f(\tau,y(\tau))\right] \right\|$$
$$\times \sum_{j=0}^{n} \left\{(n+1-j)^{\alpha}(n-j+\alpha+2) - (n-j)^{\alpha}(n-j+2+2\alpha)\right\}. \quad (4.9)$$

We know that

$$\left((n-j+1)^{\alpha} - \alpha(n-j)^{\alpha}\right) \leq ((n+1)^{\alpha} - \alpha n^{\alpha}), \quad (4.10)$$

also,

$$\sum_{j=0}^{n}(n-j+2+2\alpha) = \frac{n(n+2\alpha+4)}{2}. \quad (4.11)$$

So that

$$\|R(\alpha,n,\beta)\| < \frac{\beta\alpha(\Delta t)^{\alpha+2}}{2\Gamma(\alpha+2)} \max_{\tau \in [0, t_{n+1}]} \left\| \frac{\partial^2}{\partial \tau^2}\left[\tau^{\beta-1}f(\tau,y(\tau))\right] \right\|$$
$$\times \left\{(n+1)^{\alpha} - \alpha n^{\alpha}\right\} \frac{n(n+2\alpha+4)}{2}. \quad (4.12)$$

4.3 ERROR ANALYSIS FOR FRACTAL-FRACTIONAL CF CAUCHY PROBLEM

In this section, we consider a general Cauchy problem where the differential operator is the Caputo-Fabrizio fractional derivative:

$$_{0}^{FFE}D_{t}^{\alpha,\beta}y(t) = f(t,y(t)). \quad (4.13)$$

We first transform the above equation into integral equations as

$$y(t) = y(0) + \frac{(1-\alpha)}{M(\alpha)}\tau^{\beta-1}\beta f(t,y(t)) + \frac{\alpha\beta}{M(\alpha)}\int_{0}^{t}\tau^{\beta-1}f(\tau,y(\tau))d\tau. \quad (4.14)$$

At $t_{n+1} = (n+1)\Delta t$ and $t_n = n\Delta t$, we have

$$y(t_{n+1}) = y(t_n) + \frac{(1-\alpha)}{M(\alpha)}\left[t_n^{\beta-1}f(t_n,y(t_n)) - t_{n-1}^{\beta-1}f(t_n,y(t_n))\right]$$
$$+ \frac{\alpha\beta}{M(\alpha)}\int_{t_n}^{t_{n+1}}\tau^{\beta-1}f(\tau,y(\tau))d\tau. \quad (4.15)$$

For simplicity, we let

$$F(\tau, y(\tau)) = \tau^{\beta-1} f(\tau, y(\tau)).$$

So, we write

$$\begin{aligned} y(t_{n+1}) &= y(t_n) + \frac{(1-\alpha)}{M(\alpha)} \beta \left[t_n^{\beta-1} f(t_n, y(t_n)) - t_{n-1}^{\beta-1} f(t_{n-1}, y(t_n)) \right] \\ &\quad + \frac{\alpha\beta}{M(\alpha)} \int_{t_n}^{t_{n+1}} F(\tau, y(\tau)) d\tau. \end{aligned} \quad (4.16)$$

Here within the interval $[t_n, t_{n+1}]$, we approximate the function $F(\tau, y(\tau))$ using the Lagrange interpolation polynomial,

$$F(\tau, y(\tau)) = \frac{\tau - t_{j-1}}{\Delta t} F(y^n, t_n) - \frac{t - t_j}{\Delta t} F(y^{n-1}, t_{n-1}) + R_n(\tau). \quad (4.17)$$

Thus, the error can be evaluated as

$$R(n, \alpha, \beta) = \frac{\alpha\beta}{M(\alpha)} \int_{t_n}^{t_{n+1}} R_n(\tau) d\tau, \quad (4.18)$$

where

$$R_n(\tau) = \frac{(\tau - t_n)(\tau - t_{n-1})}{2!} \frac{\partial^2}{\partial \tau^2} \left(\tau^{\beta-1} f(\tau, y(\tau)) \Big|_{\tau=\gamma_\tau} \right). \quad (4.19)$$

Indeed an interpolation given t_n and t_{n+1}, therefore

$$\begin{aligned} \|R_n(n, \alpha, \beta)\| &= \frac{\alpha\beta}{M(\alpha)} \int_{t_n}^{t_{n+1}} \frac{(\tau - t_n)(\tau - t_{n-1})}{2!} \\ &\quad \times \frac{\partial^2}{\partial \tau^2} \left(\tau^{\beta-1} f(\tau, y(\tau)) \Big|_{\tau=\gamma_\tau} \right) d\tau, \end{aligned} \quad (4.20)$$

Therefore, the remainder is bounded as

$$\begin{aligned} \|R(n, \alpha, \beta)\| &\leq \frac{\alpha\beta}{2!M(\alpha)} \sup_{t_n \leq t \leq t_{n-1}} \frac{\partial^2}{\partial \tau^2} \left(\tau^{\beta-1} f(\tau, y(\tau)) \Big|_{\tau=\gamma_\tau} \right) \\ &\quad \times \int_{t_n}^{t_{n+1}} \frac{(\tau - t_n)(\tau - t_{n-1})}{2!} d\tau. \end{aligned} \quad (4.21)$$

4.4 ERROR ANALYSIS FOR FRACTAL-FRACTIONAL CAUCHY PROBLEM WITH MITTAG-LEFFLER KERNEL

In this section, we consider a general Cauchy problem where the differential operator is with generalised Mittag-Leffler Kernel.

$$_0^{FFM}D_t^{\alpha,\beta}y(t) = f(t,y(t)). \quad (4.22)$$

Applying the associated integral yields

$$y(t_n) = \frac{1-\alpha}{AB(\alpha)}\beta t^{\beta-1}f(t,y(t))$$
$$+ \frac{\alpha\beta}{AB(\alpha)\Gamma(\alpha)}\int_0^t \tau^{\beta-1}(t-\tau)^{\alpha-1}f(\tau,y(\tau))d\tau. \quad (4.23)$$

At $t = (n+1)\Delta t$, the above can be written as

$$y(t_{n+1}) = \frac{1-\alpha}{AB(\alpha)}\beta t_n^{\beta-1}f(t_n,y(t_n))$$
$$+ \frac{\alpha\beta}{AB(\alpha)\Gamma(\alpha)}\int_0^{t_{n+1}} \tau^{\beta-1}(t_{n+1}-\tau)^{\alpha-1}f(\tau,y(\tau))d\tau. \quad (4.24)$$

Some researchers asked why we have to put $f(t_n,y(t_n))$ inside of $f(t_{n+1},y(t_{n+1}))$, here we are avoiding implicit situation

$$y(t_{n+1}) = \frac{1-\alpha}{AB(\alpha)}\beta t_n^{\beta-1}f(t_n,y(t_n))$$
$$+ \frac{\alpha\beta}{AB(\alpha)\Gamma(\alpha)}\sum_{j=0}^n \int_{t_j}^{t_{j+1}} \tau^{\beta-1}(t_{n+1}-\tau)^{\alpha-1}$$
$$\times f(\tau,y(\tau))d\tau. \quad (4.25)$$

Again within the interval $[t_j, t_{j+1}]$, we make use of the Lagrange interpolation formula to approximate the function $F(\tau,y(\tau)) = \tau^{\beta-1}f(\tau,y(\tau))$. Thus, within this interval, $F(\tau,y(\tau)) = P_j(\tau) + R_j(\tau)$. Thus, the error is given as

$$R(\alpha,n,\beta,t) = \frac{\alpha\beta}{AB(\alpha)\Gamma(\alpha)}\sum_{j=0}^n \int_{t_j}^{t_{j+1}} R_j(\tau)(t_{n+1}-\tau)^{\alpha-1}d\tau,$$
$$= \frac{\alpha\beta}{AB(\alpha)\Gamma(\alpha)}\sum_{j=0}^n \int_{t_j}^{t_{j+1}} \frac{(\tau-t_j)(\tau-t_{j-1})}{2!}$$
$$\times \frac{\partial^2}{\partial \tau^2}\left[\tau^{\beta-1}f(\tau,y(\tau))\Big|_{\tau=\gamma_\tau}\right]$$
$$\times (t_{n+1}-\tau)^{\alpha-1}d\tau. \quad (4.26)$$

Therefore,

$$\|R(\alpha,n,\beta,t)\| = \frac{\alpha\beta}{AB(\alpha)\Gamma(\alpha)} \Bigg| \sum_{j=0}^{n} \int_{t_j}^{t_{j+1}} \frac{(\tau-t_j)(\tau-t_{j-1})}{2!}$$

$$\times \frac{\partial^2}{\partial \tau^2} \left[\tau^{\beta-1} f(\tau, y(\tau)) \Big|_{\tau=\gamma_\tau} \right]$$

$$\times (t_{n+1} - \tau)^{\alpha-1} d\tau \Bigg|,$$

$$\leq \frac{\alpha\beta}{AB(\alpha)\Gamma(\alpha)} \sum_{j=0}^{n} \int_{t_j}^{t_{j+1}} \frac{(\tau-t_j)(\tau-t_{j-1})}{2!}$$

$$\times \frac{\partial^2}{\partial \tau^2} \left[\tau^{\beta-1} f(\tau, y(\tau)) \Big|_{\tau=\gamma_\tau} \right]$$

$$\times (t_{n+1} - \tau)^{\alpha-1} d\tau,$$

$$\leq \frac{\alpha\beta}{AB(\alpha)\Gamma(\alpha)} \sum_{j=1}^{n} \int_{t_j}^{t_{j+1}} \frac{(\tau-t_j)(\tau-t_{j-1})}{2!}$$

$$\times \max_{0 \leq \tau \leq t_{n+1}} \left\{ \frac{\partial^2}{\partial \tau^2} \left[\tau^{\beta-1} f(\tau, y(\tau)) \Big|_{\tau=\gamma_\tau} \right] \right\} (t_{n+1} - \tau)^{\alpha-1},$$

$$\leq \frac{\alpha\beta}{AB(\alpha)\Gamma(\alpha)} \max_{0 \leq \tau \leq t_{n+1}} \left\{ \frac{\partial^2}{\partial \tau^2} \left[\tau^{\beta-1} f(\tau, y(\tau)) \Big|_{\tau=\gamma_\tau} \right] \right\}$$

$$\times \sum_{j=1}^{n} \int_{t_j}^{t_{j+1}} \frac{(\tau-t_j)(\tau-t_{j-1})}{2!} (t_{n+1} - \tau)^{\alpha-1} d\tau,$$

$$\leq \frac{\alpha\beta}{AB(\alpha)\Gamma(\alpha)} \max_{0 \leq \tau \leq t_{n+1}} \left\{ \frac{\partial^2}{\partial \tau^2} \left[\tau^{\beta-1} f(\tau, y(\tau)) \Big|_{\tau=\gamma_\tau} \right] \right\}$$

$$\times \sum_{j=1}^{n} \Bigg\{ \int_{t_j}^{t_{j+1}} \tau^2 (t_{n+1} - \tau)^{\alpha-1} d\tau + j(j-1)(\Delta t)^2$$

$$\times \int_{t_j}^{t_{j+1}} (t_{n+1} - \tau)^{\alpha-1} d\tau$$

$$+ (2j-1)\Delta t \int_{t_j}^{t_{j+1}} \tau(t_{n+1} - \tau)^{\alpha-1} d\tau \Bigg\},$$

$$\leq \frac{\alpha\beta}{AB(\alpha)\Gamma(\alpha)} \max_{0 \leq \tau \leq t_{n+1}} \left\{ \frac{\partial^2}{\partial \tau^2} \left[\tau^{\beta-1} f(\tau, y(\tau)) \Big|_{\tau=\gamma_\tau} \right] \right\}$$

$$\times \sum_{j=1}^{n} \Bigg\{ \int_{t_j}^{t_{j+1}} t_{n+1}^2 (t_{n+1} - \tau)^{\alpha-1} d\tau$$

$$+ j(j-1)(\Delta t)^2 \int_{t_j}^{t_{j+1}} (t_{n+1} - \tau)^{\alpha-1} d\tau$$

$$+ (2j-1)\Delta t \int_{t_j}^{t_{j+1}} \tau(t_{n+1} - \tau)^{\alpha-1} d\tau \Bigg\}. \qquad (4.27)$$

Error Analysis

Since $t_{n+1} > \tau$, then the above becomes

$$\|R(\alpha,n,\beta,t)\| \leq \frac{\alpha\beta}{AB(\alpha)\Gamma(\alpha)} \max_{0\leq t\leq t_{n+1}} \left\{\frac{\partial^2}{\partial \tau^2}\left[\tau^{\beta-1}f(\tau,y(\tau))\right]\bigg|_{\tau=\gamma_\tau}\right\}$$
$$\times \sum_{j=1}^{n} \left\{(n+1)^2(\Delta t)^2 + j(j-1)(\Delta t)^2 \right.$$
$$\left. + (2j-1)(\Delta t)^2(n+1)\right\} \int_{t_j}^{t_{j+1}} (t_{n+1}-\tau)^{\alpha-1} d\tau, \quad (4.28)$$

whereas $(t_{n+1}-\tau)^{\alpha-1}$ is positive and $t_j > 0$, therefore

$$\int_0^{t_{j+1}} (t_{n+1}-\tau)^{\alpha-1} d\tau > \int_{t_j}^{t_{j+1}} (t_{n+1}-\tau)^{\alpha-1} d\tau. \quad (4.29)$$

It follows that

$$\|R(\alpha,n,\beta,t)\| \leq \frac{\alpha\beta}{AB(\alpha)\Gamma(\alpha)} \max_{0\leq t\leq t_{n+1}} \left\{\frac{\partial^2}{\partial \tau^2}\left[\tau^{\beta-1}f(\tau,y(\tau))\right]\bigg|_{\tau=\gamma_\tau}\right\}$$
$$\times \sum_{j=1}^{n} \left\{(n+1)^2(\Delta t)^2 + j(j-1)(\Delta t)^2 \right.$$
$$\left. + (2j-1)(\Delta t)^2(n+1)\right\} \int_{t_j}^{t_{j+1}} (t_{n+1}-\tau)^{\alpha-1} d\tau,$$
$$\leq \frac{\alpha\beta(\Delta t)^{2+\alpha}}{AB(\alpha)\Gamma(\alpha+1)} \max_{0\leq t\leq t_{n+1}} \left\{\frac{\partial^2}{\partial \tau^2}\left[\tau^{\beta-1}f(\tau,y(\tau))\right]\bigg|_{\tau=\gamma_\tau}\right\}$$
$$\times \sum_{j=1}^{n} \left\{(n+1)^2 + j(j-1) + (2j-1)(n+1)\right\}$$
$$(n-j), \quad (4.30)$$

whereas $\forall j \in \{1,...,n\}$, $n > n-j$, thus

$$\|R(\alpha,n,\beta,t)\| \leq \frac{\alpha\beta(\Delta t)^{2+\alpha}}{AB(\alpha)\Gamma(\alpha+1)} \max_{0\leq t\leq t_{n+1}} \left\{\frac{\partial^2}{\partial \tau^2}\left[\tau^{\beta-1}f(\tau,y(\tau))\right]\bigg|_{\tau=\gamma_\tau}\right\}$$
$$\times \sum_{j=1}^{n} \left\{n(n+1)^2 + \frac{1}{3}n^3 + \frac{1}{2}n^2 + \frac{n}{6} - \frac{1}{2}n^2 \right.$$
$$\left. -\frac{1}{2}n + (n^2+n-n+1)n\right\},$$
$$\leq \frac{\alpha\beta(\Delta t)^{2+\alpha}}{AB(\alpha)\Gamma(\alpha+1)} \max_{0\leq t\leq t_{n+1}} \left\{\frac{\partial^2}{\partial \tau^2}\left[\tau^{\beta-1}f(t,y(t))\right]\bigg|_{t=\gamma_t}\right\}$$
$$\left\{2n^3 + 3n^2 + \frac{5n}{3} + 1\right\}. \quad (4.31)$$

5 Existence and Uniqueness of Fractal Fractional Differential Equations

5.1 INTRODUCTION

Any differential or integral equation becomes extremely complex when using the fractal-fractional differential and integral operators. Compared to well-known fractional and fractal differential and integral operators, they offer more memory. Novel classes of ordinary differential equations were created as a result of these new differential and integral operators. It should be noted that ordinary differential equations appear frequently in the social, natural, and mathematical sciences. Differentials and derivatives are used in mathematical representations of change. Differential equations, which explain dynamically changing events, evolution, and variation, are created when various differentials, derivatives, and functions are coupled by equations. Frequently, the definition of a quantity is the rate of change of another quantity. Geometry and analytical mechanics are examples of specific mathematical disciplines. Physicists and astronomers make up a large portion of the scientific fields. These differential equations are sometimes nonlinear, thus, including the fractal-fractional differential operators render these equations more complex and nonlinear. We shall note, however, that, a nonlinear system is one in which the change in the output is not proportional to the change in the input in mathematics and science. Because most systems are intrinsically nonlinear, nonlinear problems are of interest to engineers, biologists, physicists, mathematicians, and many other scientists. In contrast to considerably simpler linear systems, nonlinear dynamical systems, which describe changes in variables over time, may appear chaotic, unexpected, or illogical. A nonlinear system of equations, which is a collection of simultaneous equations in which the unknowns appear as variables of a polynomial of degree greater than one or in the argument of a function that is not a polynomial of degree one, is typically used to describe the behaviour of a nonlinear system in mathematics. Nonlinear systems are frequently approximated by linear equations because nonlinear dynamical equations are challenging to solve. This works well up to a certain accuracy and range for the input values, but linearisation hides several intriguing phenomena like solitons, chaos, and singularities. As a result, some features of a nonlinear system's dynamic behaviour may seem illogical, unpredictable, or even chaotic. Despite the fact that such chaotic activity may seem random, it is not. For instance, some meteorological phenomena are thought to be chaotic, with minor changes in one area of the system leading to complex impacts everywhere else. With existing technology, it is hard to make precise long-term forecasts because of this nonlinearity. Thus researchers mostly rely on numerical schemes to provide numerical solutions to these equations. However, they at least show that these equations admit unique solutions under some conditions which are established using fixed point theories. For example, split feasibility issues, variational inequality issues, nonlinear optimisation issues, equilibrium issues, complementarity issues, selection, and matching issues, and issues proving the existence of solutions to integral and differential equations can all be resolved using the fundamental tools provided by fixed point theory. For single-valued or set-valued mappings of abstract metric spaces, fixed point theorems are given. The fixed-point theorems for set-valued mappings are particularly useful in optimal control theory and have been widely applied to numerous issues in game theory and economics. The aforementioned equation does not, however, inevitably have a fixed point if F is not self-mapping. In this chapter using some conditions, we present a detailed analysis of the uniqueness and existing solutions of fractal-fractional ordinary differential equations. This chapter discusses the existence and uniqueness of the fractal-fractional ordinary differential equations. We focus to study the fractal-fractional operators in the sense of Caputo, Caputo-Fabrizio, and the Atangana-Baleanu derivative.

5.2   EXISTENCE AND UNIQUENESS FOR POWER LAW CASE

In this case, we consider the following ordinary differential equation

$$\begin{aligned}{}^{FFP}_{0}D^{\alpha,\beta}_{t} y(t) &= f(t,y(t)), \text{ if } t>0, \\ y(0) &= y_0, \text{ if } t=0. \end{aligned} \quad (5.1)$$

The following conditions are suggested:

1. $f(t,y(t))$ is bounded
2. $\forall t \in D_f$, $|f(t,y(t))|^2 < K(1+|y|^2)$
3. $\forall y_1, y_2 \in C^1[0,T]$, $|f(t,y_1(t)) - f(t,y_2(t))|^2 < \overline{K}|y_1 - y_2|^2$

Our system of equations is converted to an integral equation,

$$\begin{aligned}{}^{RL}_{0}D^{\alpha,\beta}_{t} y(t) &= \beta t^{\beta-1} f(t,y(t)) \text{ if } t>0 \\ y(0) &= y_0 \text{ if } t=0 \end{aligned}$$

Later, we apply the Riemann-Liouville integral to have

$$y(t) = \frac{\beta}{\Gamma(\alpha)} \int_0^t \tau^{\beta-1} f(\tau,y(\tau))(t-\tau)^{\alpha-1} d\tau, \ y(0) = y_0.$$

We construct the following mapping

$$\Pi y(t) = \frac{\beta}{\Gamma(\alpha)} \int_0^t \tau^{\beta-1} f(\tau,y(\tau))(t-\tau)^{\alpha-1} d\tau.$$

Existence and Uniqueness of Fractal Fractional Differential Equations

We first show that $\Pi y(t)$ is bounded

$$|\Pi y(t)| = \frac{\beta}{\Gamma(\alpha)}\left|\int_0^t \tau^{\beta-1} f(\tau,y(\tau))(t-\tau)^{\alpha-1} d\tau\right|.$$

Using the Holder inequality, we have

$$\begin{aligned}
|\Pi y(t)| &= \frac{\beta}{\Gamma(\alpha)}\left|\int_0^t \tau^{\beta-1}(t-\tau)^{\alpha-1} d\tau \int_0^t |f(\tau,y(\tau))| d\tau,\right. \\
&\leq \frac{\beta}{\Gamma(\alpha)} t^{\beta+\alpha-1} B(\beta,\alpha) \int_0^t |f(\tau,y(\tau))| d\tau,
\end{aligned}$$

$$|\Pi y(t)| \leq \sup_{t \in D_y} |y(t)| \leq \frac{\beta}{\Gamma(\alpha)} t^{\beta+\alpha-1} B(\beta,\alpha) \int_0^t \sup_{l \in [0,T]} |f(l,y(l))| d\tau$$

Since $f(t,y(t))$ is bounded, we have $|f(l,y(l))| < M$. Thus,

$$\begin{aligned}
|\Pi y(t)| &\leq \|\|\Pi y\|\|_\infty \leq \frac{\beta}{\Gamma(\alpha)} T^{\beta+\alpha-1} B(\alpha,\beta) M T, \\
&\leq \frac{\beta T^{\beta+\alpha} \Gamma(\alpha)\Gamma(\beta) M}{\Gamma(\alpha)\Gamma(\alpha+\beta)}, \\
&< \frac{M T^{\beta+\alpha} \Gamma(\beta+1)}{\Gamma(\alpha+\beta)}.
\end{aligned}$$

Therefore, the mapping is bounded. We now evaluate the linear growth of the following mapping:

$$|\Pi y(t)|^2 = \frac{\beta}{\Gamma(\alpha)} \int_0^t (t-\tau)^{\alpha-1} \tau^{\beta-1} f(\tau,y(\tau)) d\tau.$$

At this point, we shall use the Cauchy inequality to obtain

$$|\Pi y(t)|^2 < \frac{2\beta^2}{\Gamma^2(\alpha)}\left|\int_0^t \tau^{\beta-1}(t-\tau)^{\alpha-1}(f(\tau,y_1(\tau)) - f(\tau,y_2(\tau))) d\tau\right|^2.$$

Applying the Cauchy inequality leads to

$$\begin{aligned}
|\Pi y_1 - \Pi y_2|^2 &\leq \frac{2\beta^2}{\Gamma^2(\alpha)} \int_0^t \tau^{2\beta-2}(t-\tau)^{2\alpha-2} d\tau \int_0^t \left|f(\tau,y_1(\tau)) - f(\tau,y_2(\tau))\right| d\tau, \\
&\leq \frac{2\beta^2}{\Gamma^2(\alpha)} t^{2\beta+2\alpha-3} B(2\beta-1, 2\alpha-1) \overline{K} \int_0^t |y_1(\tau) - y_2(\tau)|^2 d\tau, \\
&\leq \frac{2\beta^2 T^{2\beta+2\alpha-2}}{\Gamma^2(\alpha)} B(2\beta-1, 2\alpha-1) \overline{K} \|y_1 - y_2\|_\infty^2, \\
&< \overline{K}_1 \|y_1 - y_2\|_\infty^2,
\end{aligned}$$

$$\overline{K}_1 = \frac{2\beta^2 T^{2\beta+2\alpha-2}}{\Gamma^2(\alpha)} \overline{K} B(2\alpha-1, 2\beta-1).$$

This shows that mapping also verifies the Lipschitz condition. Conclusion is that the mapping admits a unique solution; thus under the conditions the Cauchy problem also admits a unique solution.

5.3 EXISTENCE AND UNIQUENESS FOR MITTAG-LEFFLER CASE

We consider in this section the following fractal-fractional differential equation:

$$_0^{FFM}D_t^{\alpha,\beta}y(t) = f(t,y(t)), \text{ if } t > 0,$$

$$y(0) = y_0, \text{ if } t = 0.$$

$$_0^{FFM}D_t^{\alpha,\beta}y(t) = \beta t^{\beta-1}f(t,y(t)), \text{ if } t > 0,$$

$$y(0) = y_0, \text{ if } t = 0.$$

Applying the AB-integral yields

$$y(t) = (1-\alpha)\beta t^{\beta-1}f(t,y(t))$$
$$+ \frac{\alpha\beta}{\Gamma(\alpha)}\int_0^t \tau^{\beta-1}f(\tau,y(\tau))(t-\tau)^{\alpha-1}d\tau, \ y(0) = y_0.$$

We define the following mapping:

$$\Pi y(t) = (1-\alpha)\beta t^{\beta-1}f(t,y(t))$$
$$+ \frac{\alpha\beta}{\Gamma(\alpha)}\int_0^t \tau^{\beta-1}(t-\tau)^{\alpha-1}f(\tau,y(\tau))d\tau.$$

Since t^β is a positive function such that for all $t \in (0,1]$ $t^{\beta-1} \leq t^\beta$, then

$$\Pi y(t) \leq (1-\alpha)\beta t^\beta f(t,y(t)) + \frac{\alpha\beta}{\Gamma(\alpha)}\int_0^t \tau^{\beta-1}(t-\tau)^{\alpha-1}f(\tau,y(\tau))d\tau.$$

If we assume that the function $f(t,y(t))$ is a non-decreasing function, then under this condition using the Gronwall inequality, we obtain

$$\Pi y(t) \leq (1-\alpha)\beta t^\beta f(t,y(t)) exp\left(\int_0^t \frac{\alpha\beta}{\Gamma(\alpha)} \tau^{\beta-1}(t-\tau)^{\alpha-1}f(\tau,y(\tau))d\tau\right).$$

Owing the fact that $f(\tau,y(\tau))$ is bounded, we have in addition that

$$\Pi y(t) < (1-\alpha)\beta T^\beta f(T,y(T)) exp\left(M\frac{\alpha\beta}{\Gamma(\alpha)} T^{\alpha+\beta-1} B(\alpha,\beta)\right).$$

Existence and Uniqueness of Fractal Fractional Differential Equations

This shows Π is bounded under the prescribed conditions. Beyond the prescribed conditions, we have

$$|\Pi y(t)|^2 = \left|(1-\alpha)\beta t^{\beta-1} f(t,y(t))\right.$$
$$\left. + \frac{\alpha\beta}{\Gamma(\alpha)} \int_0^t \tau^{\beta-1}(t-\tau)^{\alpha-1} f(\tau,y(\tau)) d\tau\right|^2,$$
$$< 2(1-\alpha)^2 \beta^2 t^{2\beta-2} |f(t,y(t))|^2$$
$$+ \frac{2\alpha^2\beta^2}{\Gamma^2(\alpha)} \left|\int_0^t \tau^{\beta-1}(t-\tau)^{\alpha-1} f(\tau,y(\tau)) d\tau\right|^2,$$
$$< 2(1-\alpha)^2 \beta^2 t^{2\beta-2} K(1+|y|^2)$$
$$+ \frac{\psi\alpha^2\beta^2}{\Gamma^2(\alpha)} t^{2\beta+2\alpha-2} B(2\alpha-1, 2\beta-1) K(1+|y|^2),$$
$$< \left\{ 2(1-\alpha)^2 \beta^2 T^{2\beta-2} K \right.$$
$$\left. + \frac{\psi\alpha^2\beta^2}{\Gamma^2(\alpha)} T^{2\beta+2\alpha-2} B(2\alpha-1, 2\beta-1) \right\} (1+\|y\|^2).$$

$$|\Pi y(t)|^2 \leq \|\Pi y\|_\infty^2 \leq \overline{K}(1+\|y\|_\infty^2).$$

This shows that the mapping verifies the growth linear condition, so we shall prove the Liptschits condition also

$$|\Pi y_1 - \Pi y_2|^2 \leq 2(1-\alpha)^2 \beta^2 t^{2\beta-2} \left|f(t,y_1) - f(t,y_2)\right|^2$$
$$+ \frac{2\beta^2\alpha^2}{\Gamma^2(\alpha)} K_1 T^{2\beta-2+2\alpha} B(2\beta-1, 2\alpha-1) \|y_1 - y_2\|_\infty^2.$$

$$\|\Pi y_1 - \Pi y_2\|_\infty^2 \leq \left(2(1-\alpha)^2 \beta^2 T^{2\beta-2} K_1\right.$$
$$\left. + \frac{2\beta^2\alpha^2}{\Gamma^2(\alpha)} K_1 T^{2\beta+2\alpha-2} B(2\beta-1, 2\alpha-1)\right) \|y_1 - y_2\|_\infty^2,$$
$$\leq \overline{K}_1 \|y_1 - y_2\|_\infty^2.$$

where

$$\overline{K}_1 = 2(1-\alpha)^2 \beta^2 T^{2\beta-2} K_1 + \frac{2\beta^2\alpha^2}{\Gamma^2(\alpha)} K_1 T^{2\beta+2\alpha-2} B(2\beta-1, 2\alpha-1)).$$

With the above result, we conclude that the mapping has a unique solution which implies that the Cauchy problem has a unique solution.

5.4 EXISTENCE AND UNIQUENESS FOR EXPONENTIAL CASE

In this section, we consider the following fractal-fractional ordinary differential equation,

$$\begin{aligned}{}^{FFE}_0 D_t^{\alpha,\beta} y(t) &= f(t,y(t)), \text{ if } t > 0, \\ y(0) &= y_0, \text{ if } t = 0. \end{aligned} \quad (5.2)$$

$$\begin{aligned}{}^{CF}_0 D_t^{\alpha,\beta} y(t) &= \beta t^{\beta-1} f(t,y(t)), \text{ if } t > 0, \\ y(0) &= y_0, \text{ if } t = 0. \end{aligned} \quad (5.3)$$

Applying the Caputo-Fabrizio integral yields

$$y(t) = (1-\alpha)\beta t^{\beta-1} f(t,y(t)) + \alpha\beta \int_0^t \tau^{\beta-1} f(\tau,y(\tau))d\tau, \text{ if } t > 0,$$

$$y(0) = y_0, \text{ if } t = 0.$$

We define the following mapping:

$$\Pi y(t) = (1-\alpha)\beta t^{\beta-1} f(t,y(t)) + \alpha\beta \int_0^t \tau^{\beta-1} f(\tau,y(\tau))d\tau.$$

First, we assume that if the function $f(t,y(t))$ is increasing, then we have

$$\Pi y(t) \leq (1-\alpha)(\beta+1) t^{\beta-1} f(t,y(t)) + \alpha\beta \int_0^t \tau^{\beta-1} f(\tau,y(\tau))d\tau.$$

$$|\Pi y(t)|^2 \leq 2(1-\alpha)^2 (\beta+1)^2 t^{2\beta-2} K(1+|y|^2) + 2\alpha^2 \beta^2 K \int_0^t (1+|y|^2)d\tau.$$

$$\begin{aligned} \sup_{t \in [0,T]} |\Pi y(t)|^2 &\leq \left(2(1-\alpha)^2(\beta+1)^2 T^{2\beta-2} K + 2\alpha^2 \beta^2 T K\right)(1+|y|_\infty^2), \\ &< K(1+|y|_\infty^2), \end{aligned}$$

where

$$K = \left(2(1-\alpha)^2(\beta+1)^2 T^{2\beta-2} K + 2\alpha^2 \beta^2 T K\right).$$

On a similar way, we prove that Πy satisfies the lipschitz condition. Thus, the mapping has a unique solution.

5.5 EXISTENCE AND UNIQUENESS FOR THE CASE WITH DELTA-DIRAC KERNEL

The Dirac-Delta function plays a significant role in mathematics, applied mathematics, and others fields of science, technology, and engineering. We recall that the Dirac-Delta function is defined as

$$\delta(t) = \begin{cases} +\infty & \text{if } x=0 \\ 0 & \text{if } x \neq 0 \end{cases} \tag{5.4}$$

It is constrained to have the following condition

$$\int_{-\infty}^{+\infty} \delta(t)dt = 1. \tag{5.5}$$

One of its important property is the translation, which implies

$$\int_{-\infty}^{+\infty} f(t)\delta(t-\tau)d\tau = f(\tau). \tag{5.6}$$

Thus, in calculus, the classical derivative can be expressed in terms of the Delta-Dirac formula as follows:

$$\int_0^t \frac{df(\tau)}{d\tau}\delta(t-\tau)d\tau = \frac{df(t)}{dt}. \tag{5.7}$$

We also have that

$$\int_0^\infty e^{-st}\delta(t-a)dt = e^{-sa}. \tag{5.8}$$

Indeed a fractal-fractional derivative with Dirac-Delta function can be expressed as

$$\begin{aligned} {}_0^{FFD}D_t^{\alpha,\beta}y(t) &= \lim_{t_1 \to t} \frac{f(t_1)-f(t)}{t_1^\beta - t^\beta}, \\ &= \frac{d}{dt^\beta}\int_0^t f(\tau)\delta(t-\tau)d\tau, \\ &= \lim_{t_1 \to t} \frac{f(t_1)-f(t)}{t_1^\beta - t^\beta} = \frac{df(t)}{dt^\beta} \end{aligned} \tag{5.9}$$

which is the well-known fractal derivative. We shall therefore present an analysis for existence and uniqueness of a Cauchy problem with this differential operator. We shall assume that $y(t)$ is classically differentiable such that

$$ {}_0^{FF}D_t^{\alpha,\beta}y(t) = \frac{d}{dt}y(t)\frac{1}{\beta t^{\beta-1}} \tag{5.10}$$

at this exists.

$$ {}_0^{FFD}D_t^{\alpha,\beta}y(t) = f(t,y(t)), \text{ if } t > 0, \tag{5.11}$$
$$y(0) = y_0, \text{ if } t = 0. \tag{5.12}$$

$$\frac{d}{dt}y(t) = \beta t^{\beta-1} f(t, y(t)), \text{ if } t > 0, \tag{5.13}$$

$$y(0) = y_0, \text{ if } t = 0. \tag{5.14}$$

$$y(t) - y_0 = \beta \int_0^t \tau^{\beta-1} f(\tau, y(\tau)) d\tau \tag{5.15}$$

$$y(0) = y_0. \tag{5.16}$$

We define the following mapping:

$$\Pi y(t) = y(0) + \beta \int_0^t \tau^{\beta-1} f(\tau, y(\tau)) d\tau \tag{5.17}$$

It is assumed that $f(\tau, y(\tau))$ is bounded,

$$C_{a,b} = \overline{C}_a[t_0] \times \overline{C}_b[y_0] \tag{5.18}$$

where

$$\overline{C}_a[t_0] = [t_0 - a, t_0 + a], \overline{C}_b[y_0] = [y_0 - b, y_0 + b]. \tag{5.19}$$

The function f satisfies the Lipschitz condition

$$\left| f(t, y_1) - f(t, y_2) \right| < K |y_1 - y_2|. \tag{5.20}$$

$$\begin{aligned}
\left| \Pi y(t) - y(0) \right| &= \left| \beta \int_0^t \tau^{\beta-1} f(\tau, y(\tau)) d\tau \right|, \\
&\leq \beta \int_0^t \left| \tau^{\beta-1} f(\tau, y(\tau)) \right| d\tau, \\
&\leq \beta \int_0^t \tau^{\beta-1} \left| f(\tau, y(\tau)) \right| d\tau.
\end{aligned} \tag{5.21}$$

Since f is bounded, we have

$$\begin{aligned}
\left| \Pi y(t) - y(0) \right| &\leq \beta M \int_{t_0}^t \tau^{\beta-1} d\tau, \\
&\leq M(t^\beta - t_0^\beta), \\
&\leq Mt^\beta \leq Ma^\beta \leq b.
\end{aligned} \tag{5.22}$$

The above is true if it is imposed that

$$a < \left(\frac{b}{M}\right)^{\frac{1}{\beta}}. \tag{5.23}$$

Existence and Uniqueness of Fractal Fractional Differential Equations

This result is obtained since we have that

$$\left|\Pi y(t) - y(0)\right| \leq \sup_{t \in D_y} \left|\Pi y(t) - y(0)\right|,$$
$$= \left|\Pi y(t) - y(0)\right|_\infty. \qquad (5.24)$$

Now we evaluate the Lipschitz condition

$$\begin{aligned}
\left|\Pi y_1 - \Pi y_2\right| &= \beta \left|\int_0^t \tau^{\beta-1}\left(f(t,y_1) - f(t,y_2)\right)d\tau\right|, \\
&\leq \beta \int_0^t \tau^{\beta-1}\left|f(t,y_1) - f(t,y_2)\right|d\tau, \\
&\leq \beta K \int_0^t \tau^{\beta-1}\left|y_1 - y_2\right|d\tau, \\
&\leq \beta K \int_0^t t^{\beta-1} \sup_{l \in [0,\tau]}\left|y_1 - y_2\right|d\tau, \\
&\leq \beta K \left|y_1 - y_2\right|_\infty \int_0^t t^{\beta-1} d\tau, \\
&\leq K a^\beta \left|y_1 - y_2\right|_\infty. \qquad (5.25)
\end{aligned}$$

This will be a contraction if

$$a < \left(\frac{1}{K}\right)^{\frac{1}{\beta}}.$$

We therefore search fixed points if $a < \min\left\{\left(\frac{b}{M}\right)^{\frac{1}{\beta}}, \left(\frac{1}{K}\right)^{\frac{1}{\beta}}\right\}$
which conclude the first proof. Secondly, we consider the following hypothesis,

$$\left|f(t,y(t))\right|^2 < K\left|1 + |y|^2\right| \qquad (5.26)$$

$$\left|f(t,y_1) - f(t,y_2)\right|^2 < \overline{K}\left|y_1 - y_2\right|^2. \qquad (5.27)$$

Thus,

$$\begin{aligned}
\left|\Pi y(t)\right|^2 &= \beta^2 \left|\int_0^t \tau^{\beta-1} f(\tau, y(\tau)) d\tau\right|^2, \\
&\leq \frac{2\beta^2 t^{2\beta-1}}{2\beta-1} \int_0^t \left|f(\tau, y(\tau))\right|^2 d\tau, \\
&\leq \frac{2K\beta^2 a^{2\beta-1}}{2\beta-1}\left(1 + \|y\|_\infty^2\right)a, \\
&\leq \frac{2K\beta^2 a^{2\beta}}{2\beta-1}\left(1 + \|y\|_\infty^2\right). \qquad (5.28)
\end{aligned}$$

$$\left\|\Pi y\right\|_\infty^2 \leq K_1\left(1+\|y\|_\infty^2\right), \tag{5.29}$$

where

$$K_1 = \frac{2K\beta^2 a^{2\beta}}{2\beta - 1}.$$

We also have

$$\left|\Pi y_1 - \Pi y_2\right|_\infty^2 \leq \overline{K}_1 \left\|y_1 - y_2\right\|_\infty^2 \tag{5.30}$$

where

$$\overline{K}_1 = \frac{2\overline{K}\beta^2 a^{2\beta}}{2\beta - 1} \beta > 1/2.$$

We conclude that the defined mapping satisfies the linear growth and the Lipschitz condition under the prescribed condition. Therefore, the Cauchy problem at fractal-fractional derivative has a unique solution.

6 A Numerical Solution of Fractal-Fractional ODE with Linear Interpolation

6.1 INTRODUCTION

In mathematics, linear interpolation is a technique for curve fitting that creates new data points within the bounds of a discrete collection of known data points using linear polynomials. Since ancient times, tables have been filled in with gaps using linear interpolation. Consider the situation if one wants to estimate the population in 1994 and had a table of a country's population over the previous 30 years. A simple method for doing this is linear interpolation. It is thought that the Seleucid Empire and the Greek mathematician and astronomer Hipparchus both employed it. In computer graphics, the fundamental technique of linear interpolation between two variables is frequently utilised. It is sometimes referred to as a lerp from linear interpolation in that field's lingo. This chapter discusses the numerical approximations of the fractal-fractional differential equations using interpolations. We consider the cases with Dirac-Delta function, power law function, exponential decay kernel, and generalised Mittag-Leffler Kernel.

Fractal-fractional ordinary differential equation are very complex that there exists solutions sometimes very difficult to obtain using existing analytical methods. Thus, in this chapter, we use the linear interpolation to derive a numerical scheme that could be used to solve these equations. As background, we have that the interpolation is often used to approximate a value of some function f using two known values.

$$R_T = f(t) - p(t), \tag{6.1}$$

is called the error, $p(t)$ is the interpolation polynomials defined by

$$p(t) = f(t_j) + \frac{f(t_{j+1}) - f(t_j)}{t_{j+1} - t_j}(x - t_j). \tag{6.2}$$

The literature shows that

$$|R_T| \leq \frac{(t_{j+1} - t_j)^2}{8} \max_{t_j \leq t \leq t_{j+1}} \left\| f''(t) \right\| \tag{6.3}$$

The condition is that f is at least two times differentiable.

DOI: 10.1201/9781003359258-6

6.2 CASE WITH THE DELTA-DIRAC KERNEL

We consider the following Cauchy problem:

$$_0^{FFD}D_t^{\alpha,\beta}y(t) = f(t,y(t)), \text{ if } t>0,$$

$$y(0) = y_0, \text{ if } t=0. \quad (6.4)$$

It is assumed that $f(.)$ is twice differentiable, $y(t)$ is classically differentiable such that

$$y(t) = y(0) + \beta \int_0^t \tau^{\beta-1} f(\tau,y(\tau))d\tau, \quad (6.5)$$

at $t = t_{n+1} = (n+1)\Delta t$, where $\Delta t = t_{n+1} - t_n$, we have

$$y(t_{n+1}) = y(0) + \beta \int_0^{t_{n+1}} \tau^{\beta-1} f(\tau,y(\tau))d\tau, \quad (6.6)$$

at $t = t_n$, we have

$$y(t_n) = y(0) + \beta \int_0^{t_n} \tau^{\beta-1} f(\tau,y(\tau))d\tau. \quad (6.7)$$

Thus,

$$y(t_{n+1}) - y(t_n) = \beta \int_{t_n}^{t_{n+1}} \tau^{\beta-1} f(\tau,y(\tau))d\tau. \quad (6.8)$$

Within t_n, t_{n+1}, we approximate the function

$$f(\tau,y(\tau)) \approx P_n(\tau) = f(t_n,y(t_n)) + (\tau - t_n)\left(\frac{f(t_{n+1},y_{(t_{n+1})}) - f(t_n,y(t_n))}{\Delta t}\right). \quad (6.9)$$

Replacing, yields

$$y_{n+1} = y_n + \beta \int_{t_n}^{t_{n+1}} \tau^{\beta-1} P_n(\tau)d\tau, \quad (6.10)$$

So,

$$\begin{aligned} y_{n+1} &= y_n + \beta \int_{t_n}^{t_{n+1}} \tau^{\beta-1} \bigg(f(t_n,y(t_n)) + (\tau - t_n) \\ &\quad \times \bigg(\frac{f(t_{n+1},y_{(t_{n+1})}) - f(t_n,y(t_n))}{\Delta t}\bigg)\bigg) d\tau, \\ &= y_n + f(t_n,y_n)(t_{n+1}^\beta - t_n^\beta) + \frac{f(t_{n+1},y_{n+1}) - f(t_n,y_n)}{\Delta t} \\ &\quad \times \int_{t_n}^{t_{n+1}} \beta(\tau - t_n)\tau^{\beta-1} d\tau. \end{aligned} \quad (6.11)$$

A Numerical Solution of Fractal-Fractional ODE with Linear Interpolation

Note that

$$\int_{t_n}^{t_{n+1}} \beta(\tau - t_n)\tau^{\beta-1}d\tau = \beta \int_{t_n}^{t_{n+1}} \tau^\beta d\tau - t_n\beta \int_{t_n}^{t_{n+1}} \tau^{\beta-1}d\tau,$$

$$= \frac{\beta}{\beta+1}\left(t_{n+1}^{\beta+1} - t_n^{\beta+1}\right) - t_n\left(t_{n+1}^\beta - t_n^\beta\right). \quad (6.12)$$

Replacing, we obtain

$$y_{n+1} = y_n + f(t_n, y_n)\left(t_{n+1}^\beta - t_n^\beta\right) + \frac{f(t_{n+1}, y_{n+1}^p) - f(t_n, y_n)}{\Delta t}$$

$$\times \left\{\frac{\beta}{\beta+1}\left(t_{n+1}^{\beta+1} - t_n^{\beta+1}\right) - t_n\left(t_{n+1}^\beta - t_n^\beta\right)\right\}, \quad (6.13)$$

where

$$y_{n+1}^p = y_n + (t_{n+1}^\beta - t_n^\beta)f(t_n, y_n). \quad (6.14)$$

$$y_{n+1} = y_n + (\Delta t)^\beta f(t_n, y_n)\{(n+1)^\beta - n^\beta\}$$

$$+ (\Delta t)^\beta \{f(t_{n+1}, y_{n+1}^p) - f(t_n, y_n)\}$$

$$\times \left\{\frac{\beta}{\beta+1}\left((n+1)^{\beta+1} - n^{\beta+1}\right) - n\left((n+1)^\beta - n^\beta\right)\right\}$$

where

$$y_{n+1}^p = y_n + (\Delta t)^\beta \left\{(n+1)^\beta - n^\beta\right\} f(t_n, y_n).$$

If $\beta = 1$, we have

$$y_{n+1} = y_n + (\Delta t)f(t_n, y_n)$$

$$+ \Delta t\left\{f(t_{n+1}, y_{n+1}^p) - f(t_n, y_n)\right\}$$

$$\times \left\{\frac{1}{2}((n+1)^2 - n^2) - n\right\}.$$

$$y_{n+1} = y_n + \frac{\Delta t}{2}f(t_n, y_n) + \frac{\Delta t}{2}f(t_{n+1}, y_{n+1}^p),$$

where

$$y_{n+1}^p = y_n + (\Delta t)f(t_n, y_n).$$

Therefore,

$$y_{n+1} = y_n + \frac{\Delta t}{2}\Big(f(t_n, y_n) + f(t_n + h, y_n + (\Delta t)f(t_n, y_n))\Big).$$

6.2.1 EXAMPLES OF FRACTAL DIFFERENTIAL EQUATIONS

$$\frac{dy}{dt^\beta} = -\lambda y, \; y(0) = y_0. \tag{6.15}$$

We get

$$y(t) = y(0)\exp(-\lambda t^\beta), \; y(0) = y_0. \tag{6.16}$$

We consider the following examples:

$$\frac{dy}{dt^\beta} = t^n, \; n > 1, \; y(0) = y_0. \tag{6.17}$$

$$\frac{dy}{dt} = \beta t^{n+\beta-1}, \; y(0) = y_0. \tag{6.18}$$

$$y(t) = \frac{\beta}{\beta+n} t^{n+\beta}, \; y(0) = y_0. \tag{6.19}$$

We consider the following examples:

$$\frac{dy}{dt^\beta} = \exp(t^\beta), \; y(0) = y_0. \tag{6.20}$$

$$\frac{dy}{dt} = \beta t^{\beta-1} \exp(t^\beta), \; y(0) = y_0. \tag{6.21}$$

$$y(t) = \exp(t^\beta)\, y(0) = y_0 = 1.$$

We present the test of the suggested numerical schemes on Figure 6.1 below.

6.3 THE CASE OF POWER LAW KERNEL

In this case, we consider the following fractal-fractional Cauchy problem:

$$\begin{aligned} {}^{FFP}_0 D^{\alpha,\beta}_t y(t) &= f(t,y(t)), \; \text{if } t > 0, \\ y(0) &= y_0, \; \text{if } t = 0. \end{aligned} \tag{6.22}$$

It is assumed that $f(t,y(t))$ is twice differentiable and that $y(t)$ is at least continuous

$$\begin{aligned} {}^{RL}_0 D^{\alpha}_t y(t) &= \beta t^{\beta-1} f(t,y(t)), \; \text{if } t > 0, \\ y(0) &= y_0, \; \text{if } t = 0. \end{aligned} \tag{6.23}$$

We write further

$$y(t) = \frac{\beta}{\Gamma(\alpha)} \int_0^t \tau^{\beta-1} f(\tau, y(\tau))(t-\tau)^{\alpha-1} d\tau, \; \text{if } t > 0,$$

A Numerical Solution of Fractal-Fractional ODE with Linear Interpolation

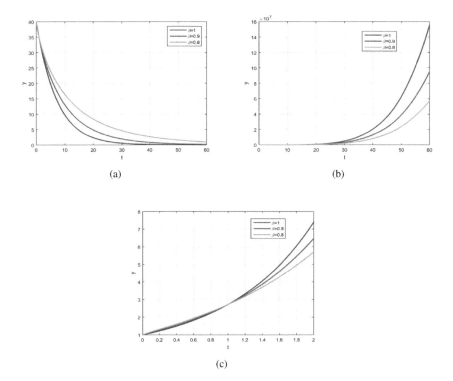

Figure 6.1 Plot of the exact solutions for different β, (a) $y(t) = y(0)\exp(-\lambda t^\beta)$, where $y(0) = 40$ and $\lambda = 0.141517$, (b) $y(t) = \frac{\beta}{\beta+n} t^{n+\beta}$, where, $n = 4$, (c) $y(t) = \exp(t^\beta)$, $y(0) = 1$.

$$y(0) = y_0, \text{ if } t = 0, \tag{6.24}$$

at $t = t_{n+1}$, we have

$$y(t_{n+1}) = \frac{\beta}{\Gamma(\alpha)} \int_0^{t_{n+1}} \tau^{\beta-1} f(\tau, y(\tau))(t_{n+1} - \tau)^{\alpha-1} d\tau. \tag{6.25}$$

Also, we write

$$y(t_{n+1}) = \frac{\beta}{\Gamma(\alpha)} \sum_{j=0}^{n} \int_{t_j}^{t_{j+1}} \tau^{\beta-1} f(\tau, y(\tau))(t_{n+1} - \tau)^{\alpha-1} d\tau. \tag{6.26}$$

Thus, within $[t_j, t_{j+1}]$, we approximate the function

$$f(\tau, y(\tau)) \approx P_j(\tau) = f(t_j, y_j) + (t - t_j) \frac{f(t_{j+1}, y_{j+1}) - f(t_j, y_j)}{\Delta t}, \tag{6.27}$$

replacing yields

$$y(t_{n+1}) \approx y_{n+1} = \frac{\beta}{\Gamma(\alpha)} \sum_{j=0}^{n} \int_{t_j}^{t_{j+1}} \tau^{\beta-1}(t_{n+1}-\tau)^{\alpha-1}$$
$$\times \left(f(t_j, y_j) + (t-t_j)\frac{f(t_{j+1}, y_{j+1}) - f(t_j, y_j)}{\Delta t} \right) d\tau,$$
$$= \frac{\beta}{\Gamma(\alpha)} \sum_{j=0}^{n} f(t_j, y_j) \int_{t_j}^{t_{j+1}} \tau^{\beta-1}(t_{n+1}-\tau)^{\alpha-1} d\tau$$
$$+ \frac{\beta}{\Gamma(\alpha)} \sum_{j=0}^{n} \frac{f(t_{j+1}, y_{j+1}) - f(t_j, y_j)}{\Delta t}$$
$$\times \int_{t_j}^{t_{j+1}} (t_{n+1}-\tau)^{\alpha-1}(t-t_j) d\tau. \quad (6.28)$$

Noting that

$$\int_{t_j}^{t_{j+1}} \tau^{\beta-1}(t_{n+1}-\tau)^{\alpha-1} d\tau = \int_{0}^{t_{j+1}} \tau^{\beta-1}(t_{n+1}-\tau)^{\alpha-1} d\tau$$
$$- \int_{0}^{t_j} \tau^{\beta-1}(t_{n+1}-\tau)^{\alpha-1} d\tau, \quad (6.29)$$

where

$$\int_{0}^{t_{j+1}} \tau^{\beta-1}(t_{n+1}-\tau)^{\alpha-1} d\tau = t_{n+1}^{\alpha+\beta-1} \int_{0}^{\frac{t_{j+1}}{t_{n+1}}} y^{\beta-1}(1-y)^{\alpha} y^{-1} dy,$$
$$= t_{n+1}^{\alpha+\beta-1} B\left(\frac{t_{j+1}}{t_{n+1}}, \beta, \alpha\right). \quad (6.30)$$

$$\int_{0}^{t_j} \tau^{\beta-1}(t_{n+1}-\tau)^{\alpha-1} d\tau = t_{n+1}^{\alpha+\beta-1} B\left(\frac{t_j}{t_{n+1}}, \beta, \alpha\right). \quad (6.31)$$

Therefore,

$$\int_{t_j}^{t_{j+1}} \tau^{\beta-1}(t_{n+1}-\tau)^{\alpha-1}(\tau-t_j) d\tau = \int_{t_j}^{t_{j+1}} \tau^{\beta}(t_{n+1}-\tau)^{\alpha-1} d\tau$$
$$-t_j \int_{t_j}^{t_{j+1}} \tau^{\beta-1}(t_{n+1}-\tau)^{\alpha-1} d\tau,$$
$$= t_{n+1}^{\alpha+\beta}\left\{ B\left(\frac{t_{j+1}}{t_{n+1}}, \beta+1, \alpha\right)\right.$$
$$\left. -B\left(\frac{t_j}{t_{n+1}}, \beta+1, \alpha\right)\right\}$$
$$-t_j\left\{ B\left(\frac{t_{j+1}}{t_{n+1}}, \beta, \alpha\right) - B\left(\frac{t_j}{t_{n+1}}, \beta, \alpha\right)\right\}$$
$$t_{n+1}^{\alpha+\beta-1}.$$

A Numerical Solution of Fractal-Fractional ODE with Linear Interpolation

Replacing yields

$$
\begin{aligned}
y_{n+1} &= \frac{\beta}{\Gamma(\alpha)} \sum_{j=0}^{n} f(t_j, y_j) \left\{ t_{n+1}^{\alpha+\beta-1} B\left(\frac{t_{j+1}}{t_{n+1}}, \beta, \alpha\right) \right. \\
&\quad \left. - t_{n+1}^{\alpha+\beta-1} B\left(\frac{t_j}{t_{n+1}}, \beta, \alpha\right) \right\} \\
&\quad + \frac{\beta}{\Gamma(\alpha)} \sum_{j=0}^{n} \frac{f(t_{j+1}, y_{j+1}) - f(t_j, y_j)}{\Delta t} \\
&\quad \times \left\{ t_{n+1}^{\alpha+\beta} \left[B\left(\frac{t_{j+1}}{t_{n+1}}, \beta+1, \alpha\right) - B\left(\frac{t_j}{t_{n+1}}, \beta+1, \alpha\right) \right] \right. \\
&\quad \left. - t_j \left[B\left(\frac{t_j}{t_{n+1}}, \beta, \alpha\right) - B\left(\frac{t_j}{t_{n+1}}, \beta, \alpha\right) \right] t_{n+1}^{\alpha+\beta-1} \right\}. \quad (6.32)
\end{aligned}
$$

$$
\begin{aligned}
y_{n+1} &= \frac{\beta}{\Gamma(\alpha)} (\Delta t)^{\alpha+\beta-1} \sum_{j=0}^{n} f(t_j, y_j)(n+1)^{\alpha+\beta-1} \\
&\quad \times \left\{ B\left(\frac{t_{j+1}}{t_{n+1}}, \beta, \alpha\right) - B\left(\frac{t_j}{t_{n+1}}, \beta, \alpha\right) \right\} \\
&\quad + \frac{\beta}{\Gamma(\alpha)} (\Delta t)^{\alpha+\beta-1} \sum_{j=0}^{n-1} \left[f(t_{j+1}, y_{j+1}) - f(t_j, y_j) \right] \\
&\quad \times \left\{ (n+1)^{\alpha+\beta} B\left(\frac{t_{j+1}}{t_{n+1}}, \beta+1, \alpha\right) - n^{\alpha+\beta} B\left(\frac{t_j}{t_{n+1}}, \beta+1, \alpha\right) \right. \\
&\quad \left. - j\left[B\left(\frac{t_{j+1}}{t_{n+1}}, \beta, \alpha\right) - B\left(\frac{t_j}{t_{n+1}}, \beta, \alpha\right) \right] (n+1)^{\alpha+\beta-1} \right\} \\
&\quad + \frac{\beta(\Delta t)^{\alpha+\beta-1}}{\Gamma(\alpha)} \left[f(t_n + h, y_{n+1}^p) - f(t_n, y_n) \right] \\
&\quad \times \left\{ (n+1)^{\alpha+\beta} \left(B(1, \beta+1, \alpha) - B\left(\frac{t_n}{t_{n+1}}, \beta+1, \alpha\right) \right) \right. \\
&\quad \left. - n \left(B(1, \beta, \alpha) - B\left(\frac{t_n}{t_{n+1}}, \beta, \alpha\right) \right) (n+1)^{\alpha+\beta-1} \right\} \quad (6.33)
\end{aligned}
$$

where

$$
\begin{aligned}
y_{n+1}^p &= y_0 + \frac{\beta}{\Gamma(\alpha+1)} (\Delta t)^{\alpha+\beta-1} \sum_{j=0}^{n} f(t_j, y_j)(n+1)^{\alpha+\beta-1} \\
&\quad \times \left\{ B\left(\frac{t_{j+1}}{t_{n+1}}, \beta, \alpha\right) - B\left(\frac{t_j}{t_{n+1}}, \beta, \alpha\right) \right\}
\end{aligned}
$$

where $B(.,.,.)$ is the well-known incomplete beta function defined as

$$
B(x, \alpha, \beta) = \int_0^x u^{\alpha-1}(1-u)^{\beta-1} d\tau. \quad (6.34)
$$

6.4 CASE WITH EXPONENTIAL DECAY KERNEL

In this section, we consider the fractal-fractional Cauchy problem with exponential kernel

$$\begin{aligned}{}^{FFE}_{0}D_t^{\alpha,\beta}y(t) &= f(t,y(t)), \text{ if } t>0,\\ y(0) &= y_0, \text{ if } t=0.\end{aligned} \quad (6.35)$$

$$\begin{aligned}{}^{CF}_{0}D_t^{\alpha}y(t) &= \beta t^{\beta-1}f(t,y(t)), \text{ if } t>0,\\ y(0) &= y_0, \text{ if } t=0.\end{aligned} \quad (6.36)$$

$$\begin{aligned}y(t) &= (1-\alpha)\beta t^{\beta-1}f(t,y(t))\\ &+ \alpha\beta \int_0^t \tau^{\beta-1}f(\tau,y(\tau))d\tau, \ y(0)=y_0,\end{aligned} \quad (6.37)$$

at $t = t_{n+1}$, we have

$$\begin{aligned}y(t_{n+1}) &= (1-\alpha)\beta t_{n+1}^{\beta-1}f(t_{n+1},y_{n+1})\\ &+ \alpha\beta \int_0^{t_{n+1}} \tau^{\beta-1}f(\tau,y(\tau))d\tau,\end{aligned} \quad (6.38)$$

at $t = t_n$, we have

$$\begin{aligned}y(t_n) &= (1-\alpha)\beta t_n^{\beta-1}f(t_n,y_n)\\ &+ \alpha\beta \int_0^{t_n} \tau^{\beta-1}f(\tau,y(\tau))d\tau.\end{aligned} \quad (6.39)$$

Therefore,

$$\begin{aligned}y(t_{n+1}) &= y(t_n) + (1-\alpha)\beta\left[t_{n+1}^{\beta-1}f(t_{n+1},y_{n+1}) - t_n^{\beta-1}f(t_n,y(t_n))\right]\\ &+ \alpha\beta \int_{t_n}^{t_{n+1}} \tau^{\beta-1}f(\tau,y(\tau))d\tau.\end{aligned} \quad (6.40)$$

Using the polynomial approximations of $f(\tau,y(\tau))$ within $[t_n, t_{n+1}]$ yields,

$$\begin{aligned}y_{n+1} &= y_n + (1-\alpha)\beta\left[t_{n+1}^{\beta-1}f(t_{n+1},y_{n+1}^p) - t_n^{\beta-1}f(t_n,y_n)\right]\\ &+ \alpha\left[f(t_n,y_n)(t_{n+1}^\beta - t_n^\beta) + \frac{f(t_{n+1},y_{n+1}^p) - f(t_n,y_n)}{\Delta t}\right.\\ &\left.\left\{\frac{\beta}{\beta+1}(t_{n+1}^{\beta+1} - t_n^{\beta+1}) - t_n(t_{n+1}^\beta - t_n^\beta)\right\}\right],\end{aligned} \quad (6.41)$$

A Numerical Solution of Fractal-Fractional ODE with Linear Interpolation

$$y_{n+1} = y_n + (1-\alpha)\beta(\Delta t)^{\beta-1}$$
$$\left\{(n+1)^{\beta-1}f(t_n+h, y_{n+1}^p) - n^{\beta-1}f(t_n, y_n)\right\}$$
$$+\alpha(\Delta t)^\beta \left\{f(t_n, y_n)((n+1)^\beta - n^\beta) + (f(t_n+h, y_{n+1}^p) - f(t_n, y_n))\right\}$$
$$\times \left\{\frac{\beta}{\beta+1}[(n+1)^{\beta+1} - n^{\beta+1}] - n\left((n+1)^\beta - n^\beta\right)\right\},$$

where

$$y_{n+1}^p = y_0 + (1-\alpha)\beta t_n^{\beta-1} f(t_n, y_n)$$
$$+\alpha \sum_{j=0}^{n} f(t_j, y_j) \beta \int_{t_j}^{t_{j+1}} \tau^{\beta-1}(\tau - t_j)d\tau.$$

$$y_{n+1}^p = y_0 + (1-\alpha)\beta t_n^{\beta-1} f(t_n, y_n)$$
$$+\alpha \sum_{j=0}^{n} f(t_j, y_j) \left[\frac{\beta}{\beta+1}t_{j+1}^{\beta+1} - \frac{\beta}{\beta+1}t_j^{\beta+1} - t_j(t_{j+1}^\beta - t_j^\beta)\right].$$

$$y_{n+1}^p = y_0 + (1-\alpha)\beta(\Delta t)^{\beta-1} n^{\beta-1} f(t_n, y_n)$$
$$+\alpha \sum_{j=0}^{n} f(t_j, y_j) \left[\frac{\beta}{\beta+1}[t_{j+1}^{\beta+1} - t_j^{\beta+1}] - t_j(t_{j+1}^\beta - t_j^\beta)\right].$$

6.4.1 EXAMPLES OF FRACTAL-FRACTIONAL WITH EXPONENTIAL DECAY FUNCTION

$$_0^{FFE}D_t^{\alpha,\beta} y(t) = t^n, \text{ if } t > 0,$$
$$y(0) = y_0 = 0. \quad (6.42)$$

$$_0^{CFR}D_t^\alpha y(t) = \beta t^{\beta+n-1},$$
$$y(0) = y_0 = 0. \quad (6.43)$$

$$y(t) = (1-\alpha)\beta t^{\alpha+\beta-1} + \frac{\alpha\beta}{\beta+n}t^{n+\beta},$$
$$y(0) = y_0. \quad (6.44)$$

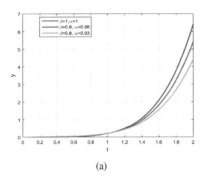
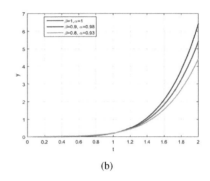

Figure 6.2 Plot of the exact solutions for different β, (a) $y(t) = (1-\alpha)\beta t^{\alpha+\beta-1} + \frac{\alpha+\beta}{n}t^{\beta+n}$, (b)$y(t) = (1-\alpha)\beta t^{\beta-1} Sin(t^\beta) - \alpha Cos(t^\beta) + \alpha$.

Example, we consider the following fractal-fractional Cauchy problem:

$$_0^{FFE}D_t^{\alpha,\beta}y(t) = Sin(t^\beta),$$

$$y(0) = y_0 = 0. \tag{6.45}$$

$$_0^{CFR}D_t^\alpha y(t) = \beta t^{\beta-1} Sin(t^\beta),$$

$$y(0) = y_0 = 0. \tag{6.46}$$

$$y(t) = (1-\alpha)\beta t^{\beta-1} Sin(t^\beta) - \alpha Cos(t^\beta) + \alpha,$$

$$y(0) = y_0 = 0. \tag{6.47}$$

The graphical solution to the above exact solution is given in Figure 6.2.

6.5 CASE WITH GENERALISED MITTAG-LEFFLER KERNEL

We consider in this section a fractal-fractional Cauchy problem with the generalised Mittag-Leffler Kernel:

$$_0^{FFM}D_t^{\alpha,\beta}y(t) = f(t,y(t)), \text{ if } t > 0,$$

$$y(0) = y_0, \text{ if } t = 0. \tag{6.48}$$

We have for AB-RL case

$$_0^{ABR}D_t^\alpha y(t) = \beta t^{\beta-1} f(t,y(t)), \text{ if } t > 0,$$

A Numerical Solution of Fractal-Fractional ODE with Linear Interpolation

$$y(0) = y_0, \text{ if } t = 0. \tag{6.49}$$

$$y(t) = (1-\alpha)\beta t^{\beta-1} f(t, y(t))$$
$$+ \frac{\alpha\beta}{\Gamma(\alpha)} \int_0^t \tau^{\beta-1}(t-\tau)^{\alpha-1} f(\tau, y(\tau)) d\tau, \ y(0) = y_0. \tag{6.50}$$

At $t = t_{n+1}$, we have

$$y(t_{n+1}) = (1-\alpha)\beta t_{n+1}^{\beta-1} f(t_{n+1}, y(t_{n+1}))$$
$$+ \frac{\alpha\beta}{\Gamma(\alpha)} \int_0^{t_{n+1}} \tau^{\beta-1}(t_{n+1}-\tau)^{\alpha-1} f(\tau, y(\tau)) d\tau. \tag{6.51}$$

$$y(t_{n+1}) = (1-\alpha)\beta t_{n+1}^{\beta-1} f(t_{n+1}, y(t_{n+1}))$$
$$+ \frac{\alpha\beta}{\Gamma(\alpha)} \sum_{j=0}^{n} \int_{t_j}^{t_{j+1}} (t_{n+1}-\tau)^{\alpha-1} \tau^{\beta-1} f(\tau, y(\tau)) d\tau. \tag{6.52}$$

Approximating the function $f(\tau, y(\tau))$ by polynomial approximations, we have

$$y_{n+1} = \beta t_{n+1}^{\beta}(1-\alpha) f(t_{n+1}, y_{n+1}^p)$$
$$+ \frac{\alpha\beta}{\Gamma(\alpha)} \sum_{j=0}^{n} f(t_j, y_j) \left\{ t_{n+1}^{\alpha+\beta-1} \left[B\left(\frac{t_{j+1}}{t_{n+1}}, \beta, \alpha\right) \right.\right.$$
$$\left.\left. - B\left(\frac{t_j}{t_{n+1}}, \beta, \alpha\right) \right] \right\}$$
$$+ \frac{\alpha\beta}{\Gamma(\alpha)} \sum_{j=0}^{n-1} \left(\frac{f(t_{j+1}, y_{j+1}) - f(t_j, y_j)}{\Delta t} \right)$$
$$\left\{ t_{n+1}^{\alpha+\beta} \left[B\left(\frac{t_{j+1}}{t_{n+1}}, \beta+1, \alpha\right) - B\left(\frac{t_j}{t_{n+1}}, \beta+1, \alpha\right) \right] \right.$$
$$\left. - t_j t_{n+1}^{\alpha+\beta-1} \left[B\left(\frac{t_{j+1}}{t_{n+1}}, \beta, \alpha\right) - B\left(\frac{t_j}{t_{n+1}}, \beta, \alpha\right) \right] \right\}$$
$$+ \frac{\alpha\beta}{\Gamma(\alpha)} \left(\frac{f(t_n+h, y_{n+1}^p) - f(t_n, y_n)}{\Delta t} \right)$$
$$\times (\Delta t)^{\alpha+\beta} \left\{ (n+1)^{\alpha+\beta} \{ B(1, \beta+1, \alpha) - B(\frac{t_n}{t_{n+1}}, \beta+1, \alpha) \} \right.$$
$$\left. - n(n+1)^{\alpha+\beta-1} \{ B(1, \beta, \alpha) - B(\frac{t_n}{t_{n+1}}, \beta, \alpha) \} \right\} \tag{6.53}$$

$$y_{n+1}^p = (1-\alpha)\beta f(t_n, y_n)$$

$$+ (\Delta t)^{\alpha+\beta-1} \frac{\alpha\beta}{\Gamma(\alpha)} \sum_{j=0}^{n} f(t_j, y_j)(n+1)^{\alpha+\beta-1}$$
$$\times \left[B\left(\frac{t_{j+1}}{t_{n+1}}, \beta, \alpha\right) - B\left(\frac{t_j}{t_{n+1}}, \beta, \alpha\right) \right]. \tag{6.54}$$

7 Numerical Scheme of Fractal-Fractional ODE with Middle Point Interpolation

7.1 INTRODUCTION

The centre of a line segment is known as the midpoint. It is the centroid of the segment and of the ends, and it is equally distant from both of them. It cuts the section in half. By initially building a lens out of circular arcs of equal radii centred at the two endpoints and joining the cusps of the lens, one can determine the midpoint of a line segment immersed in a plane. The midpoint of the segment is then the place where the line joining the cusps intersects the segment. According to the Mohr-Mascheroni theorem, it is more difficult to find the midpoint when using only a compass, but it is still doable. The numerical scheme for ordinary differential equations has been derived using this concept. The midpoint technique is a one-step procedure for solving the differential equation numerically in the field of numerical analysis, a subfield of applied mathematics. The implicit technique is the most straightforward collocation approach, and when used with Hamiltonian dynamics, it is referred to as a symplectic integrator. The explicit midpoint method is also occasionally referred to as the modified Euler method. For further clarification, see list of Runge-Kutta methods. Take note that the modified Euler method might also refer to Heun's method. On the other hand, because fractal-fractional differential equations are difficult to solve analytically, numerical methods are used to generate approximations of the solutions. The middle point concept will be employed in this chapter to create a numerical method for solving fractal-fractional differential equations. This chapter discusses the approximation of the fractal-fractional differential equations using the middle point interpolation. We consider the cases for the Dirac-Delta, Power law, exponential, and Mitag-Leffler law. We derive the approximations for the above-mentioned cases in details.

7.2 NUMERICAL SCHEME FOR DELTA-DIRAC CASE

In this section, we consider the following fractal-fractional Cauchy problem:

$$\frac{dy}{dt^\beta} = f(t, y(t)) \text{ if } t > 0$$

$$y(0) = y_0, \text{ if } t = 0.$$

We assume that $f(t,y(t))$ is differentiable twice and that $y(t)$ is classically differentiable

$$\frac{dy}{dt} = \beta t^{\beta-1} f(t,y(t)) \text{ if } t > 0$$

$$y(0) = y_0, \text{ if } t = 0.$$

Then, we can write

$$y(t) = y_0 + \int_0^t \beta t^{\beta-1} f(t,y(t)) d\tau,$$

$$y(0) = y_0.$$

For $t = t_{n+1}$, and $t = t_n$, we have

$$y(t_{n+1}) = y_0 + \int_0^{t_{n+1}} \beta \tau^{\beta-1} f(\tau,y(\tau)) d\tau,$$

and

$$y(t_n) = y_0 + \int_0^{t_n} \beta \tau^{\beta-1} f(\tau,y(\tau)) d\tau,$$

Further, we write

$$y(t_{n+1}) = y_{(t_n)} + \int_{t_n}^{t_{n+1}} \beta \tau^{\beta-1} f(\tau,y(\tau)) d\tau.$$

Within $[t_n, t_{n+1}]$, we use the middle point approximation for the function $f(t,y(y))$. Thus,

$$y(t_{n+1}) = y_{(t_n)} + \beta \int_{t_n}^{t_{n+1}} \tau^{\beta-1} f\left(t_n + \frac{\Delta t}{2}, \frac{y_n + y_{n+1}}{2}\right).$$

$$y(t_{n+1}) = y_n + \beta f\left(t_n + \frac{\Delta t}{2}, \frac{y_n + y_{n+1}^p}{2}\right) \left\{\frac{t_{n+1}^\beta}{\beta} - \frac{t_n^\beta}{\beta}\right\}.$$

$$y(t_{n+1}) = y_n + (\Delta t)^\beta f\left(t_n + \frac{\Delta t}{2}, \frac{y_n + y_{n+1}^p}{2}\right) \left\{(n+1)^\beta - n^\beta\right\},$$

Nothing that

$$y_{n+1}^p = y_n + (\Delta t)^\beta f(t_n, y_n)\{(n+1)^\beta - n^\beta\}.$$

Replacing yields

$$y(t_{n+1}) = y_n + (\Delta t)^\beta f\left(t_n + \frac{\Delta t}{2}, y_n + (\Delta t)^\beta f(t_n, y_n)\left\{(n+1)^\beta - n^\beta\right\}\right)$$

Numerical Scheme of Fractal-Fractional ODE with Middle Point Interpolation 75

$$\times \left\{(n+1)^\beta - n^\beta\right\}.$$

Therefore, the numerical approximation of the Cauchy problem for fractal-fractional with Delta-Dirac kernel with middle point method is given by

$$y_{n+1} = y_n + (\Delta t)^\beta f\left(t_n + \frac{\Delta t}{2}, y_n + (\Delta t)^\beta f(t_n, y_n)\right.$$
$$\left.\times \left\{(n+1)^\beta - n^\beta\right\} + y_n\right). \tag{7.1}$$

Noting that when $\beta = 1$,

$$y_{n+1} = y_n + \Delta t f\left(t_n + \frac{\Delta t}{2}, y_n + \Delta t f(t_n, y_n)\right), \tag{7.2}$$

which is the middle point for the classical differentiation.

7.3 NUMERICAL SCHEME FOR EXPONENTIAL CASE

We consider in this section the following fractal-fractional Cauchy problem:

$$^{FFE}_0 D_t^{\alpha,\beta} y(t) = f(t, y(t)), \text{ if } t > 0,$$

$$y(0) = y_0, \text{ if } t = 0.$$

We can write further

$$^{CFR}_0 D_t^\alpha y(t) = \beta t^{\beta-1} f(t, y(t)), \text{ if } t > 0,$$

$$y(0) = y_0, \text{ if } t = 0.$$

We have further the following:

$$y(t) = \beta t^{\beta-1}(1-\alpha)f(t,y(t))t^{\beta-1} + \alpha\beta \int_0^t \tau^{\beta-1} f(\tau, y(\tau))d\tau, \text{ if } t > 0,$$

$$y(0) = y_0, \text{ if } t = 0.$$

At $t = t_{n+1}$, and $t = t_n$, we have

$$y(t_{n+1}) = (1-\alpha)\beta t_{n+1}^{\beta-1} f(t_{n+1}, y_{n+1}^p) + \alpha\beta \int_0^{t_{n+1}} \tau^{\beta-1} f(\tau, y(\tau))d\tau,$$

and

$$y(t_n) = (1-\alpha)\beta t_n^{\beta-1} f(t_n, y_n) + \alpha\beta \int_0^{t_n} \tau^{\beta-1} f(\tau, y(\tau))d\tau.$$

Therefore,

$$y(t_{n+1}) = y_n + (1-\alpha)\beta\left(t_{n+1}^{\beta-1}f(t_{n+1},y_{n+1}^p) - t_n^{\beta-1}f(t_n,y_n)\right)$$
$$+\alpha\beta\int_{t_n}^{t_{n+1}}\tau^{\beta-1}f(\tau,y(\tau))d\tau,$$

$$y_{n+1} = y_n + (1-\alpha)\beta\left[t_{n+1}^{\beta-1}f(t_{n+1},y_{n+1}^p) - t_n^{\beta-1}f(t_n,y_n)\right]$$
$$+\alpha(\Delta t)^\beta f\left(t_n + \frac{\Delta t}{2}, y_n + (\Delta t)^\beta f(t_n,y_n)\{(n+1)^\beta - n^\beta\}\right),$$

where

$$y_{n+1}^p = y_n + \alpha(\Delta t)^\beta f\left(t_n + \frac{\Delta t}{2}, y_n + (\Delta t)^\beta f(t_n,y_n)\{(n+1)^\beta - n^\beta\}\right).$$

Noting that when $\beta = 1$, we have

$$y_{n+1} = y_n + (1-\alpha)f(t_{n+1},y_{n+1}^p) - f(t_n,y_n)\Big]$$
$$+\alpha\Delta t f\left(t_n + \frac{\Delta t}{2}, y_n + \Delta t f(t_n,y_n)\right).$$

7.4 NUMERICAL SCHEME FOR POWER LAW CASE

In this section, we consider the fractal-fractional Cauchy problem for the power law case

$$_0^{FFP}D_t^{\alpha,\beta}y(t) = f(t,y(t)), \text{ if } t > 0,$$
$$y(0) = y_0, \text{ if } t = 0.$$

We can write further

$$_0^{RL}D_t^\alpha y(t) = \beta t^{\beta-1}f(t,y(t)), \text{ if } t > 0,$$
$$y(0) = y_0, \text{ if } t = 0.$$

Further, we write

$$y(t) = \frac{\beta}{\Gamma(\alpha)}\int_0^t \tau^{\beta-1}f(\tau,y(\tau))(t-\tau)^{\alpha-1}d\tau,$$
$$y(0) = y_0.$$

At $t = t_{n+1}$, we have

$$y(t_{n+1}) = \frac{\beta}{\Gamma(\alpha)}\int_0^{t_{n+1}}\tau^{\beta-1}f(\tau,y(\tau))(t_{n+1}-\tau)^{\alpha-1}d\tau,$$

Numerical Scheme of Fractal-Fractional ODE with Middle Point Interpolation

$$y(t_{n+1}) = \frac{\beta}{\Gamma(\alpha)} \sum_{j=0}^{n} \int_{t_j}^{t_{j+1}} \tau^{\beta-1} f(\tau, y(\tau))(t_{n+1} - \tau)^{\alpha-1} d\tau.$$

Within $[t_j, t_{j+1}]$, we approximate the function $f(\tau, y(\tau))$ using the middle point method. Thus,

$$y_{n+1} = \frac{\beta}{\Gamma(\alpha)} \sum_{j=0}^{n} \int_{t_j}^{t_{j+1}} (t_{n+1} - \tau)^{\alpha-1} \tau^{\beta-1} f\left(t_j + \frac{\Delta t}{2}, \frac{y_j + y_{j+1}}{2}\right) d\tau.$$

Further, we have

$$y_{n+1} = \frac{\beta}{\Gamma(\alpha)} \sum_{j=0}^{n} f\left(t_j + \frac{\Delta t}{2}, \frac{y_j + y_{j+1}}{2}\right) \int_{t_j}^{t_{j+1}} \tau^{\beta-1} (t_{n+1} - \tau)^{\alpha-1} d\tau.$$

Noting that

$$\int_{t_j}^{t_{j+1}} \tau^{\beta-1} (t_{n+1} - \tau)^{\alpha-1} d\tau = \int_{0}^{t_{j+1}} \tau^{\beta-1} (t_{n+1} - \tau)^{\alpha-1} d\tau$$
$$- \int_{0}^{t_j} \tau^{\beta-1} (t_{n+1} - \tau)^{\alpha-1} d\tau,$$

where

$$\int_{0}^{t_{j+1}} \tau^{\beta-1} (t_{n+1} - \tau)^{\alpha-1} d\tau = t_{n+1}^{\alpha+\beta-1} B\left(\frac{t_{j+1}}{t_{n+1}}, \beta, \alpha\right),$$

$$\int_{0}^{t_j} \tau^{\beta-1} (t_{n+1} - \tau)^{\alpha-1} d\tau = t_{n+1}^{\alpha+\beta-1} B\left(\frac{t_j}{t_{n+1}}, \beta, \alpha\right).$$

Therefore,

$$y_{n+1} = \frac{\beta}{\Gamma(\alpha)} t_{n+1}^{\alpha+\beta-1} \sum_{j=0}^{n-1} f\left(t_j + \frac{\Delta t}{2}, \frac{y_j + y_{j+1}}{2}\right) \left\{ B\left(\frac{t_{j+1}}{t_{n+1}}, \beta, \alpha\right) \right.$$
$$\left. - B\left(\frac{t_j}{t_{n+1}}, \beta, \alpha\right) \right\}$$
$$+ \frac{\beta}{\Gamma(\alpha)} t_{n+1}^{\alpha+\beta-1} f\left(t_n + \frac{\Delta t}{2}, \frac{y_n + y_{n+1}^p}{2}\right) \left\{ B(1, \beta, \alpha) - B\left(\frac{t_n}{t_{n+1}}, \beta, \alpha\right) \right\}.$$

7.5 NUMERICAL SCHEME FOR THE MITTAG-LEFFLER CASE

In this section, we consider the following fractal-fractional Cauchy problem for the Mittag-Leffler case

$$_{0}^{FFM}D_{t}^{\alpha,\beta} y(t) = f(t, y(t)), \text{ if } t > 0,$$

$$y(0) = y_0, \text{ if } t = 0.$$

We can write further

$$_{0}^{ABR}D_{t}^{\alpha}y(t) = \beta t^{\beta-1}f(t,y(t)), \text{ if } t > 0,$$

$$y(0) = y_0, \text{ if } t = 0.$$

Further, we write

$$y(t) = (1-\alpha)\beta t^{\beta-1}f(t,y(t)) + \frac{\alpha\beta}{\Gamma(\alpha)}\int_0^t \tau^{\beta-1}f(\tau,y(\tau))(t-\tau)^{\alpha-1}d\tau,$$

$$y(0) = y_0.$$

For $t = t_{n+1}$, we have

$$y(t_{n+1}) = (1-\alpha)\beta t_{n+1}^{\beta-1}f(t_{n+1},y^p(t_{n+1}))$$

$$+ \frac{\alpha\beta}{\Gamma(\alpha)}\sum_{j=0}^{n}\int_{t_j}^{t_{j+1}} \tau^{\beta-1}f(\tau,y(\tau))(t_{n+1}-\tau)^{\alpha-1}d\tau,$$

whereas

$$y^p(t_{n+1}) = \frac{\beta}{\Gamma(\alpha)}\int_0^{t_{n+1}} \tau^{\beta-1}(t_{n+1}-\tau)^{\alpha-1}f(\tau,y(\tau))d\tau,$$

$$= \frac{\beta}{\Gamma(\alpha)}\sum_{j=0}^{n}\int_{t_j}^{t_{j+1}} \tau^{\beta-1}(t_{n+1}-\tau)^{\alpha-1}f(\tau,y(\tau))d\tau.$$

Hence, we use the Euler approximations

$$y_{n+1}^p = \frac{\beta}{\Gamma(\alpha)}\sum_{j=0}^{n}\int_{t_j}^{t_{j+1}} f(t_j,y_j)(t_{n+1}-\tau)^{\alpha-1}\tau^{\beta-1}d\tau,$$

$$= \frac{\beta}{\Gamma(\alpha)}\sum_{j=0}^{n} f(t_j,y_j)t_{n+1}^{\alpha+\beta-1}\left[B\left(\frac{t_{j+1}}{t_{n+1}},\beta,\alpha\right) - B\left(\frac{t_j}{t_{n+1}},\beta,\alpha\right)\right].$$

Therefore, the numerical scheme is given as

$$y_{n+1} = \frac{\beta}{\Gamma(\alpha)}t_{n+1}^{\alpha+\beta-1}\sum_{j=0}^{n-1} f\left(t_j + \frac{\Delta t}{2}, \frac{y_j+y_{j+1}}{2}\right)$$

$$\times \left[B\left(\frac{t_{j+1}}{t_{n+1}},\beta,\alpha\right) - B\left(\frac{t_j}{t_{n+1}},\beta,\alpha\right)\right]$$

$$+ \frac{\beta}{\Gamma(\alpha)}t_{n+1}^{\alpha+\beta-1}f\left(t_n + \frac{\Delta t}{2}, \frac{y_n}{2} + \frac{y_{n+1}^p}{2}\right)$$

$$\times \left\{B(1,\beta,\alpha) - B\left(\frac{t_n}{t_{n+1}},\beta,\alpha\right)\right\},$$

and

$$y_{(t_{n+1})} \approx y_{n+1} = (1-\alpha)\beta t_{n+1}^{\beta-1} f(t_{n+1}, y_{n+1}^p)$$
$$+ \frac{\alpha\beta}{\Gamma(\alpha)} \sum_{j=0}^{n-1} f\left(t_j + \frac{h}{2}, \frac{y_j + y_{j+1}}{2}\right) t_{n+1}^{\alpha+\beta-1}$$
$$\times \left[B\left(\frac{t_{j+1}}{t_{n+1}}, \beta, \alpha\right) - B\left(\frac{t_j}{t_{n+1}}, \beta, \alpha\right) \right]$$
$$+ \frac{\alpha\beta}{\Gamma(\alpha)} t_{n+1}^{\alpha+\beta-1} \left\{ B(1, \beta, \alpha) - B\left(\frac{t_n}{t_{n+1}}, \beta, \alpha\right) \right\},$$

where

$$y_{n+1}^p = \frac{\beta}{\Gamma(\alpha)} \sum_{j=0}^{n} f(t_j, y_j) t_{n+1}^{\alpha+\beta-1} \left[B\left(\frac{t_{j+1}}{t_{n+1}}, \beta, \alpha\right) - B\left(\frac{t_j}{t_{n+1}}, \beta, \alpha\right) \right].$$

8 Fractal-Fractional Euler Method

8.1 INTRODUCTION

The Euler method, sometimes known as the forward Euler method, is a first-order numerical approach for resolving ordinary differential equations with a specified beginning value. It is the simplest Runge-Kutta method and the most fundamental explicit method for numerical integration of ordinary differential equations. Leonhard Euler is honoured by having the Euler technique named after him. Since the Euler method is a first-order approach, the local error is proportional to the square of the step size, and the global error is proportional to the step size. The Euler method frequently serves as the foundation for creating more complicated procedures. This approach has been broadened to encompass the idea of fractional calculus, and it has proven to be an effective numerical method for handling linear fractional differential equations. In this chapter, we will develop a numerical method for fractal-fractional Cauchy problems using the power law, the generalised Mittag-Leffler kernel, the exponential decay kernel, and the Delta-Dirac kernel. Predictor-corrector methods are a class of algorithms used in numerical analysis to integrate ordinary differential equations in order to identify an unknown function that satisfies a given differential equation. This notion will be applied in several cases. Such algorithms all follow a two-step process: A function is fitted to the function values and derivative values at a prior set of points in the first prediction phase in order to extrapolate ahead and predict the value of this function at a subsequent, new point. The second phase corrects the original approximation by utilising the function's expected value and another technique to interpolate the value of the unknown function at the same subsequent point. The purpose of this chapter to give new numerical schemes for the solution of fractal-fractional differential equations using the euler method. We give the approximations of Euler method for the Dirac-delta function and then show the consistency and the stability of the obtained scheme. We then present the numerical scheme for fractal-fractional differential equation with exponential kernel, power law kernel, and with the generalised Mittag-Leffler kernel. Also, the consistency and stability with the above cases are given.

8.2 EULER METHOD WITH DIRAC-DELTA

In this section, we consider the following fractal-fractional Cauchy problem:

$$\frac{dy}{dt^\beta} = f(t, y(t)) \text{ if } t > 0,$$
$$= y(0) = y_0, \text{ if } t = 0. \qquad (8.1)$$

Assumptions:

1. $y(t)$ is differentiable in a classical way,
2. $f(t,y(t))$ differentiable or at least continuous,
3. $f(t,y(t))$ is bounded and stratify Lipschitz condition,

$$\frac{dy}{dt} = \beta t^{\beta-1} f(t,y(t)) \text{ if } t > 0,$$
$$= y(0) = y_0, \text{ if } t = 0. \quad (8.2)$$

Thanks to the differentiability of $y(t)$. Now applying the integral both sides, yields

$$y(t) = y(0) + \beta \int_0^t \tau^{\beta-1} f(\tau,y(\tau)) d\tau,$$
$$y(0) = y_0. \quad (8.3)$$

At $t = t_{n+1}$, we have

$$y(t_{n+1}) = y(0) + \beta \int_0^{t_{n+1}} \tau^{\beta-1} f(\tau,y(\tau)) d\tau,$$
$$y(0) = y_0. \quad (8.4)$$

At $t = t_n$, we have

$$y(t_n) = y(0) + \beta \int_0^{t_n} \tau^{\beta-1} f(\tau,y(\tau)) d\tau,$$
$$y(0) = y_0. \quad (8.5)$$

Subtracting fields

$$y(t_{n+1}) = y(t_n) + \beta \int_{t_n}^{t_{n+1}} \tau^{\beta-1} f(\tau,y(\tau)) d\tau. \quad (8.6)$$

We approximate $f(\tau,y(\tau))$ within $[t_n, t_{n+1}]$

$$y_{n+1} = y_n + \beta f(t_n,y_n) \int_{t_n}^{t_{n+1}} \tau^{\beta-1} d\tau,$$
$$= y_n + \beta f(t_n,y_n) \left[\frac{t_{n+1}^\beta}{\beta} - \frac{t_n^\beta}{\beta}\right],$$
$$= y_n + f(t_n,y_n)(\Delta t)^\beta \left((n+1)^\beta - n^\beta\right). \quad (8.7)$$

Therefore, the recursive formula is given by

$$y_{n+1} = y_n + (\Delta t)^\beta \left((n+1)^\beta - n^\beta\right) f(t_n,y_n), \quad (8.8)$$

Fractal-Fractional Euler Method

of course, when $\beta = 1$, we have

$$y_{n+1} = y_n + \Delta t f(t_n, y_n). \tag{8.9}$$

We consider verify the consistency and the stability of the obtained scheme

$$\left| y(t_{n+1}) - y_{n+1} \right| = \left| y(t_n) + \beta \int_{t_n}^{t_{n+1}} f(\tau, y(\tau)) \tau^{\beta-1} d\tau \right. \\ \left. - y_n - (\Delta t)^\beta f(t_n, y_n)\left((n+1)^\beta - n^\beta\right) \right|, \tag{8.10}$$

$$\left| y(t_{n+1}) - y_{n+1} \right| \leq \left| y(t_n) - y_n \right| + \left| \int_{t_n}^{t_{n+1}} \beta \tau^{\beta-1} f(\tau, y(\tau)) \right. \\ \left. - f(t_n, y(t_n)) \right| d\tau. \tag{8.11}$$

Since the function $f(t, y(t))$ is differentiable, thanks to the mean value theorem, we have

$$\left| y(t_{n+1}) - y_{n+1} \right| \leq \left| y(t_n) - y_n \right| + \left| \int_{t_n}^{t_{n+1}} \beta \tau^{\beta-1} \left| f'(c, y(c) \right| d\tau, \tag{8.12}$$

where $c \in [t_n, t_{n+1}]$,

$$\left| y(t_{n+1}) - y_{n+1} \right| \leq \left| y(t_n) - y_n \right| + |f'(c, y(c)| \left\{ t_{n+1}^\beta - t_n^\beta \right\},$$
$$\leq \left| y(t_n) - y_n \right| + |f'(c, y(c)|(\Delta t)^\beta \left\{ t_{n+1}^\beta - t_n^\beta \right\}. \tag{8.13}$$

$$\lim_{\Delta t \to 0} \left| y(t_{n+1}) - y_{n+1} \right| \leq \lim_{\Delta t \to 0} \left| y(t_n) - y_n \right|, \tag{8.14}$$

$$\lim_{\Delta t \to 0, 1 \leq n \leq N} \frac{\left| y(t_{n+1}) - y_{n+1} \right|}{\left| y(t_n) - y_n \right|} \leq 1. \tag{8.15}$$

$$\lim_{\Delta t \to 0, n \to \infty} \left| y(t_{n+1}) - y_{n+1} \right| \leq q^n \left| y(t_0) - y_0 \right| = 0, \tag{8.16}$$

since $q \leq 1$. The perturbation scheme is given as

$$\widetilde{y}_{n+1} + y_{n+1} = \widetilde{y}_n + y_n + (\Delta t)^\beta \left\{ (n+1)^\beta - n^\beta \right\} f(t_n, y_n + \widetilde{y}_n). \tag{8.17}$$

Subtraction from non-perturbation scheme gives

$$\widetilde{y}_{n+1} = \widetilde{y}_n + (\Delta t)^\beta \left\{ (n+1)^\beta - n^\beta \right\} \left(f(t_n, y_n + \widetilde{y}_n) - f(t_n, y_n) \right). \tag{8.18}$$

$$|\tilde{y}_{n+1}| \leq |\tilde{y}_n| + (\Delta t)^\beta \{(n+1)^\beta - n^\beta\} \left|\left(f(t_n, y_n + \tilde{y}_n) - f(t_n, y_n)\right)\right|,$$
$$\leq |\tilde{y}_n| + (\Delta t)^\beta \{(n+1)^\beta - n^\beta\} K |\tilde{y}_n|. \quad (8.19)$$

Since $f(t,y)$ is Lipschitz respect to y, therefore,

$$|\tilde{y}_{n+1}| \leq |\tilde{y}_n| \left(1 + (\Delta t)^\beta K \{(n+1)^\beta - n^\beta\}\right). \quad (8.20)$$

Further,

$$\frac{|\tilde{y}_n|}{|\tilde{y}_{n+1}|} \leq \frac{1}{\left(1 + (\Delta t)^\beta K \{(n+1)^\beta - n^\beta\}\right)}, \quad (8.21)$$

but $\{(n+1)^\beta - n^\beta\} \leq 1$, which shows the stability of the scheme.

8.3 FRACTAL-FRACTIONAL EULER METHOD WITH THE EXPONENTIAL KERNEL

In this section, we consider the following Cauchy problem:

$$\begin{aligned} {}_0^{FFF}D_t^{\alpha,\beta} y(t) &= f(t, y(t)), \text{ if } t > 0, \\ y(0) &= y_0, \text{ if } t = 0. \end{aligned} \quad (8.22)$$

$$\begin{aligned} {}_0^{CFR}D_t^\alpha y(t) &= \beta t^{\beta-1} f(t, y(t)) \text{ if } t > 0 \\ y(0) &= y_0 \text{ if } t = 0 \end{aligned} \quad (8.23)$$

$$\begin{aligned} y(t) &= (1-\alpha)\beta t^{\beta-1} f(t, y(t)) \\ &\quad + \alpha\beta \int_0^t \tau^{\beta-1} f(\tau, y(\tau))d\tau, \text{ if } t > 0, \\ y(0) &= y_0, \text{ if } t = 0, \end{aligned} \quad (8.24)$$

at $t = t_{n+1}$, we have

$$\begin{aligned} y(t_{n+1}) &= \beta t_{n+1}^{\beta-1} f(t_{n+1}, y(t_{n+1})) \\ &\quad + \alpha\beta \int_0^{t_{n+1}} \tau^{\beta-1} f(\tau, y(\tau))d\tau, \\ &= \beta t_{n+1}^{\beta-1} f(t_{n+1}, y^P(t_{n+1})) \end{aligned}$$

Fractal-Fractional Euler Method

$$+\alpha\beta \sum_{j=0}^{n} \int_{t_j}^{t_{j+1}} f(\tau,y(\tau))\tau^{\beta-1}d\tau \tag{8.25}$$

$$\begin{aligned}
y_{n+1} &= \beta t_{n+1}^{\beta-1} f(t_{n+1},y^p(t_{n+1})) + \alpha\beta \sum_{j=0}^{n} \int_{t_j}^{t_{j+1}} \tau^{\beta-1} d\tau, \\
&= \beta t_{n+1}^{\beta-1} f(t_{n+1},y^p(t_{n+1})) + \alpha\beta \sum_{j=0}^{n} f(t_j,y_j)\left\{\frac{t_{j+1}^{\beta}}{\beta} - \frac{t_j^{\beta}}{\beta}\right\}, \\
&= \beta t_{n+1}^{\beta-1} f(t_{n+1},y^p(t_{n+1})) + \alpha \sum_{j=0}^{n} f(t_j,y_j)\left\{(j+1)^{\beta} - j^{\beta}\right\}, \\
&= \beta t_{n+1}^{\beta-1} f(t_{n+1},y^p(t_{n+1})) + \alpha(\Delta t)^{\beta} \sum_{j=0}^{n} f(t_j,y_j) \\
&\quad \times \left\{(j+1)^{\beta} - j^{\beta}\right\},
\end{aligned} \tag{8.26}$$

where

$$y^p(t_{n+1}) = y_n + \alpha(\Delta t)^{\beta}\{(n+1)^{\beta} - n^{\beta}\} f(t_n,y_n). \tag{8.27}$$

Alternatively, we have

$$\begin{aligned}
y_{n+1} &= y_n + \alpha\beta\left(t_{n+1}^{\beta-1} f(t_n,y_{n+1}^p) - t_n^{\beta-1} f(t_n,y_n)\right) \\
&\quad + \alpha(\Delta t)^{\beta}\{(n+1)^{\beta} - n^{\beta}\} f(t_n,y_n),
\end{aligned} \tag{8.28}$$

where

$$y_{n+1}^p = y_n + \alpha(\Delta t)^{\beta}\{(n+1)^{\beta} - n^{\beta}\} f(t_n,y_n). \tag{8.29}$$

The stability and consistency of the above method can be deduced from the one presented earlier.

8.4 FRACTAL-FRACTIONAL EULER METHOD FOR POWER LAW KERNEL

In this section, we consider the fractal-fractional Cauchy problem with power law kernel.

$$\begin{aligned}
{}_0^{FFP}D_t^{\alpha,\beta} y(t) &= f(t,y(t)), \text{ if } t > 0, \\
y(0) &= y_0, \text{ if } t = 0.
\end{aligned} \tag{8.30}$$

$${}_0^{RL}D_t^{\alpha} y(t) = \beta t^{\beta-1} f(t,y(t)) \text{ if } t > 0$$

We write

$$y(t) = \frac{\beta}{\Gamma(\alpha)} \int_0^t \tau^{\beta-1} f(\tau, y(\tau))(t-\tau)^{\alpha-1} d\tau, \; y(0) = y_0, \quad (8.32)$$

at $t = t_{n+1}$, we have

$$y(t_{n+1}) = \frac{\beta}{\Gamma(\alpha)} \int_0^{t_{n+1}} \tau^{\beta-1} f(\tau, y(\tau))(t_{n+1}-\tau)^{\alpha-1} d\tau, \quad (8.33)$$

$$\begin{aligned} y(t_{n+1}) &= \frac{\beta}{\Gamma(\alpha)} \sum_{j=0}^{n} \int_{t_j}^{t_{j+1}} \tau^{\beta-1} f(\tau, y(\tau))(t_{n+1}-\tau)^{\alpha-1} d\tau, \\ &\approx \frac{\beta}{\Gamma(\alpha)} \sum_{j=0}^{n} f(t_j, y_j) \int_{t_j}^{t_{j+1}} \tau^{\beta-1} (t_{n+1}-\tau)^{\alpha-1} d\tau, \\ &\approx \frac{\beta}{\Gamma(\alpha)} \sum_{j=0}^{n} f(t_j, y_j) t_{n+1}^{\alpha+\beta-1} \\ &\quad \times \left[B\left(\frac{t_{j+1}}{t_{n+1}}, \beta, \alpha\right) - B\left(\frac{t_j}{t_{n+1}}, \beta, \alpha\right) \right]. \end{aligned} \quad (8.34)$$

Of course that if $\beta = 1$, we have

$$y_{n+1} = \frac{(\Delta t)^\alpha}{\Gamma(\alpha+1)} \sum_{j=0}^{n} f(t_j, y_j) \delta_{n,j}^{\alpha,1}, \quad (8.35)$$

where

$$\delta_{n,j}^{\alpha,1} = (n-j+1)^\alpha - (n-j)^\alpha, \quad (8.36)$$

if in addition, $\alpha = 1$, then we have

$$\delta_{n,j}^{1,1} = 1. \quad (8.37)$$

Thus,

$$y(t_{n+1}) \approx y_{n+1} = (\Delta t) \sum_{j=0}^{n} f(t_j, y_j), \quad (8.38)$$

$$y_{n+1} = (\Delta t) \sum_{j=0}^{n} f(t_j, y_j), \quad (8.39)$$

$$\left| y(t_{n+1}) - y_{n+1} \right| = \left| \frac{\beta}{\Gamma(\alpha)} \sum_{j=0}^{n} \int_{t_j}^{t_{j+1}} \tau^{\beta-1} f(\tau, y(\tau))(t_{n+1}-\tau)^{\alpha-1} d\tau \right.$$

Fractal-Fractional Euler Method

$$-\frac{\beta}{\Gamma(\alpha)}\sum_{j=0}^{n}f(t_j,y_j)t_{n+1}^{\beta+\alpha-1}\left[B\left(\frac{t_{j+1}}{t_{n+1}},\beta,\alpha\right)-\left(\frac{t_j}{t_{n+1}},\beta,\alpha\right)\right]\Big|,$$

$$=\Big|\frac{\beta}{\Gamma(\alpha)}\sum_{j=0}^{n}\int_{t_j}^{t_{j+1}}\tau^{\beta-1}f(\tau,y(\tau))(t_{n+1}-\tau)^{\alpha-1}d\tau$$

$$-\frac{\beta}{\Gamma(\alpha)}\sum_{j=0}^{n}\int_{t_j}^{t_{j+1}}f(t_j,y_j)\tau^{\beta-1}(t_{n+1}-\tau)^{\beta-1}d\tau\Big|,$$

$$\leq \frac{\beta}{\Gamma(\alpha)}\sum_{j=0}^{n}\int_{t_j}^{t_{j+1}}\tau^{\beta-1}\Big|f(\tau,y(\tau))-f(t_j,y_j)\Big|$$

$$(t_{n+1}-\tau)^{\alpha-1}d\tau. \tag{8.40}$$

Since $f(t,y(t))$ is differentiable with respect to y, we have

$$\Big|y(t_{n+1})-y_{n+1}\Big| \leq \frac{\beta}{\Gamma(\alpha)}\sum_{j=0}^{n}\int_{t_j}^{t_{j+1}}\tau^{\beta-1}|f'(c,y(c))|(\tau-t_j)(t_{n+1}-\tau)^{\alpha-1}d\tau.$$

$$\Big|y(t_{n+1})-y_{n+1}\Big| \leq \frac{\beta}{\Gamma(\alpha)}\sum_{j=0}^{n}|f'(c,y(c))|\int_{t_j}^{t_{j+1}}(\tau-t_j)(t_{n+1}-\tau)^{\alpha-1}d\tau.$$

Noting that

$$\int_{t_j}^{t_{j+1}}\tau^{\beta-1}(\tau-t_j)(t_{n+1}-\tau)^{\alpha-1}d\tau = \int_{t_j}^{t_{j+1}}\tau^{\beta}(t_{n+1}-\tau)^{\alpha-1}d\tau$$

$$-t_j\int_{t_j}^{t_{j+1}}\tau^{\beta-1}(t_{n+1}-\tau)^{\alpha-1}d\tau,$$

where

$$\int_{t_j}^{t_{j+1}}\tau^{\beta}(t_{n+1}-\tau)^{\alpha-1}d\tau = t_{n+1}^{\alpha+\beta}\left[B\left(\frac{t_{j+1}}{t_{n+1}},\beta+1,\alpha\right)\right.$$

$$\left.-B\left(\frac{t_j}{t_{n+1}},\beta+1,\alpha\right)\right], \tag{8.41}$$

and

$$\int_{t_j}^{t_{j+1}}\tau^{\beta-1}(t_{n+1}-\tau)^{\alpha-1}d\tau = t_{n+1}^{\alpha+\beta-1}\left[B\left(\frac{t_{j+1}}{t_{n+1}},\beta,\alpha\right)\right.$$

$$\left.-B\left(\frac{t_j}{t_{n+1}},\beta,\alpha\right)\right]. \tag{8.42}$$

Replacing yields

$$\Big|y(t_{n+1})-y_{n+1}\Big| \leq \frac{\beta}{\Gamma(\alpha)}|f'(c,y(c))|\sum_{j=0}^{n}t_{n+1}^{\alpha+\beta-1}\left[t_{n+1}B\left(\frac{t_{j+1}}{t_{n+1}},\beta+1,\alpha\right)\right.$$

$$-t_{n+1}B\left(\frac{t_j}{t_{n+1}}, \beta+1, \alpha\right)$$
$$-t_jB\left(\frac{t_{j+1}}{t_{n+1}}, \beta, \alpha\right) + t_jB\left(\frac{t_j}{t_{n+1}}, \beta, \alpha\right)\bigg]. \tag{8.43}$$

$$\left|y(t_{n+1}) - y_{n+1}\right| \leq \frac{\beta}{\Gamma(\alpha)}(\Delta t)^{\alpha+\beta}|f'(c,y(c))|\sum_{j=0}^{n}(n+1)^{\alpha+\beta-1}$$
$$\left[(n+1)B\left(\frac{t_{j+1}}{t_{n+1}}, \beta+1, \alpha\right) - (n+1)B\left(\frac{t_j}{t_{n+1}}, \beta+1, \alpha\right)\right.$$
$$\left. - jB\left(\frac{t_{j+1}}{t_{n+1}}, \beta, \alpha\right) + jB\left(\frac{t_j}{t_{n+1}}, \beta, \alpha\right)\right]. \tag{8.44}$$

Therefore,

$$\lim_{\Delta t \to 0}\left|y(t_{n+1}) - y_{n+1}\right| \leq \lim_{\Delta t \to 0}\frac{\beta}{\Gamma(\alpha)}(\Delta t)^{\alpha+\beta}|f'(c,y(c))|\sum_{j=0}^{n}(n+1)^{\alpha+\beta-1}$$
$$\left[(n+1)B\left(\frac{t_{j+1}}{t_{n+1}}, \beta+1, \alpha\right) - (n+1)B\left(\frac{t_j}{t_{n+1}}, \beta+1, \alpha\right)\right.$$
$$\left. - jB\left(\frac{t_{j+1}}{t_{n+1}}, \beta, \alpha\right) + jB\left(\frac{t_j}{t_{n+1}}, \beta, \alpha\right)\right] = 0. \tag{8.45}$$

Therefore,

$$\lim_{\Delta t \to 0}\left|y(t_{n+1}) - y_{n+1}\right| = 0, \tag{8.46}$$

which shows that the scheme is consistent.

8.5 FRACTAL-FRACTIONAL EULER METHOD WITH THE GENERALISED MITTAG-LEFFLER

In this section, a fractal-fractional Cauchy problem with the generalised Mittag-Leffler function is considered

$$\begin{aligned}{}_{0}^{FFM}D_t^{\alpha,\beta}y(t) &= f(t,y(t)), \text{ if } t > 0, \\ y(0) &= y_0, \text{ if } t = 0. \end{aligned} \tag{8.47}$$

$$\begin{aligned}{}_{0}^{ABR}D_t^{\alpha}y(t) &= \beta t^{\beta-1}f(t,y(t)) \text{ if } t > 0 \\ y(0) &= y_0 \text{ if } t = 0. \end{aligned} \tag{8.48}$$

$$y(t) = \beta t^{\beta-1}f(t,y(t))(1-\alpha) + \frac{\alpha\beta}{\Gamma(\alpha)}\int_0^t$$

Fractal-Fractional Euler Method

$$\times \tau^{\beta-1}(t-\tau)^{\alpha-1}f(\tau,y(\tau))d\tau, \text{ if } t > 0$$

$$y(0) = y_0 \text{ if } t = 0. \tag{8.49}$$

At $t = t_{n+1}$, we have

$$y(t_{n+1}) = \beta t_{n+1}^{\beta-1} f(t_{n+1}, y^p(t_{n+1}))(1-\alpha)$$
$$+ \frac{\alpha\beta}{\Gamma(\alpha)} \int_0^{t_{n+1}} \tau^{\beta-1}(t_{n+1}-\tau)^{\alpha-1}f(\tau,y(\tau))d\tau. \tag{8.50}$$

$$y(t_{n+1}) = \beta t_{n+1}^{\beta-1} f(t_{n+1}, y^p(t_{n+1}))(1-\alpha)$$
$$+ \frac{\alpha\beta}{\Gamma(\alpha)} \sum_{j=0}^{n} \int_{t_j}^{t_{j+1}} \tau^{\beta-1}(t_{n+1}-\tau)^{\alpha-1}f(\tau,y(\tau))d\tau. \tag{8.51}$$

$$y(t_{n+1}) \approx y_{n+1} = \beta t_{n+1}^{\beta-1} f(t_{n+1}, y_{n+1}^p)(1-\alpha)$$
$$+ \frac{\alpha\beta}{\Gamma(\alpha)} \sum_{j=0}^{n} \int_{t_j}^{t_{j+1}} \tau^{\beta-1}(t_{n+1}-\tau)^{\alpha-1}f(\tau,y(\tau))d\tau. \tag{8.52}$$

$$y_{n+1} = \beta t_{n+1}^{\beta-1} f(t_{n+1}, y_{n+1}^p)(1-\alpha)$$
$$+ \frac{\alpha\beta}{\Gamma(\alpha)} \sum_{j=0}^{n} f(t_j, y_j) \int_{t_j}^{t_{j+1}} (t_{n+1}-\tau)^{\alpha-1}\tau^{\beta-1}d\tau. \tag{8.53}$$

$$y_{n+1} = \beta t_{n+1}^{\beta-1} f(t_{n+1}, y_{n+1}^p)(1-\alpha) + \frac{\alpha\beta}{\Gamma(\alpha)} \sum_{j=0}^{n} f(t_j, y_j)$$
$$\times t_{n+1}^{\alpha+\beta-1} \left[B\left(\frac{t_{j+1}}{t_{n+1}}, \beta, \alpha\right) - B\left(\frac{t_j}{t_{n+1}}, \beta, \alpha\right) \right]. \tag{8.54}$$

$$y_{n+1} = \beta t_{n+1}^{\beta-1} f(t_{n+1}, y_{n+1}^p)(1-\alpha) + \frac{\alpha\beta}{\Gamma(\alpha)} (\Delta t)^{\alpha+\beta-1}(n+1)^{\alpha+\beta}$$
$$\sum_{j=0}^{n} f(t_j, y_j) \left[B\left(\frac{t_{j+1}}{t_{n+1}}, \beta, \alpha\right) - B\left(\frac{t_j}{t_{n+1}}, \beta, \alpha\right) \right], \tag{8.55}$$

where

$$y_{n+1}^p = \frac{\alpha\beta}{\Gamma(\alpha)} (\Delta t)^{\alpha+\beta-1}(n+1)^{\alpha+\beta-1} \sum_{j=0}^{n} f(t_j, y_j)$$
$$\times \left[B\left(\frac{t_{j+1}}{t_{n+1}}, \beta, \alpha\right) - B\left(\frac{t_j}{t_{n+1}}, \beta, \alpha\right) \right]. \tag{8.56}$$

9 Application of Fractal-Fractional Operators to a Chaotic Model

9.1 INTRODUCTION

The concept of local differentiation and integration is known to be the classical differentiation. The classical differentiation is further extended into the concept of fractional derivative with Caputo type with local and singular kernel. Further, the idea was extended in the sense of Caputo-Fabrizio operator with non-local and singular kernel. Moreover, the fractional calculus was extended to have operator with non-local and non-singular kernel known as Atangana-Baleanu operator. The beginning of the fractional calculus was the use of the power law kernel for the classical derivative. This concept has been used by the researchers working in different branch of science and engineering and achieved useful results. With the new concept of differentiation and integration with singular and son-singular kernel have been used in literature extensively by the researchers around the globe, see [75, 95, 110, 169, 181, 211].

The literature related to fractional operators, and the most important integral and operators that attain much interest from the readers and researchers around the world are Riemann-Liouville and Caputo derivatives, Caputo-Fabrizio fractional derivative, and the Atangana-Baleanu. These fractional derivative are based on some of the fundamental laws such as power law, exponential decay law with Delta Dirac, and the generalised Mittag-Leffler function and the first derivative. Some recent related work with these defined operators and their kernel is given in [74, 76, 111, 112, 195, 242, 279]. This discussion comes from the exchange of ideas among L Hopital and Leibniz, while another innovative results established by the authors for classical differentiations known as fractals [47, 48, 97, 227]. This operators has been used in many areas but not get popularity among researchers working in science and related fields. However, this fractal derivative appears in many problems in nature such as ground water transport problem, etc. [10]. The aims of this chapter is to investigate the dynamics of a chaotic model in different fractal-fractional operators. The newly operators we use to apply on the model are fractal-fractional Atangana-Baleanu, fractal-fractional Caputo-Fabrizio, and the fractal-fractional Caputo derivative. We apply each operator on the model and provide a novel numerical procedure for their solution. The numerical results with various values of the fractal and fractional operators are shown. We suggest that these new operators in the sense

DOI: 10.1201/9781003359258-9

of Atangana-Baleanu, Caputo-Fabrizio, and Caputo provide better understanding of the chaotic dynamics.

9.2 MODEL

The aim of the present section is to present the novel procedure for solution of the chaotic problem in different fractal-fractional operators. We consider three different newly defined operators known as fractal-fractional Caputo, fractal-fractional Caputo-Fabrizio, and the fractal-fractional Atangana-Baleanu. The chaotic model we consider here is given by the following equations:

$$\begin{cases} \frac{dx(t)}{dt} = Ay(t), \\ \frac{dy(t)}{dt} = x(t) - x^3(t) - Ay(t) + B\cos(Ct), \\ \frac{dz(t)}{dt} = By(t) + \sin(y(t)) + x(t), \end{cases} \quad (9.1)$$

where $x(t)$, $y(t)$, and $z(t)$ are the state variables while A, B, and C are the parameters. This system is chaotic for the given values of the parameters $A = 0.25$, $B = 0.3$, $C = 1$ with the initial conditions on state variables $x(0) = 0.01$, $y(0) = 0.5$, and $z(0) = -0.6$. The fractal-fractional representation of the model (9.1) is given by

$$\begin{cases} {}^{FF}D_{0,t}^{p,q}\bigl(x(t)\bigr) = Ay(t), \\ {}^{FF}D_{0,t}^{p,q}\bigl(y(t)\bigr) = x(t) - x^3(t) - Ay(t) + B\cos(Ct), \\ {}^{FF}D_{0,t}^{p,q}\bigl(z(t)\bigr) = By(t) + \sin(y(t)) + x(t), \end{cases} \quad (9.2)$$

where p represents the fractional order while q is fractal order. The equilibrium points associated with the model (9.2) are discussed in the following subsection.

9.2.1 FIXED POINTS

Here, we determine the equilibrium points of the model (9.2) in fractal-fractional derivatives by setting the time rate of change equal to zero

$$\begin{cases} 0 = Ay(t), \\ 0 = x(t) - x^3(t) - Ay(t) + B\cos(Ct), \\ 0 = By(t) + \sin(y(t)) + x(t), \end{cases} \quad (9.3)$$

The solution of the equation (9.3) give one equilibrium point say $E_1 = (0,0,0)$. At the given equilibrium point E_1 in the model (9.2), we determine their asymptotical stability. The related stability of the given system is shown in the following form:

$$J(E_1) = \begin{pmatrix} 0 & A & 0 \\ 1 & -A & 0 \\ 1 & B+\sin'(0) & 0 \end{pmatrix}.$$

Now, we have the eigenvalues of the matrix $J(E_1)$ which are given by $\lambda_1 = 0$, $\lambda_2 = \frac{1}{2}\left(-A - \sqrt{A+4}\sqrt{A}\right)$, and $\lambda_3 = \frac{1}{2}\left(\sqrt{A}\sqrt{A+4} - A\right)$. It can be observed that the first eigenvalues are zero with no imaginary part, and the other two have the negative real parts. So, we can say that the model at the given equilibrium point is unstable.

9.3 EXISTENCE AND UNIQUENESS

Due to the highly nonlinear nature of the model under discussion, finding the precise solution is quite difficult. In order to obtain the numerical solution to the system, one will therefore rely on the current numerical scheme. However, demonstrating that the system has a singular system solution under certain circumstances is frequently a mathematical exercise. To meet the Lipchitz criterion and linear growth, these conditions must be met. This section of the model's existence and uniqueness system is presented. We begin by assuming that all of the solutions are continuous and bounded.

$$\frac{dx(t)}{dt} = Ay(t),$$
$$\frac{dy(t)}{dt} = x(t) - x^3(t) - Ay(t) + B\cos(Ct),$$
$$\frac{dz(t)}{dt} = By(t) + \sin(y(t)) + x(t). \tag{9.4}$$

We write

$$\frac{dx(t)}{dt} = f_x(t,x,y,z),$$
$$\frac{dy(t)}{dt} = f_y(t,x,y,z),$$
$$\frac{dz(t)}{dt} = f_z(t,x,y,z). \tag{9.5}$$

We shall find the following conditions

$$\left\| f_x(t,x,y,z) \right\|_\infty^2 < K_x \left(1 + \|x\|_\infty^2 \right) \tag{9.6}$$

$$\left\| f_x(t,x_1,y,z) - f_x(t,x_2,y,z) \right\|_\infty^2 \leq \overline{K}_x \left\| x_1 - x_2 \right\|_\infty^2. \tag{9.7}$$

This will be shown for all f_y and f_z.

$$\left\| f_x(t,x,y,z) \right\|^2 = A^2 \|y\|^2. \tag{9.8}$$

$$\left\| f_x(t,x,y,z) \right\|^2 \leq \sup_{t \in [0,T]} \left\| f_x(t,x,y,z) \right\|^2,$$
$$\leq A^2 \sup_{t \in [0,T]} |y(t)|^2,$$
$$\leq A^2 \|y\|_\infty^2 (1 + \|x\|_\infty^2),$$

$$\leq K_x\left(1+\|x\|_\infty^2\right), \tag{9.9}$$

where $K_x = A^2\|y\|_\infty^2$. With $f_y(t,x,y,z)$ we have

$$\begin{aligned}\left\|f_y(t,x,y,z)\right\|_\infty^2 &\leq 4\|x\|_\infty^2 + 4\|x\|_\infty^6 + 4A^2\|y\|_\infty^2 B^2, \\ &\leq 4\left(\|x\|_\infty^2 + \|x\|_\infty^6 + B^2\right) \\ &\quad \left(1 + \frac{A^2}{\|x\|_\infty^2 + \|x\|_\infty^6 + B^2}\|y\|_\infty^2\right).\end{aligned} \tag{9.10}$$

If $\frac{A^2}{\|x\|_\infty^2 + \|x\|_\infty^6 + B^2} < 1$, then

$$\left\|f_y(t,x,y,z)\right\|_\infty^2 \leq K_y\left(1+\|y\|_\infty^2\right). \tag{9.11}$$

For $f_z(t,x,y,z)$, we have the following result

$$\begin{aligned}\left\|f_z(t,x,y,z)\right\|_\infty^2 &\leq 3B^2\|y\|_\infty^2 + 3 + 3\|x\|_\infty^2, \\ &\leq \left(3B^2\|y\|_\infty^2 + 3 + 3\|x\|_\infty^2\right)\left(1+\|z\|_\infty^2\right), \\ &\leq K_z\left(1+\|z\|_\infty^2\right).\end{aligned} \tag{9.12}$$

Therefore, if $\frac{A^2}{\|x\|_\infty^2 + \|x\|_\infty^6 + B^2} < 1$. The system satisfies the linear growth. We now check the Lipschitz condition:

$$\left\|f_x(t,x_1,y,z) - f_x(t,x_2,y,z)\right\|_\infty^2 = 0 \leq \|x_1 - x_2\|_\infty^2 \tag{9.13}$$

$$\begin{aligned}\left\|f_y(t,x,y_1,z) - f_y(t,x,y_2,z)\right\|_\infty^2 &= A^2\|y_1 - y_2\|_\infty^2, \\ &\leq \frac{3}{2}A^2\|y_1 - y_2\|_\infty^2, \\ &\leq \overline{K}_y\|y_1 - y_2\|_\infty^2.\end{aligned} \tag{9.14}$$

Finally,

$$\left\|f_z(t,x,y,z_1) - f_x(t,x_2,y,z_2)\right\|_\infty^2 = 0 \leq \left(1+\|z\|_\infty^2\right). \tag{9.15}$$

The system also satisfies the Lipschitz condition. We conclude that if $\frac{A^2}{\|x\|_\infty^2 + \|x\|_\infty^6 + B^2} < 1$, then under the condition that $\|x\|_\infty^2 < M_x$, $\|y\|_\infty^2 < M_y$, and $\|z\|_\infty^2 < M_z$. The system admits a unique system of solution.

Application of Fractal-Fractional Operators to a Chaotic Model   95

9.4 STABILITY OF THE USED NUMERICAL SCHEME

Numerical stability is a widely desired quality of numerical algorithms in the mathematical discipline of numerical analysis. Depending on the situation, stability might mean several things. The first is numerical linear algebra, and the second are strategies for approximating discrete solutions to ordinary and partial differential equations. Instabilities brought on by proximity to singularities of various types, such as very small or virtually colliding eigenvalues, are the main issue in numerical linear algebra. The worry with numerical methods for differential equations, on the other hand, is the rise of round-off errors and/or tiny changes in initial data that could result in a significant departure from the exact solution in the final answer. We present in this section the stability of the used numerical scheme. We recall that the used method was obtained by approximation of the nonlinear function with Lagrange interpolation. We shall note that between points t_n and t_{n+1}, the Lagrange interpolates the value near function $f(t, y(t))$ as

$$f(t, y(t)) \approx \frac{\tau - t_{n-1}}{\Delta t} f(t_n, y_n) - \frac{\tau - t_n}{\Delta t} f(t_{n-1}, y_{n-1}). \tag{9.16}$$

We shall present the stability of this method for all cases. For the classical derivative, we have the following approximations after using the Lagrange:

$$y_{n+1} = y_n + \frac{3}{2} \Delta t f(t_n, y_n) - \Delta t f(t_{n-1} y_{n-1}), \tag{9.17}$$

which is the approximate solution, whereas the exact solution is given as

$$y_{(t_{n+1})} = y_{(t_n)} + \int_{t_n}^{t_{n+1}} f(\tau, y(\tau)) d\tau. \tag{9.18}$$

For stability, we consider three perturbations terms, $\tilde{y}_{n+1}$, $\tilde{y}_n$, and $\tilde{y}_{n-1}$, which represent the perturbations of y_{n+1}, y_n and y_{n-1}, respectively. The perturbed equation becomes

$$y_{n+1} + \tilde{y}_{n+1} = y_n + \tilde{y}_n + \frac{3}{2} \Delta t f(t_n, y_n + \tilde{y}_n) - \frac{\Delta t}{2} f(t_n, y_{n-1} + \tilde{y}_{n-1}). \tag{9.19}$$

Subtracting the perturbed equation from approximated yields

$$\begin{aligned} \tilde{y}_{n+1} &= \tilde{y}_n + \frac{3}{2} \Delta t f(t_n, y_n + \tilde{y}_n) - \frac{3}{2} \Delta t f(t_n, y_n) \\ &\quad + \frac{\Delta t}{2} f(t_{n-1}, y_{n-1}) - \frac{\Delta t}{2} f(t_{n-1}, y_{n-1} + \tilde{y}_{n-1}). \end{aligned} \tag{9.20}$$

Of course, we have that the nonlinear function $f(t, y)$ satisfies the Lipschitz conditions with respect to y. Therefore,

$$|\tilde{y}_{n+1}| \leq |\tilde{y}_n| + \frac{3}{2} \Delta t \left| f(t_n, y_n + \tilde{y}_n) - f(t_{n-1}, y_{n-1} + \tilde{y}_{n-1}) \right|$$

$$+\frac{\Delta t}{2}\left|f(t_{n-1},y_{n-1}+\tilde{y}_{n-1})-f(t_n,y_n)\right|. \tag{9.21}$$

Using the Lipschitz condition of f with respect to y, then

$$|\tilde{y}_{n+1}| \leq |\tilde{y}_n| + \frac{3}{2}\Delta t K|\tilde{y}_n| + \frac{\Delta t}{2}K|\tilde{y}_{n-1}|. \tag{9.22}$$

We choose the

$$\max_{0 \leq n \leq N}\{\tilde{y}_{n-1},\tilde{y}_n\},$$

then,

$$|\tilde{y}_{n+1}| = \max_{0 \leq n \leq N}\{\tilde{y}_n,\tilde{y}_{n-1}\}\left\{1+\frac{3}{2}\Delta t K + \frac{\Delta t K}{2}\right\}. \tag{9.23}$$

$$|\tilde{y}_{n+1}| \leq (1+2\Delta t K)\max_{0 \leq n \leq N}\{\tilde{y}_n,\tilde{y}_{n-1}\},$$

$$\leq C \max_{0 \leq n \leq N}\{\tilde{y}_n,\tilde{y}_{n-1}\}, \tag{9.24}$$

which proves that the scheme is stable in the case of classical derivative. The case with $\beta \neq 1$ can be deduced. Since

$$y_{n+1} = y_n + \frac{3}{2}\Delta t \beta t_n^{\beta-1} f(t_n,y_n) - \frac{\Delta t}{2}\beta t_{n-1}^{\beta-1} f(t_{n-1},y_{n-1}), \tag{9.25}$$

in this case following the routine presented above, we obtain

$$|\tilde{y}_{n+1}| \leq (1+2\Delta t^\beta \beta n^{\beta-1})\max_{1 \leq n \leq N}\{\tilde{y}_{n-1},\tilde{y}_n\}. \tag{9.26}$$

9.5 CASE FOR POWER LAW

With the power law function using the Lagrange polynomial interpolation, the Cauchy problem with fractal-fractional derivative can be approximated as follows:

$$\begin{aligned} y_{n+1} &= \frac{\beta(\Delta t)^\alpha}{\Gamma(\alpha)}\sum_{j=1}^n t_j^{\beta-1} f(t_j,y_j) \\ &\quad \times \left((n-j+1)^\alpha(n-j+2+\alpha)-(n-j)^\alpha(n-j+2+2\alpha)\right) \\ &\quad -\frac{\beta(\Delta t)^\alpha}{\Gamma(\alpha)}\sum_{j=1}^n f(t_{j-1},y_{j-1}) \\ &\quad \times t_{j-1}^{\beta-1}\left\{(n-j+1)^{\alpha+1}-(n-j)^\alpha(n-j+1+\alpha)\right\}. \end{aligned} \tag{9.27}$$

We consider the approximate perturbed values $\tilde{y}_{n+1}, \tilde{y}_n, \tilde{y}_j$, and $\tilde{y}_{j-1}$ for y_{n+1}, y_n, y_j, and y_{j-1}, respectively; thus the perturbed scheme becomes.

Application of Fractal-Fractional Operators to a Chaotic Model

$$\tilde{y}_{n+1} + y_{n+1} = \frac{\beta(\Delta t)^\alpha}{\Gamma(\alpha)} \sum_{j=1}^{n} t_j^{\beta-1} f(t_j, y_j + \tilde{y}_j) \delta_{n,j}^{\alpha,1}$$
$$- \frac{\beta(\Delta t)^\alpha}{\Gamma(\alpha)} \sum_{j=1}^{n} t_{j-1}^{\beta-1} f(t_{j-1}, y_{j-1} + \tilde{y}_{j-1}) \delta_{n,j}^{\alpha,2}. \quad (9.28)$$

Subtracting from the approximate solutions yields

$$\tilde{y}_{n+1} = \frac{\beta(\Delta t)^\alpha}{\Gamma(\alpha)} \sum_{j=1}^{n} t_j^{\beta-1} f(t_j, y_j + \tilde{y}_j) \delta_{n,j}^{\alpha,1}$$
$$- \frac{\beta(\Delta t)^\alpha}{\Gamma(\alpha)} \sum_{j=1}^{n} t_{j-1}^{\beta-1} f(t_j, y_j) \delta_{n,j}^{\alpha,1}$$
$$- \frac{\beta(\Delta t)^\alpha}{\Gamma(\alpha)} \sum_{j=1}^{n} t_{j-1}^{\beta-1} f(t_{j-1}, y_{j-1} + \tilde{y}_{j-1}) \delta_{n,j}^{\alpha,2}$$
$$+ \frac{\beta(\Delta t)^\alpha}{\Gamma(\alpha)} \sum_{j=1}^{n} t_{j-1}^{\beta-1} f(t_{j-1}, y_{j-1}) \delta_{n,j}^{\alpha,2}. \quad (9.29)$$

Taking absolute value on both sides yields

$$|\tilde{y}_{n+1}| \leq \frac{(\Delta t)^\alpha \beta}{\Gamma(\alpha)} \sum_{j=1}^{n} t_j^{\beta-1} |\delta_{n,j}^{\alpha,1}| \left| f(t_j, y_j + \tilde{y}_j) - f(t_j, y_j) \right|$$
$$+ \frac{(\Delta t)^\alpha \beta}{\Gamma(\alpha)} \sum_{j=1}^{n} t_{j-1}^{\beta-1} |\delta_{n,j}^{\alpha,2}| \left| f(t_{j-1}, y_{j-1} + \tilde{y}_{j-1}) \right.$$
$$\left. - f(t_{j-1}, y_{j-1}) \right|. \quad (9.30)$$

Using the Lipschitz condition of $f(t, y(t))$ with respect to y, we obtain

$$|\tilde{y}_{n+1}| \leq \frac{(\Delta t)^\alpha \beta}{\Gamma(\alpha)} K \sum_{j=1}^{n} t_j^{\beta-1} |\delta_{n,j}^{\alpha,1}| |\tilde{y}_j|$$
$$+ \frac{(\Delta t)^\alpha \beta}{\Gamma(\alpha)} K \sum_{j=1}^{n} t_{j-1}^{\beta-1} |\delta_{n,j}^{\alpha,2}| |\tilde{y}_{j-1}|. \quad (9.31)$$

Noting that $t_{j-1}^{\beta-1} < t_j^{\beta-1}$, we have

$$|\tilde{y}_{n+1}| \leq \frac{(\Delta t)^\alpha \beta}{\Gamma(\alpha)} K \sum_{j=1}^{n} t_j^{\beta-1} \{|\delta_{n,j}^{\alpha,1}| |\tilde{y}_j| + |\delta_{n,j}^{\alpha,2}| |\tilde{y}_{j-1}|\}. \quad (9.32)$$

But, for all $t_{n+1} \geq t_j$, we choose $\tilde{z}_j = \max_{i \leq j \leq n}\{\tilde{y}_{j-1}, \tilde{y}_j\}$,

$$|\tilde{y}_{n+1}| \leq \frac{(\Delta t)^\alpha \beta}{\Gamma(\alpha)} K (\Delta t)^{\beta-1} n^{\beta-1} \sum_{j=1}^{n} |\tilde{z}_j| \{|\delta_{n,j}^{\alpha,1}| + |\delta_{n,j}^{\alpha,2}|\}. \quad (9.33)$$

We now evaluate $|\delta_{n,j}^{\alpha,1}|$ and $|\delta_{n,j}^{\alpha,2}|$

$$\begin{aligned}|\delta_{n,j}^{\alpha,1}| &= \left\|(n-j+1)^{\alpha}(n-j+2+\alpha)-(n-j)^{\alpha}(n-j+2+2\alpha)\right\|,\\ &\leq (n-j+1)^{\alpha}(n-j+2+\alpha)+(n-j)^{\alpha}(n-j+2+2\alpha),\\ &\leq (n-j+1)^{\alpha}\{n-j+2+\alpha+n-j+2+2\alpha\},\\ &\leq (n-j+1)^{\alpha}\{2n+4+3\alpha-2j\},\\ &\leq (n+1)^{\alpha}(2n+7).\end{aligned}$$ (9.34)

Since $4+3\alpha-2j<7$

$$\begin{aligned}\delta_{n,j}^{\alpha,2} &= \left\|(n-j+1)^{\alpha+1}-(n-j)^{\alpha}(n-j+1+\alpha)\right\|,\\ &\leq (n-j+1)^{\alpha+1}+(n-j)^{\alpha}(n-j+1+\alpha),\\ &\leq (n-j+1)^{\alpha+1}(1+n-j+1+\alpha),\\ &\leq (n-j+1)^{\alpha+1}(3+n-j),\\ &\leq (n+1)^{\alpha+1}(n+3).\end{aligned}$$ (9.35)

Therefore,

$$\begin{aligned}|\tilde{y}_{n+1}| &\leq \frac{(\Delta t)^{\alpha+\beta-1}}{\Gamma(\alpha)}\beta K n^{\beta-1}\sum_{j=1}^{n}|\tilde{z}_j|\{(n+1)^{\alpha}(2n+7)+(n+1)^{\alpha+1}(n+3)\},\\ &\leq \frac{2(\Delta t)^{\alpha+\beta-1}}{\Gamma(\alpha)}BK(n+1)^{\alpha+1}(2n+7)\sum_{j=1}^{n}|\tilde{z}_j|,\\ &\leq C(n+1)^{\alpha+1}(2n+7)\sum_{j=1}^{n}|\tilde{z}_j|,\end{aligned}$$ (9.36)

The case of exponential decay and the generalised Mittag-Leffler functions can be deduced from the case of classical derivative and power law case, respectively. The conclusion is due to the fact that for the case of exponential decay law, we have

$$\begin{aligned}y_{n+1} &= y_n+(1-\alpha)\beta\left(t_n^{\beta-1}f(t_n,y_n)-t_{n-1}^{\beta-1}f(t_{n-1},y_{n-1})\right.\\ &\quad \left.+\alpha\beta\int_{t_n}^{t_{n+1}}\tau^{\beta-1}f(\tau,y(\tau))d\tau\right),\end{aligned}$$ (9.37)

$$y_{n+1} = y_n+(1-\alpha)\beta\left\{t_n^{\beta-1}f(t_n,y_n)-t_{n-1}^{\beta-1}f(t_{n-1},y_{n-1})\right\}$$

Application of Fractal-Fractional Operators to a Chaotic Model 99

$$+\alpha\beta\left\{\frac{3}{2}t_n^{\beta-1}f(t_n,y_n)\Delta t - \frac{t_{n-1}^{\beta-1}}{2}\Delta t f(t_{n-1},y_{n-1})\right\}. \quad (9.38)$$

Thus,

$$\begin{aligned}\widetilde{y}_{n+1} + y_{n+1} &= y_n + \widetilde{y}_n \\ &+ (1-\alpha)\beta\left(t_n^{\beta-1}f(t_n,y_n+\widetilde{y}_n) - t_{n-1}f(t_{n-1},y_{n-1}+\widetilde{y}_{n-1})\right) \\ &+ \alpha\beta\left\{\frac{3}{2}\Delta t t_n^{\beta-1}f(t_n,y_n+\widetilde{y}_n)\right. \\ &\left. - \frac{\Delta t}{2}t_{n-1}^{\beta-1}f(t_{n-1},y_{n-1}+\widetilde{y}_{n-1})\right\}. \quad (9.39)\end{aligned}$$

For the generalised Mittag-Leffler, we have that

$$\begin{aligned}y_{n+1} &= t_n^{\beta-1}\beta(1-\alpha)f(t_n,y_n) + \frac{\alpha\beta}{\Gamma(\alpha+2)}(\Delta t)^\alpha \sum_{j=1}^n t_j^{\beta-1}f(t_j,y_j)\delta_{n,j}^{\alpha,1} \\ &- \frac{\alpha\beta}{\Gamma(\alpha+2)}(\Delta t)^\alpha \sum_{j=1}^n f(t_{j-1},y_{j-1})\delta_{n,j}^{\alpha,2}. \quad (9.40)\end{aligned}$$

Therefore, the perturbed term is given as

$$\begin{aligned}\widetilde{y}_{n+1} + y_{n+1} &= t_n^{\beta-1}\beta(1-\alpha)f(t_n,y_n+\widetilde{y}_n) + \frac{\beta\alpha(\Delta t)^\alpha}{\Gamma(\alpha+2)}\sum_{j=1}^n t_j^{\beta-1}f(t_j,y_j+\widetilde{y}_j) \\ &- \frac{\beta\alpha(\Delta t)^\alpha}{\Gamma(\alpha+2)}\sum_{j=1}^n t_{j-1}f(t_{j-1},y_{j-1}+\widetilde{y}_{j_1}). \quad (9.41)\end{aligned}$$

The subtraction gives

$$\begin{aligned}\widetilde{y}_{n+1} &= t_n^{\beta-1}\beta(1-\alpha)\left\{f(t_n,y_n+\widetilde{y}_n) - f(t_n,y_n)\right\} \\ &+ \frac{\beta\alpha(\Delta t)^\alpha}{\Gamma(\alpha+2)}\sum_{j=1}^n t_j^{\beta-1}\left(f(t_j,y_j+\widetilde{y}_j) - f(t_j,y_j)\right)\delta_{n,j}^{\alpha,1} \\ &- \frac{\beta\alpha(\Delta t)^\alpha}{\Gamma(\alpha+2)}\sum_{j=1}^n t_j^\beta\left(f(t_{j-1},y_{j-1}+\widetilde{y}_{j-1})\right. \\ &\left. - f(t_{j-1},y_{j-1})\right)\delta_{n,j}^{\alpha,2}. \quad (9.42)\end{aligned}$$

Thus, applying the absolute value on both sides produces

$$\begin{aligned}|\widetilde{y}_{n+1}| &\leq n^{\beta-1}\Delta t^{\beta-1}(1-\alpha)K|\widetilde{y}_n| + \frac{(\Delta t)^{\alpha+\beta-1}}{\Gamma(\alpha)}\beta K n^{\beta-1} \\ &\times \sum_{j=1}^n |\widetilde{z}_j|\left\{(n+1)^\alpha(2n+7) + (n+1)^{\alpha+1}(n+3)\right\},\end{aligned}$$

$$\leq n^{\beta-1}\Delta t^{\beta-1}(1-\alpha)K|\tilde{y}_n|$$
$$+\frac{2(\Delta t)^{\alpha+\beta-1}}{\Gamma(\alpha)}BK(n+1)^{\alpha+1}(2n+7)\sum_{j=1}^{n}|\tilde{z}_j|,$$
$$\leq n^{\beta-1}\Delta t^{\beta-1}(1-\alpha)K|\tilde{y}_n|+C(n+1)^{\alpha+1}(2n+7)\sum_{j=1}^{n}|\tilde{z}_j|. \quad (9.43)$$

The next subsections deal the numerical procedure for the chaotic model (9.2) in the sense of fractal-fractional Caputo, CF, and AB.

9.6 NUMERICAL SCHEMES AND ITS SIMULATIONS

9.6.1 NUMERICAL PROCEDURE IN THE SENSE OF FRACTAL-FRACTIONAL-CAPUTO OPERATOR

This subsection determines the numerical procedure for the solution of the fractal-fractional model given by (9.2) in fractal-fractional-Caputo operator. We convert the model (9.2) to volterra form as the fractional integral is differentiable, so in the Riemann-Liouville sense, we have

$$^{FFP}D_{0,t}^{p,q}f(t)=\frac{1}{\Gamma(1-p)}\frac{d}{dt}\int_{0}^{t}(t-q)^{-p}f(q)dq\frac{1}{qt^{q-1}}, \quad (9.44)$$

we have the following result:

$$^{RL}D_{0,t}^{p}\left(x(t)\right) = qt^{q-1}\left[Ay(t)\right],$$
$$^{RL}D_{0,t}^{p}\left(y(t)\right) = qt^{q-1}\left[x(t)-x^3(t)-Ay(t)+B\cos(Ct)\right],$$
$$^{RL}D_{0,t}^{p}\left(z(t)\right) = qt^{q-1}\left[By(t)+\sin(y(t))+x(t)\right]. \quad (9.45)$$

The above equations are represented in the form of RL derivative, to change it into Caputo derivative, for the purpose to make sure to have the initial conditions in integer case. So, we have to apply both sides the Riemann-Liouville fractional integral and obtain the result below:

$$x(t) = \frac{q}{\Gamma(p)}\int_{0}^{t}\lambda^{q-1}(t-\lambda)^{p-1}f_1(x,y,z,\lambda)d\lambda,$$
$$y(t) = \frac{q}{\Gamma(p)}\int_{0}^{t}\lambda^{q-1}(t-\lambda)^{p-1}f_2(x,y,z,\lambda)d\lambda,$$
$$z(t) = \frac{q}{\Gamma(p)}\int_{0}^{t}\lambda^{q-1}(t-\lambda)^{p-1}f_3(x,y,z,\lambda)d\lambda, \quad (9.46)$$

where

$$f_1(x,y,z,\lambda) = Ay(t),$$

Application of Fractal-Fractional Operators to a Chaotic Model

$$f_2(x,y,z,\lambda) = x(t) - x^3(t) - Ay(t) + B\cos(Ct),$$

$$f_3(x,y,z,\lambda) = By(t) + \sin(y(t)) + x(t).$$

We now describe here a novel procedure for the model given above using the approach at t_{n+1}. Then, our model becomes

$$x^{n+1} = \frac{q}{\Gamma(p)} \int_0^t \lambda^{q-1}(t_{n+1}-\lambda)^{p-1} f_1(x,y,z,\lambda) d\lambda,$$

$$y^{n+1} = \frac{q}{\Gamma(p)} \int_0^t \lambda^{q-1}(t_{n+1}-\lambda)^{p-1} f_2(x,y,z,\lambda) d\lambda,$$

$$z^{n+1} = \frac{q}{\Gamma(p)} \int_0^t \lambda^{q-1}(t_{n+1}-\lambda)^{p-1} f_3(x,y,z,\lambda) d\lambda. \quad (9.47)$$

The approximations of the above integral lead to the following:

$$x^{n+1} = \frac{q}{\Gamma(p)} \sum_{j=0}^{n} \int_{t_j}^{t_{j+1}} \lambda^{q-1}(t_{n+1}-\lambda)^{p-1} f_1(x,y,z,\lambda) d\lambda,$$

$$y^{n+1} = \frac{q}{\Gamma(p)} \sum_{j=0}^{n} \int_{t_j}^{t_{j+1}} \lambda^{q-1}(t_{n+1}-\lambda)^{p-1} f_2(x,y,z,\lambda) d\lambda,$$

$$z^{n+1} = \frac{q}{\Gamma(p)} \sum_{j=0}^{n} \int_{t_j}^{t_{j+1}} \lambda^{q-1}(t_{n+1}-\lambda)^{p-1} f_3(x,y,z,\lambda) d\lambda. \quad (9.48)$$

Now approximating the function $\lambda^{q-1} f_i(x,y,z,\lambda)$ for $i=1,2,3$ in expression above in the interval $[t_j, t_{j+1}]$ through the Lagrangian piece-wise interpolation, we get

$$P_j(\lambda) = \frac{\lambda - t_{j-1}}{t_j - t_{j-1}} t_j^{q-1} f_1(x^j, y^j, z^j, t_j) - \frac{\lambda - t_j}{t_j - t_{j-1}} t_{j-1}^{q-1} f_1(x^{j-1}, y^{j-1}, z^{j-1}, t_{j-1}),$$

$$Q_j(\lambda) = \frac{\lambda - t_{j-1}}{t_j - t_{j-1}} t_j^{q-1} f_2(x^j, y^j, z^j, t_j) - \frac{\lambda - t_j}{t_j - t_{j-1}} t_{j-1}^{q-1} f_2(x^{j-1}, y^{j-1}, z^{j-1}, t_{j-1}),$$

$$R_j(\lambda) = \frac{\lambda - t_{j-1}}{t_j - t_{j-1}} t_j^{q-1} f_3(x^j, y^j, z^j, t_j)$$
$$- \frac{\lambda - t_j}{t_j - t_{j-1}} t_{j-1}^{q-1} f_3(x^{j-1}, y^{j-1}, z^{j-1}, t_{j-1}). \quad (9.49)$$

We get the following

$$x^{n+1} = \frac{q}{\Gamma(p)} \sum_{j=0}^{n} \int_{t_j}^{t_{j+1}} \lambda^{q-1}(t_{n+1}-\lambda)^{p-1} P_j(\lambda) d\lambda,$$

$$y^{n+1} = \frac{q}{\Gamma(p)} \sum_{j=0}^{n} \int_{t_j}^{t_{j+1}} \lambda^{\kappa-1}(t_{n+1}-\lambda)^{p-1} Q_j(\lambda) d\lambda,$$

$$z^{n+1} = \frac{q}{\Gamma(p)} \sum_{j=0}^{n} \int_{t_j}^{t_{j+1}} \lambda^{\kappa-1}(t_{n+1}-\lambda)^{p-1} R_j(\lambda) d\lambda. \quad (9.50)$$

Solving (9.50), we obtain the following:

$$\begin{aligned}
x^{n+1} &= \frac{qh^p}{\Gamma(p+2)} \sum_{j=1}^{n} \left[t_j^{q-1} f_1(x^j, y^j, z^j, t_j) \right. \\
&\quad \times \left((n+1-j)^p(n-j+2+p) - (n-j)^p(n-j+2+2p) \right) \\
&\quad - t_{j-1}^{q-1} f_1(x^{j-1}, y^{j-1}, z^{j-1}, t_{j-1}) \\
&\quad \left. \times \left((n-j+1)^{p+1} - (n-j)^p(n-j+1+p) \right) \right], \\
y^{n+1} &= \frac{qh^p}{\Gamma(p+2)} \sum_{j=1}^{n} \left[t_j^{q-1} f_2(x^j, y^j, z^j, t_j) \right. \\
&\quad \times \left((n+1-j)^p(n-j+2+p) - (n-j)^p(n-j+2+2p) \right) \\
&\quad - t_{j-1}^{q-1} f_2(x^{j-1}, y^{j-1}, z^{j-1}, t_{j-1}) \\
&\quad \left. \times \left((n-j+1)^{p+1} - (n-j)^p(n-j+1+p) \right) \right], \\
z^{n+1} &= \frac{qh^p}{\Gamma(p+2)} \sum_{j=1}^{n} \left[t_j^{q-1} f_3(x^j, y^j, z^j, t_j) \right. \\
&\quad \times \left((n+1-j)^p(n-j+2+p) - (n-j)^p(n-j+2+2p) \right) \\
&\quad - t_{j-1}^{q-1} f_3(x^{j-1}, y^{j-1}, z^{j-1}, t_{j-1}) \\
&\quad \left. \times \left((n-j+1)^{p+1} - (n-j)^p(n-j+1+p) \right) \right]. \quad (9.51)
\end{aligned}$$

Next, in the following subsection, we present the numerical procedure for the fractal-fractional model in the sense of Caputo-Fabrizio operator.

9.6.2 NUMERICAL PROCEDURE FOR FRACTAL-FRACTIONAL CAPUTO-FABRIZIO OPERATOR

The aim of this subsection is to present a numerical procedure for the solution of the given model in fractal-fractional Caputo-Fabrizio sense. In order to do this, we first write the model in the following form:

$$\begin{aligned}
{}^{CF}D_{0,t}^p \left(x(t) \right) &= qt^{q-1} f_1(x, y, z, t), \\
{}^{CF}D_{0,t}^p \left(y(t) \right) &= qt^{q-1} f_2(x, y, z, t), \\
{}^{CF}D_{0,t}^p \left(z(t) \right) &= qt^{q-1} f_3(x, y, z, t). \quad (9.52)
\end{aligned}$$

Application of Fractal-Fractional Operators to a Chaotic Model

We obtain the following by using the Caputo-Fabrizio integral:

$$x(t) = x^0 + \frac{qt^{q-1}(1-p)}{M(p)}f_1(x,y,z,t) + \frac{pq}{M(p)}\int_0^t \lambda^{q-1}f_1(x,y,z,\lambda)d\lambda,$$

$$y(t) = y^0 + \frac{qt^{q-1}(1-p)}{M(p)}f_2(x,y,z,t) + \frac{pq}{M(p)}\int_0^t \lambda^{q-1}f_2(x,y,z,\lambda)d\lambda,$$

$$z(t) = z^0 + \frac{qt^{q-1}(1-p)}{M(p)}f_3(x,y,z,t)$$
$$+ \frac{pq}{M(p)}\int_0^t \lambda^{q-1}f_3(x,y,z,\lambda)d\lambda, \tag{9.53}$$

The following is presented at t_{n+1}:

$$x^{n+1}(t) = x^n + \frac{qt^{q-1}(1-p)}{M(p)}f_1(x^n,y^n,z^n,t_n) + \frac{pq}{M(p)}\int_0^{t_{n+1}} \lambda^{q-1}f_1(x,y,z,\lambda)d\lambda,$$

$$y^{n+1}(t) = y^n + \frac{qt^{q-1}(1-p)}{M(p)}f_2(x^n,y^n,z^n,t_n) + \frac{pq}{M(p)}\int_0^{t_{n+1}} \lambda^{q-1}f_2(x,y,z,\lambda)d\lambda,$$

$$z^{n+1}(t) = z^n + \frac{qt^{q-1}(1-p)}{M(p)}f_3(x^n,y^n,z^n,t_n)$$
$$+ \frac{pq}{M(p)}\int_0^{t_{n+1}} \lambda^{q-1}f_3(x,y,z,\lambda)d\lambda. \tag{9.54}$$

Further, we have the following:

$$x^{n+1}(t) = x^n + \frac{qt^{q-1}(1-p)}{M(p)}$$
$$\times f_1(x^n,y^n,z^n,t_n) - \frac{qt_{n-1}^{q-1}(1-p)}{M(p)}f_1(x^{n-1},y^{n-1},z^{n-1},t_{n-1})$$
$$+ \frac{pq}{M(p)}\int_{t_n}^{t_{n+1}} \lambda^{q-1}f_1(x,y,z,\lambda)d\lambda,$$

$$y^{n+1}(t) = y^n + \frac{qt^{q-1}(1-p)}{M(p)}$$
$$\times f_2(x^n,y^n,z^n,t_n) - \frac{qt_{n-1}^{q-1}(1-p)}{M(p)}f_2(x^{n-1},y^{n-1},z^{n-1},t_{n-1})$$
$$+ \frac{pq}{M(p)}\int_{t_n}^{t_{n+1}} \lambda^{q-1}f_2(x,y,z,\lambda)d\lambda,$$

$$z^{n+1}(t) = z^n + \frac{qt^{q-1}(1-p)}{M(p)}$$
$$\times f_3(x^n,y^n,z^n,t_n) - \frac{qt_{n-1}^{q-1}(1-p)}{M(p)}f_3(x^{n-1},y^{n-1},z^{n-1},t_{n-1})$$
$$+ \frac{pq}{M(p)}\int_{t_n}^{t_{n+1}} \lambda^{q-1}f_3(x,y,z,\lambda)d\lambda. \tag{9.55}$$

Now using the Lagrange polynomial piece-wise interpolation and integrating, we obtain

$$x^{n+1}(t) = x^n + \frac{qt_n^{q-1}(1-p)}{M(p)}$$
$$\times f_1(x^n, y^n, z^n, t_n) - \frac{qt_{n-1}^{q-1}(1-p)}{M(p)} f_1(x^{n-1}, y^{n-1}, z^{n-1}, t_{n-1})$$
$$+ \frac{pq}{M(p)} \left[\frac{3h}{2} t_n^{q-1} f_1(x^n, y^n, z_2^n, t_n) - \frac{h}{2} t_{n-1}^{q-1} f_1(x^{n-1}, y^{n-1}, z^{n-1}, t_{n-1}) \right],$$

$$y^{n+1}(t) = y^n + \frac{qt_n^{q-1}(1-p)}{M(p)}$$
$$\times f_2(x^n, y^n, z^n, t_n) - \frac{qt_{n-1}^{q-1}(1-p)}{M(p)} f_2(x^{n-1}, y^{n-1}, z^{n-1}, t_{n-1})$$
$$+ \frac{pq}{M(p)} \left[\frac{3h}{2} t_n^{q-1} f_1(x^n, y^n, z_2^n, t_n) - \frac{h}{2} t_{n-1}^{q-1} f_2(x^{n-1}, y^{n-1}, z^{n-1}, t_{n-1}) \right],$$

$$z^{n+1}(t) = z^n + \frac{qt_n^{q-1}(1-p)}{M(p)}$$
$$\times f_3(x^n, y^n, z^n, t_n) - \frac{qt_{n-1}^{q-1}(1-p)}{M(p)} f_1(x^{n-1}, y^{n-1}, z^{n-1}, t_{n-1})$$
$$+ \frac{pq}{M(p)} \left[\frac{3h}{2} t_n^{q-1} f_3(x^n, y^n, z_2^n, t_n) \right.$$
$$\left. - \frac{h}{2} t_{n-1}^{q-1} f_3(x^{n-1}, y^{n-1}, z^{n-1}, t_{n-1}) \right]. \quad (9.56)$$

Further, we obtain the following:

$$x^{n+1} = x^n + qt_n^{q-1} \left(\frac{1-p}{M(p)} + \frac{3ph}{2M(p)} \right) f_1(x^n, y^n, z^n, t_n)$$
$$- qt_{n-1}^{q-1} \left(\frac{1-p}{M(p)} + \frac{ph}{2M(p)} \right) f_1(x^{n-1}, y^{n-1}, z^{n-1}, t_{n-1}),$$
$$y^{n+1} = y^n + qt_n^{q-1} \left(\frac{1-p}{M(p)} + \frac{3ph}{2M(p)} \right) f_2(x^n, y^n, z^n, t_n)$$
$$- qt_{n-1}^{q-1} \left(\frac{1-p}{M(p)} + \frac{ph}{2M(p)} \right) f_2(x^{n-1}, y^{n-1}, z^{n-1}, t_{n-1}),$$
$$z^{n+1} = z^n + qt_n^{q-1} \left(\frac{1-p}{M(p)} + \frac{3ph}{2M(p)} \right) f_3(x^n, y^n, z^n, t_n)$$
$$- qt_{n-1}^{q-1} \left(\frac{1-p}{M(p)} + \frac{ph}{2M(p)} \right) f_3(x^{n-1}, y^{n-1}, z^{n-1}, t_{n-1}). \quad (9.57)$$

Next, we present the numerical procedure for the fractal-fractional in the sense of Atangana-Baleanu derivative in the following subsection:

9.6.3 NUMERICAL PROCEDURE FOR FRACTAL-FRACTIONAL ATANGANA-BALEANU OPERATOR

In order to have a numerical procedure for the solution of fractal-fractional model in the sense of Atangana-Baleanu operator, we express first the model (9.2) in the following form:

$$\begin{aligned}
{}^{ABR}D_{0,t}^{\alpha}\left(x(t)\right) &= qt^{q-1}f_1(x,y,z,t), \\
{}^{ABR}D_{0,t}^{\alpha}\left(y(t)\right) &= qt^{q-1}f_2(x,y,z,t), \\
{}^{ABR}D_{0,t}^{\alpha}\left(z(t)\right) &= qt^{q-1}f_3(x,y,z,t).
\end{aligned} \qquad (9.58)$$

The following is presented on the basis of Atangana-Baleanu integral:

$$\begin{aligned}
x(t) &= \frac{qt^{q-1}(1-p)}{AB(p)}f_1(x,y,z,t) \\
&+ \frac{pq}{AB(p)\Gamma(p)}\int_0^t \lambda^{q-1}(t-\lambda)^{p-1}f_1(x,y,z,\lambda)d\lambda, \\
y(t) &= \frac{qt^{q-1}(1-p)}{AB(p)}f_2(x,y,z,t) \\
&+ \frac{pq}{AB(p)\Gamma(p)}\int_0^t \lambda^{q-1}(t-\lambda)^{p-1}f_2(x,y,z,\lambda)d\lambda, \\
z(t) &= \frac{qt^{q-1}(1-p)}{AB(p)}f_3(x,y,z,t) \\
&+ \frac{pq}{AB(p)\Gamma(p)}\int_0^t \lambda^{q-1}(t-\lambda)^{p-1}f_3(x,y,z,\lambda)d\lambda.
\end{aligned} \qquad (9.59)$$

Further, at t_{n+1}, we get

$$\begin{aligned}
x^{n+1} &= \frac{qt_n^{q-1}(1-p)}{AB(\alpha)}f_1(x^n,y^n,z^n,t_n) \\
&+ \frac{pq}{AB(p)\Gamma(p)}\int_0^{t_{n+1}} \lambda^{q-1}(t_{n+1}-\lambda)^{p-1}f_1(x,y,z,\lambda)d\lambda, \\
y^{n+1} &= \frac{qt_n^{q-1}(1-p)}{AB(\alpha)}f_2(x^n,y^n,z^n,t_n) \\
&+ \frac{pq}{AB(p)\Gamma(p)}\int_0^{t_{n+1}} \lambda^{q-1}(t_{n+1}-\lambda)^{p-1}f_2(x,y,z,\lambda)d\lambda, \\
z^{n+1} &= \frac{qt_n^{q-1}(1-p)}{AB(\alpha)}f_3(x^n,y^n,z^n,t_n) \\
&+ \frac{pq}{AB(p)\Gamma(p)}\int_0^{t_{n+1}} \lambda^{q-1}(t_{n+1}-\lambda)^{p-1}f_3(x,y,z,\lambda)d\lambda. \quad (9.60)
\end{aligned}$$

We obtain the following after simplification the integral in the above equations:

$$x^{n+1} = \frac{qt_n^{q-1}(1-p)}{AB(p)} f_1(x^n, y^n, z^n, t_n)$$
$$+ \frac{pq}{AB(p)\Gamma(p)} \sum_{j=0}^{n} \int_{t_j}^{t_{j+1}} \lambda^{q-1}(t_{n+1}-\lambda)^{p-1} f_1(x,y,z,\lambda) d\lambda,$$

$$y^{n+1} = \frac{qt_n^{q-1}(1-p)}{AB(p)} f_2(x^n, y^n, z^n, t_n)$$
$$+ \frac{pq}{AB(p)\Gamma(p)} \sum_{j=0}^{n} \int_{t_j}^{t_{j+1}} \lambda^{q-1}(t_{n+1}-\lambda)^{p-1} f_2(x,y,z,\lambda) d\lambda,$$

$$z^{n+1} = \frac{qt_n^{q-1}(1-p)}{AB(p)} f_3(x^n, y^n, z^n, t_n)$$
$$+ \frac{pq}{AB(p)\Gamma(p)} \sum_{j=0}^{n} \int_{t_j}^{t_{j+1}} \lambda^{q-1}(t_{n+1}-\lambda)^{p-1} f_3(x,y,z,\lambda) d\lambda. \quad (9.61)$$

Now approximating the expressions in (9.61) given by $\lambda^{q-1} f_1(x,y,z,\lambda)$, $\lambda^{q-1} f_2(x,y,z,\lambda)$, and $\lambda^{q-1} f_3(x,y,z,\lambda)$ in the given internal $[t_j, t_{j+1}]$, the following numerical scheme is presented:

$$x^{n+1} = \frac{qt_n^{q-1}(1-p)}{AB(p)} f_1(x^n, y^n, z^n, t_n)$$
$$+ \frac{q(\Delta t)^p}{AB(p)\Gamma(p+2)} \sum_{j=1}^{n} \Big[t_j^{q-1} f_1(x^j, y^j, z^j, t_j)$$
$$\times \Big((n+1-j)^p(n-j+2+\alpha) - (n-j)^p(n-j+2+2p)\Big)$$
$$- t_{j-1}^{q-1} f_1(x^{j-1}, y^{j-1}, z^{j-1}, t_{j-1})$$
$$\times \Big((n-j+1)^{p+1} - (n-j)^p(n-j+1+p)\Big) \Big],$$

$$y^{n+1} = \frac{qt_n^{q-1}(1-p)}{AB(p)} f_2(x^n, y^n, z^n, t_n)$$
$$+ \frac{q(\Delta t)^p}{AB(p)\Gamma(p+2)} \sum_{j=1}^{n} \Big[t_j^{q-1} f_2(x^j, y^j, z^j, t_j)$$
$$\times \Big((n+1-j)^p(n-j+2+\alpha) - (n-j)^p(n-j+2+2p)\Big)$$
$$- t_{j-1}^{q-1} f_2(x^{j-1}, y^{j-1}, z^{j-1}, t_{j-1})$$
$$\times \Big((n-j+1)^{p+1} - (n-j)^p(n-j+1+p)\Big) \Big],$$

$$z^{n+1} = \frac{qt_n^{q-1}(1-p)}{AB(p)} f_3(x^n, y^n, z^n, t_n)$$

$$+ \frac{q(\Delta t)^p}{AB(p)\Gamma(p+2)} \sum_{j=1}^{n} \left[t_j^{q-1} f_3(x^j, y^j, z^j, t_j) \right.$$
$$\times \left((n+1-j)^p(n-j+2+\alpha) - (n-j)^p(n-j+2+2p) \right)$$
$$- t_{j-1}^{q-1} f_3(x^{j-1}, y^{j-1}, z^{j-1}, t_{j-1})$$
$$\left. \times \left((n-j+1)^{p+1} - (n-j)^p(n-j+1+p) \right) \right]. \tag{9.62}$$

9.7 NUMERICAL RESULTS

In this section, we obtain the numerical solution of the fractal-fractional model (9.2). We applied above three different fractal-fractional operators in the sense of Caputo, Caputo-Fabrizio, and Atangana-Baleanu on model (9.2). We obtain many graphical illustrations for many values of the fractional order p and fractal order q. In this simulation, we consider the step size 0.01 and the initial value of t is 0.1. We obtained the graphical results for the fractal-fractional model in the sense of Caputo, Caputo-Fabrizio, and Atangana-Baleanu in Figures 9.1–9.15. Figures 9.1–9.5 represent the numerical solution of the fractal-fractional model in the sense of Caputo operator. Figures 9.6–9.9 represent the graphical results for fractal-fractional model in the sense of Caputo-Fabrizio operator while the graphical results in 9.10–9.15 are for the Atangana-Baleanu fractional fractal operators. Figures are obtained for each operator in the way by considering initially fixing the fractal order and varies the fractional order, then, we fix the fractional order and varies the fractal order, varying both fractal and fractional order parameters for the same values and then for both the arbitrary values. From these graphical results, we obtain some new interesting results, in the case of fractal-fractional operators in the sense of Caputo, Caputo-Fabrizio, and the Atangana-Baleanu. It can be seen from the graphical results that fractal-fractional operators on a chaotic model provide an interesting graphical results, while such results cannot be seen in integer order derivative.

9.8 CONCLUSION

We presented a new chaotic model and obtained the suitable parameters for which the model behave chaotically. We presented by applying the newly defined operators known as fractal-fractional in the sense of Caputo, Caputo-Fabrizio, and the Atangana-Baleanu. We apply each operator on the chaotic model and presented a novel numerical procedure for their solution. The numerical results with arbitrary values of the fractal and fractional order parameters are obtained and discussed in details. The fractal-fractional operators in the sense of Caputo, Caputo-Fabrizio, and the Atangana-Baleanu operator attract interesting results.

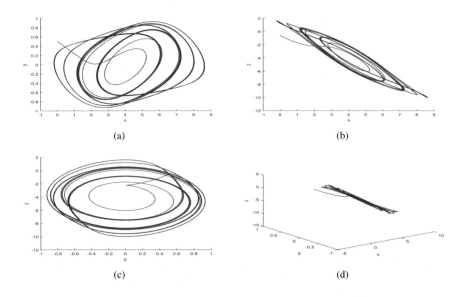

Figure 9.1 Model simulation of fractal-fractional Caputo model when $p=1, q=1$.

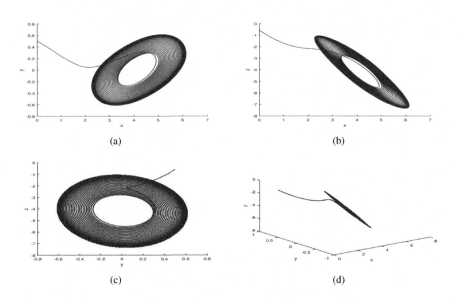

Figure 9.2 Model simulation of fractal-fractional Caputo model when $p=0.95, q=1$.

Application of Fractal-Fractional Operators to a Chaotic Model

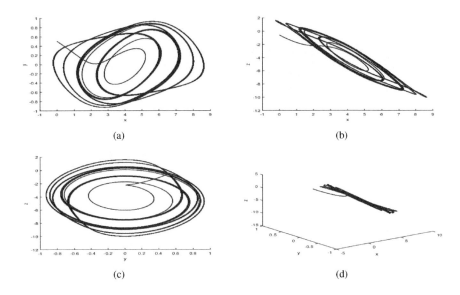

Figure 9.3 Model simulation of fractal-fractional Caputo model when $p = 01, q = 0.95$.

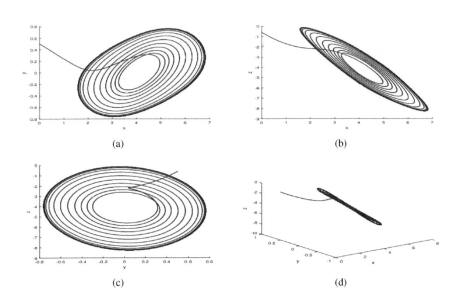

Figure 9.4 Model simulation of fractal-fractional Caputo model when $p = q = 0.96$.

110 Numerical Methods for Fractal-Fractional Differential

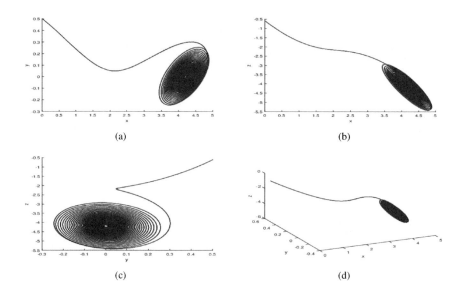

Figure 9.5 Model simulation of fractal-fractional Caputo model when $p = q = 0.94$.

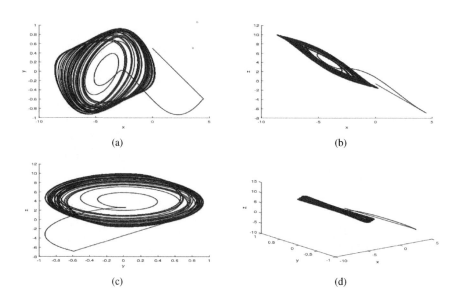

Figure 9.6 Model simulation of fractal-fractional Caputo-Fabrizio model when $p = 1$, $q = 1$.

Application of Fractal-Fractional Operators to a Chaotic Model 111

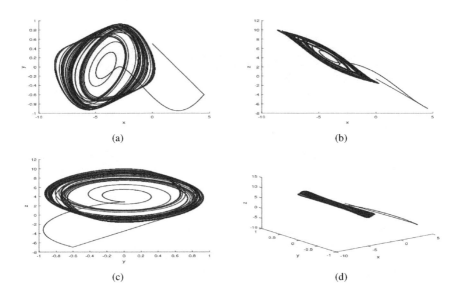

Figure 9.7 Model simulation of fractal-fractional Caputo-Fabrizio model when $p = 1$, $q = 0.99$.

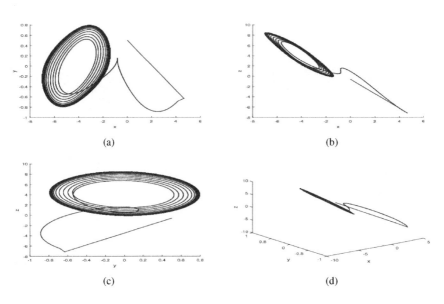

Figure 9.8 Model simulation of fractal-fractional Caputo-Fabrizio model when $p = q = 0.97$.

112 Numerical Methods for Fractal-Fractional Differential

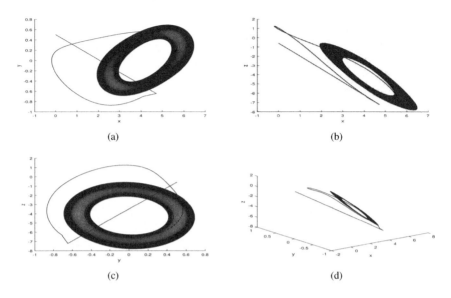

Figure 9.9 Model simulation of fractal-fractional Caputo-Fabrizio model when $p = q = 0.96$.

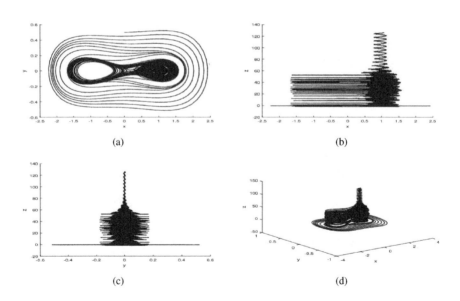

Figure 9.10 Model simulation of fractal-fractional Atangana-Baleanu model when $p = q = 1$.

Application of Fractal-Fractional Operators to a Chaotic Model

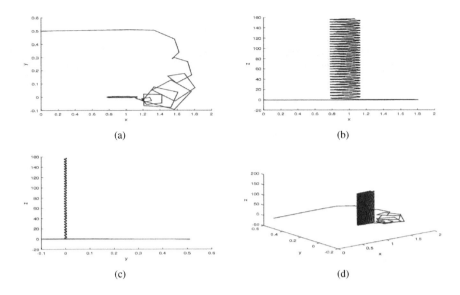

Figure 9.11 Model simulation of fractal-fractional Atangana-Baleanu model when $p = 0.95, q = 1$.

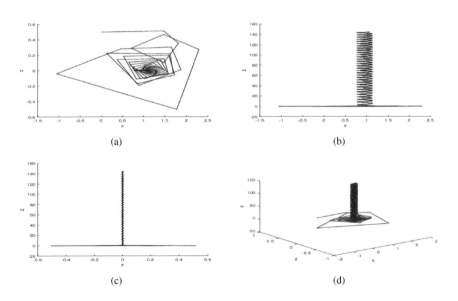

Figure 9.12 Model simulation of fractal-fractional Atangana-Baleanu model when $p = 0.93, q = 1$.

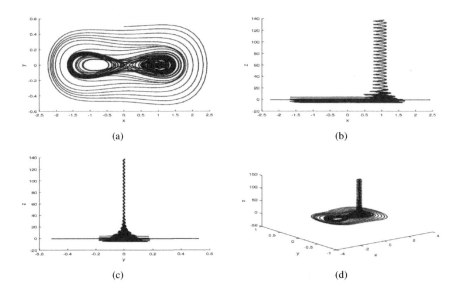

Figure 9.13 Model simulation of fractal-fractional Atangana-Baleanu model when $p = 01, q = 0.99$.

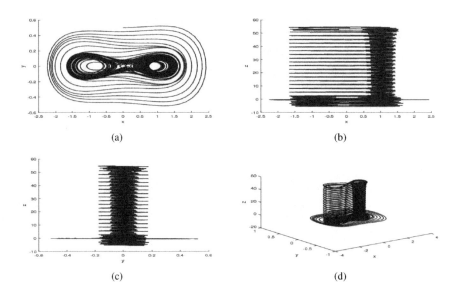

Figure 9.14 Model simulation of fractal-fractional Atangana-Baleanu model when $p = 01, q = 0.95$.

Application of Fractal-Fractional Operators to a Chaotic Model 115

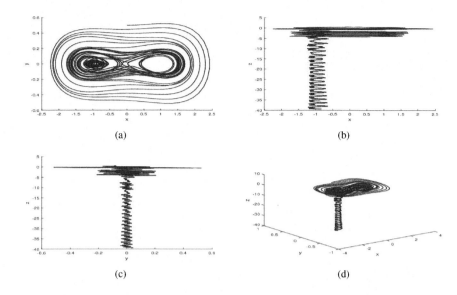

Figure 9.15 Model simulation of fractal-fractional Atangana-Baleanu model when $p = 01, q = 0.8$.

10 Fractal-Fractional Modified Chua Chaotic Attractor

10.1 INTRODUCTION

The chaotic models or systems are nonlinear extremely with the properties of highly sensitive to the initial conditions. It includes the properties such as sensitive to the initial values of the model variables, sporadic, and unusual conduct. The butterfly effect is considered the important name of the sensitivity of chaotic models [60]. After the discovery of the first chaotic system in 1963 by Lorenz [150] which consists of three equations, numerous chaotic systems have been developed and analysed [135, 146, 274]. Chaotic systems relying on the Lorenz include Lu and Chen chaotic systems [152] and Liu chaotic system [145]. The chaotic system developed in the past has three dimensional with one positive Lyapunov exponent. But there may be some other fascinating cases which include the complex factors that are not studied effectively. For example, the mind-bogglin Lorenz conditions that are used to reproduce and depict physics of detuned laser and thermal convection of fluid flows [72] while some of its dynamical properties are found contemplated [157]. The nuclear polarisation sufficiency and the electric field abundance are both complex, see for more information [197] and reference therein. Further, for application of the chaotic systems in security system, see [186], and some other complicated complexes such as Lu and Chen systems are formulated and considered as of late in [158]. The present chapter investigates the dynamics of a new chaotic model in fractal-fractional operators. The model is formulated, and the appropriate initial conditions are proposed for which the model becomes chaotic. Fractal-fractional operators in the sense of Caputo, Caputo-Fabrizio, and Atangana-Baleanu have been applied on the model and shown their results. We present a novel numerical procedure to obtain the graphical results for the fractal-fractional model with various values of the fractal and fractional orders.

10.2 MODEL FRAMEWORK

In the present section, we consider a new chaotic model given by the following system of differential equations:

$$\begin{cases} \frac{dx(t)}{dt} = \gamma(y(t) - h), \\ \frac{dy(t)}{dt} = x(t) - y(t) + z(t), \\ \frac{dz(t)}{dt} = -\beta y(t), \end{cases} \quad (10.1)$$

where $h = -B \tan\left(\frac{\pi x(t)}{2A} + d\right)$, $x(t)$, $y(t)$, and $z(t)$, are the state variables while γ, B, A, d, and β are the parameters. This system is chaotic for the given values of the parameters $\gamma = 6.91$, $B = 0.24$, $A = 2.1$, $d = 0$, $\beta = 12.57$ with the initial conditions on state variables are $x(0) = 2$, $y(0) = 1$, and $z(0) = 1$. The fractal-fractional representation of the model (10.1) is given by:

$$\begin{cases} {}^{FF}D_{0,t}^{p,q}\Big(x(t)\Big) = \gamma(y(t) - h), \\ {}^{FF}D_{0,t}^{p,q}\Big(y(t)\Big) = x(t) - y(t) + z(t), \\ {}^{FF}D_{0,t}^{p,q}\Big(z(t)\Big) = -\beta y(t), \end{cases} \tag{10.2}$$

where p represents the fractional order while q is fractal order. The equilibrium points of the model (10.2) are obtained by setting

$$\begin{cases} 0 = \gamma(y(t) - h), \\ 0 = x(t) - y(t) + z(t), \\ 0 = -\beta y(t), \end{cases} \tag{10.3}$$

we get

$$E = (x^*, y^*, z^*) = (-\frac{2Ad}{\pi}, 0, \frac{2Ad}{\pi}). \tag{10.4}$$

The Jacobian matrix evaluated at point E is given by

$$J(E) = \begin{pmatrix} \frac{B\pi\gamma}{2A} & \gamma & 0 \\ 1 & -1 & 1 \\ 0 & -\beta & 0 \end{pmatrix}. \tag{10.5}$$

For the Jacobian matrix $J(E)$, we have the following eigenvalues

$$1, -\frac{\pi\beta B\gamma}{2A}, \beta - \frac{\pi B\gamma}{2A}. \tag{10.6}$$

We see that $\lambda_1 = 1$, that is positive and hence λ_1 do not possess the negative real part. The second eigenvalue $\lambda_2 = -\frac{\pi\beta B\gamma}{2A} < 0$ and $\lambda_3 = \beta - \frac{\pi B\gamma}{2A}$, and it can be negative if $2\beta A < \pi B\gamma$. From the above discussion, we conclude that the given system is unstable at E.

10.3 EXISTENCE AND UNIQUENESS CONDITIONS

We assume that for all $t \in [0, T]$, the functions $x(t)$, $y(t)$, and $z(t)$ are bounded. For simplicity, we put

$$\frac{dx}{dt} = r(y(t) - h) = f_x(t, x, y, z),$$

$$\frac{dy}{dt} = x(t) - y(t) + z(t) = f_y(t, x, y, z),$$

$$\frac{dz}{dt} = -\beta y(t) = f_z(t,x,y,z), \tag{10.7}$$

we note that $h = -B\tan\left(\frac{\pi x(t)}{2A} + d\right)$.

$$\|x\|_\infty^2 \leq M_x, \ \|y\|_\infty^2 \leq M_y, \text{ and } \|z\|_\infty^2 \leq M_z. \tag{10.8}$$

We shall prove that $f_x, f_y,$ and f_z satisfy both the linear growth and the Lipschitz conditions. We start with the linear growth

$$\begin{aligned}\|f_x(t,x,y,z)\|^2 &= 2r^2\|y(t)\|^2 + r^2\|h\|^2,, \\ &\leq 2r^2\Big(\|y(t)\|^2 + \|h\|^2\Big). \end{aligned} \tag{10.9}$$

We shall note that

$$\tan z = \int_0^z \frac{dz}{\cos^2 \tau} = x + \frac{1}{3}x^3 + ..., \tag{10.10}$$

For the sake of our own demonstration, we consider the first term of the above expression, such that

$$h \approx B\left(\frac{\pi x(t)}{2A} + d\right) \tag{10.11}$$

since h has negative sign. Therefore, replacing yields

$$\begin{aligned}\|f_x(t,x,y,z)\|^2 &\leq 2r^2\left(\|y(t)\|^2 + \left(\frac{B\pi}{A}\right)^2\|x(t)\|^2 + \left(\frac{B\pi}{A}d\right)^2\right), \\ &\leq 2r^2\left(\sup_{t\in D_y}\|y(t)\|^2 + \left(\frac{B\pi}{A}\right)^2\|x(t)\|^2 + \left(\frac{B\pi}{A}d\right)^2\right), \\ &\leq 2r^2\left(M_y + \left(\frac{B\pi}{A}\right)^2 d^2\right)\left[1 + \frac{\left(\frac{B\pi}{A}\right)^2\|x(t)\|}{M_y + \left(\frac{\pi Bd}{A}\right)^2}\right], \\ &\leq 2r^2\left(M_y + \left(\frac{B\pi d}{A}\right)^2\right)\left[1 + \frac{\left(\frac{B\pi}{A}\right)^2\|x(t)\|^2}{M_y + \left(\frac{\pi Bd}{A}\right)^2}\right], \end{aligned} \tag{10.12}$$

if $\dfrac{\left(\frac{B\pi}{A}\right)^2}{M_y + \left(\frac{\pi Bd}{A}\right)^2} < 1$, then

$$\|f_x(t,x,y,z)\|^2 \leq 2r^2\left(M_y + \left(\frac{B\pi d}{A}\right)^2\right)\left(1 + \|x(t)\|^2\right),$$

$$\leq K_x\left(1+\|x\|^2\right), \tag{10.13}$$

where

$$K_x = 2r^2\left(M_y + \left(\frac{B\pi d}{A}\right)^2\right). \tag{10.14}$$

$$\begin{aligned}
\|f_y(t,x,y,z)\|^2 &= |x(t)-y(t)+z(t)|^2, \\
&\leq 3|x(t)|^2 + 3|y(t)|^2 + 3|z(t)|^2, \\
&\leq 3\sup_{t\in D_x}|x(t)|^2 + 3|y(t)|^2 + 3\sup_{t\in D_z}|z(t)|^2, \\
&\leq 3\left(M_x + M_z + |y(t)|^2\right), \\
&\leq 3\left(M_x + M_z\right)\left(1 + \frac{|y(t)|^2}{M_x + M_z}\right).
\end{aligned} \tag{10.15}$$

If $\frac{1}{M_x+M_z} < 1$, then

$$\begin{aligned}
|f_y(t,x,y,z)|^2 &\leq 3(M_x + M_z)(1+|y(t)|^2), \\
&\leq K_y\left(1+|y|^2\right),
\end{aligned} \tag{10.16}$$

where $K_y = 3(M_x + M_z)$

$$\begin{aligned}
\|f_z(t,x,y,z)\|^2 &= |\beta y(t)|^2, \\
&= \beta^2|y(t)|^2 \leq \beta^2\|y\|_\infty^2, \\
&\leq \beta^2\|y\|_\infty^2(1+|z|^2), \\
&\leq \beta^2 M_z(1+|z|^2), \\
&\leq K_z(1+|z|^2),
\end{aligned} \tag{10.17}$$

where $K_z = \beta^2 M_z$. The system satisfies the linear growth condition if the following inequality is reached

$$\max\left\{\frac{1}{M_x+M_z}, \frac{\left(\frac{\beta\pi}{A}\right)^2}{M_x+\left(\frac{\pi Bd}{A}\right)^2}\right\} < 1. \tag{10.18}$$

We shall now verify the Lipschitz condition of the system

$$|f_x(t,x_1,y,z) - f_x(t,x_2,y,z)|^2 = r^2|h_1 - h_2|^2,$$

$$= r^2 \left| Btan\left(\frac{\pi x_1}{2A} + d\right) \right.$$
$$\left. - Btan\left(\frac{\pi x_2}{2A} + d\right) \right|^2,$$
$$= r^2 B^2 \left| tan\left(\frac{\pi x_1}{2A} + d\right) \right.$$
$$\left. - tan\left(\frac{\pi x_2}{2A} + d\right) \right|^2. \quad (10.19)$$

Using approximation for the sake of our demonstration, we obtain

$$|f_x(t,x_1,y,z) - f_x(t,x_2,y,z)|^2 = r^2 B^2 \frac{\pi^2}{4A^2}|x_1 - x_2|^2,$$
$$\leq \frac{3r^2 B^2 \pi^2}{4A^2}|x_1 - x_2|^2,$$
$$\leq \overline{K}_x |x_1 - x_2|^2, \quad (10.20)$$

where

$$\overline{K}_x = \frac{3r^2 B^2 \pi^2}{4A^2}.$$

$$|f_y(t,x,y_1,z) - f_y(t,x,y_2,z)|^2 = |y_1 - y_2|^2,$$
$$= \frac{3}{2}|y_1 - y_2|^2, \quad (10.21)$$

where $K_y = \frac{3}{2}$

$$|f_x(t,x,y,z_1) - f_x(t,x,y,z_2)|^2 = 0 \leq (1 + |z|^2). \quad (10.22)$$

Therefore, the system verifies the Lipschitz condition. We conclude that if

$$\max\left\{\frac{1}{M_x + M_z}, \frac{\left(\frac{B\pi}{A}\right)^2}{M_z + \left(\frac{\pi Bd}{A}\right)^2}\right\} < 1, \quad (10.23)$$

then, the system admits a unique solution.

10.4 CONSISTENCY OF THE SCHEME

In this section, we present the consistency of the used method. The aim is to show that $y(t_{n+1})$ is the exact solution of a given fractal-fractional differential equation, y_{n+1} its approximated solution,

$$\lim_{h \to 0} |y(t_{n+1}) - y_{n+1}| = 0. \quad (10.24)$$

For the case with fractal-derivative, we have

$$y_{n+1} = y_n + \frac{3}{2}t_n^{\beta-1}\beta\Delta t f(t_n, y_n) - \frac{\Delta t \beta t_{n-1}^{\beta-1}}{2}f(t_{n-1}, y_{n-1}). \qquad (10.25)$$

Whereas the exact solution is

$$y(t_{n+1}) = y_{(t_n)} + \beta \int_{t_n}^{t_{n+1}} \tau^{\beta-1} f(\tau, y(\tau))d\tau. \qquad (10.26)$$

Therefore, substituting provides

$$\begin{aligned}
|y(t_{n+1}) - y_{n+1}| &= \left| y_{t_n} + \beta \int_{t_n}^{t_{n+1}} \tau^{\beta-1} f(\tau, y(\tau))d\tau - y_n - \frac{3}{2}t_n^{\beta-1}\beta\Delta t f(t_n, y_n) \right.\\
&\quad \left. + \frac{\Delta t \beta t_{n-1}^{\beta-1}}{2}f(t_{n-1}, y_{n-1}) \right|,\\
&= \left| y(t_n) - y_n + \beta \int_{t_n}^{t_{n+1}} \tau^{\beta-1} f(\tau, y(\tau))d\tau \right.\\
&\quad - \beta \int_{t_n}^{t_{n+1}} t_n^{\beta-1} f(t_n, y_n)d\tau\\
&\quad \left. - \frac{\Delta t \beta t_n^{\beta-1}}{2}f(t_n, y_n) + \frac{\beta \Delta t}{2}t_{n-1}^{\beta-1} f(t_{n-1}, y_{n-1}) \right|,\\
&\leq |y_{t_n} - y_n| + \beta \int_{t_n}^{t_{n+1}} \left| \tau^{\beta-1} f(\tau, y(\tau)) - t_n^{\beta-1} f(t_n, y_n) \right| d\tau\\
&\quad + \frac{\Delta t \beta}{2} \left| t_n^{\beta-1} f(t_n, y_n) - t_{n-1}^{\beta-1} f(t_{n-1}, y_{n-1}) \right|,\\
&\leq |y_{t_n} - y_n| + \beta \left| \left(\tau^{\beta-1} f(\tau, y(\tau)) \right)'_{\tau=c} \right| \int_{t_n}^{t_{n+1}} (\tau - t_n)d\tau\\
&\quad + \frac{(\Delta t)^2 \beta}{2} \left| \left(t^{\beta-1} f(t, y(t)) \right)'_{t=c_1} \right|, \qquad (10.27)
\end{aligned}$$

where $c_1 \in [t_n, t_{n-1}]$ and $c \in [t_n, t_{n+1}]$. Thanks to the mean value theorem,

$$|y(t_{n+1}) - y_{n+1}| \leq |y(t_n) - y_n| + \beta M_c \frac{(\Delta t)^2}{2} + \frac{(\Delta t)^2}{2}\beta M_c, \qquad (10.28)$$

where

$$\begin{aligned}
M_c &= (\beta - 1)c^{\beta-2} f(c, y(c)) + c^{\beta-1} f'(c, y(c)),\\
M_{c_1} &= (\beta - 1)c_1^{\beta-2} f(c_1, y(c_1)) + c_1^{\beta-1} f'(c_1, y(c_1)).
\end{aligned}$$

Thus,

$$\lim_{\Delta t \to 0} |y(t_{n+1}) - y_{n+1}| \leq |y(t_n) - y_n|, \qquad (10.29)$$

Fractal-Fractional Modified Chua Chaotic Attractor

for all $n \geq 0$,

$$\lim_{\Delta t \to 0} \frac{|y(t_{n+1}) - y_{n+1}|}{|y(t_n) - y_n|} \leq 1, \tag{10.30}$$

$$\lim_{\Delta t \to 0} |y(t_{n+1}) - y_{n+1}| \leq q^n |\delta(t_0) - \delta_0|, \tag{10.31}$$

where $q < 1$

$$\lim_{\Delta t \to 0, n \to \infty} |y(t_{n+1}) - y_{n+1}| = 0. \tag{10.32}$$

This shows that the method is consistent.

10.4.1 FOR THE CASE OF POWER LAW

For the case of power law, we have that the exact solution is

$$y_{(t_{n+1})} = \frac{\beta}{\Gamma(\alpha)} \int_0^{t_{n+1}} \tau^{\beta-1} (t_{n+1} - \tau)^{\alpha-1} d\tau, \tag{10.33}$$

and the approximate solution is given by

$$\begin{aligned} y_{n+1} &= \frac{\beta \alpha (\Delta t)^\alpha}{\Gamma(\alpha+2)} \sum_{j=1}^n t_j^{\beta-1} f(t_j, y_j) \delta_{n,j}^{\alpha,1} \\ &\quad - \frac{\beta \alpha (\Delta t)^\alpha}{\Gamma(\alpha+2)} \sum_{j=1}^n t_{j-1}^{\beta-1} f(t_{j-1}, y_{j-1}) \delta_{n,j}^{\alpha,2}. \end{aligned} \tag{10.34}$$

Therefore,

$$\begin{aligned} |y_{(t_{n+1})} - y_{n+1}| &= \left| \frac{\beta}{\Gamma(\alpha)} \int_0^{t_{n+1}} \tau^{\beta-1} (t_{n+1} - \tau)^{\alpha-1} d\tau \right. \\ &\quad - \frac{\beta \alpha (\Delta t)^\alpha}{\Gamma(\alpha+2)} \sum_{j=1}^n t_j^{\beta-1} f(t_j, y_j) \delta_{n,j}^{\alpha,1} \\ &\quad \left. + \frac{\beta \alpha (\Delta t)^\alpha}{\Gamma(\alpha+2)} \sum_{j=1}^n t_{j-1}^{\beta-1} f(t_{j-1}, y_{j-1}) \delta_{n,j}^{\alpha,2} \right|, \\ &= \left| \frac{\beta}{\Gamma(\alpha)} \sum_{j=0}^n \int_{t_j}^{t_{j+1}} \tau^{\beta-1} (t_{n+1} - \tau)^{\alpha-1} f(\tau, y(\tau)) d\tau \right. \\ &\quad - \frac{\beta \alpha (\Delta t)^\alpha}{\Gamma(\alpha+2)} \sum_{j=1}^n t_j^{\beta-1} f(t_j, y_j) \delta_{n,j}^{\alpha,1} \\ &\quad \left. + \frac{\beta \alpha}{\Gamma(\alpha+2)} \sum_{j=1}^n t_{j-1}^{\beta-1} f(t_{j-1}, y_{j-1}) \delta_{n,j}^{\alpha,2} \right|. \end{aligned} \tag{10.35}$$

$$\left|y_{(t_{n+1})} - y_{n+1}\right| = \left|\frac{\beta}{\Gamma(\alpha)} \sum_{j=0}^{n} \int_{t_j}^{t_{j+1}} \tau^{\beta-1}(t_{n+1} - \tau)^{\alpha-1} f(\tau, y(\tau)) d\tau \right.$$

$$-\frac{\beta}{\Gamma(\alpha)} \sum_{j=0}^{n} \int_{t_j}^{t_{j+1}} (t_{n+1} - \tau)^{\alpha-1} \left[t_j^{\beta-1} \frac{\tau - t_{j-1}}{\Delta t} f(t_j, y_j)\right.$$

$$\left.\left. + t_{j-1}^{\beta-1} \frac{\tau - t_j}{\Delta t} f(t_{j-1}, y_{j-1})\right] d\tau \right|,$$

$$\leq \frac{\beta}{\Gamma(\alpha)} \sum_{j=0}^{n} \int_{t_j}^{t_{j+1}} \tau^{\beta-1}(t_{n+1} - \tau)^{\alpha-1} \left|f(\tau, y(\tau))\right| d\tau$$

$$+ \frac{\beta}{\Gamma(\alpha)} \sum_{j=0}^{n} \int_{t_j}^{t_{j+1}} (t_{n+1} - \tau)^{\alpha-1}$$

$$\times \left[t_j^{\beta-1} f(t_j, y_j) - t_{j-1}^{\beta-1} f(t_{j-1}, y_{j-1})\right] (\tau - t_j) d\tau. \quad (10.36)$$

By the mean value theorem, we have

$$\left|y_{(t_{n+1})} - y_{n+1}\right| \leq \frac{\beta}{\Gamma(\alpha)} \sum_{j=0}^{n} \int_{t_j}^{t_{j+1}} M \tau^{\beta-1}(t_{n+1} - \tau)^{\alpha-1} d\tau$$

$$+ \frac{\beta \Delta t}{\Gamma(\alpha)} \sum_{j=0}^{n} \int_{t_j}^{t_{j+1}} \left|C^{\beta-1} f(c, y(c))\right|$$

$$(t_{n+1} - \tau)^{\alpha-1} (\tau - t_j) d\tau,$$

$$\leq \frac{\beta M}{\Gamma(\alpha)} \sum_{j=0}^{n} \int_{t_j}^{t_{j+1}} \tau^{\beta-1}(t_{n+1} - \tau)^{\alpha-1} d\tau +$$

$$\frac{\beta \Delta t}{\Gamma(\alpha)} \left|C^{\beta-1} f(c, y(c))\right| \sum_{j=0}^{n} \int_{t_j}^{t_{j+1}} (t_{n+1} - \tau)^{\alpha-1} (\tau - t_j) d\tau,$$

$$\leq \frac{\beta M}{\Gamma(\alpha)} \sum_{j=0}^{n} \left[t_{n+1}^{\beta+\alpha-1} B\left(\frac{t_{j+1}}{t_{n+1}}, \beta, \alpha\right) - t_{n+1}^{\alpha+\beta-1} B\left(\frac{t_j}{t_{n+1}}, \beta, \alpha\right)\right]$$

$$+ \frac{\beta \Delta t}{\Gamma(\alpha)} \left|C^{\beta-1} f(c, y(c))\right| \sum_{j=0}^{n} \left[t_{n+1}^{\alpha+1} B\left(\frac{t_{j+1}}{t_{n+1}}, 1, \alpha\right)\right.$$

$$\left. - t_{n+1}^{\alpha+1} B\left(\frac{t_j}{t_{n+1}}, 1, \alpha\right)\right]. \quad (10.37)$$

$$\left|y_{(t_{n+1})} - y_{n+1}\right| \leq \frac{\beta M}{\Gamma(\alpha)} (\Delta t)^{\alpha+\beta-1} \sum_{j=0}^{n} (n+1)^{\beta+\alpha-1}$$

$$\times \left[B\left(\frac{t_{j+1}}{t_{n+1}}, \beta, \alpha\right) - B\left(\frac{t_j}{t_{n+1}}, \beta, \alpha\right)\right]$$

$$+ \frac{\beta (\Delta t)^{\alpha+2}}{\Gamma(\alpha)} \left|C^{\beta-1} f(c, y(c))\right| (n+1)^{\alpha+1}$$

$$\times \left[B\left(\frac{t_{j+1}}{t_{n+1}}, 1, \alpha\right) - B\left(\frac{t_j}{t_{n+1}}, 1, \alpha\right)\right] \quad (10.38)$$

$$\lim_{\Delta t \to 0} \left| y_{(t_{n+1})} - y_{n+1} \right| = 0. \tag{10.39}$$

The cases for exponential decay and the generalised Mittag-Leffler function will be derived easily from the above.

10.5 NUMERICAL PROCEDURE FOR THE CHAOTIC MODEL

The aim of the present section is to present the novel procedure for solution of the chaotic problem in different fractal-fractional operators. We consider three different newly defined operators known as fractal-fractional Caputo, fractal-fractional Caputo-Fabrizio, and the fractal-fractional Atangana-Baleanu.

10.5.1 NUMERICAL PROCEDURE IN THE SENSE OF FRACTAL-FRACTIONAL-CAPUTO OPERATOR

This subsection determines the numerical procedure for the solution of the fractal-fractional model given by (10.2) in the sense of fractal-fractional-Caputo operator. We convert the model (10.2) to volterra form as the fractional integral is differentiable, so in the Riemann-Liouville sense, we have

$$^{FFP}D_{0,t}^{p,q}f(t) = \frac{1}{\Gamma(1-p)} \frac{d}{dt} \int_0^t (t-q)^{-p} f(q) dv \frac{1}{qt^{q-1}}, \tag{10.40}$$

we have the following result:

$$\begin{aligned}
{}^{RL}D_{0,t}^{p}\left(x(t)\right) &= qt^{q-1}\left[\gamma(y(t)-h)\right], \\
{}^{RL}D_{0,t}^{p}\left(y(t)\right) &= qt^{q-1}\left[x(t)-y(t)+z(t)\right], \\
{}^{RL}D_{0,t}^{p}\left(z(t)\right) &= qt^{q-1}\left[-\beta y(t)\right].
\end{aligned} \tag{10.41}$$

Replacing RL to Caputo derivative and applying further the application of the Riemann-Liouville fractional integral to both sides give the following result:

$$\begin{aligned}
x(t) &= \frac{q}{\Gamma(p)} \int_0^t \lambda^{q-1}(t-\lambda)^{p-1} f_1(x,y,z,\lambda) d\lambda, \\
y(t) &= \frac{q}{\Gamma(p)} \int_0^t \lambda^{q-1}(t-\lambda)^{p-1} f_2(x,y,z,\lambda) d\lambda, \\
z(t) &= \frac{q}{\Gamma(p)} \int_0^t \lambda^{q-1}(t-\lambda)^{p-1} f_3(x,y,z,\lambda) d\lambda,
\end{aligned} \tag{10.42}$$

where

$$\begin{aligned}
f_1(x,y,z,\lambda) &= \gamma(y(t)-h), \\
f_2(x,y,z,\lambda) &= x(t)-y(t)+z(t),
\end{aligned}$$

$$f_3(x,y,z,\lambda) = -\beta y(t).$$

We now describe here a novel procedure for the model given above using the approach at $t = t_{n+1}$. Then, our model becomes

$$x^{n+1} = \frac{q}{\Gamma(p)} \int_0^t \lambda^{q-1}(t_{n+1}-\lambda)^{p-1} f_1(x,y,z,\lambda) d\lambda,$$

$$y^{n+1} = \frac{q}{\Gamma(p)} \int_0^t \lambda^{q-1}(t_{n+1}-\lambda)^{p-1} f_2(x,y,z,\lambda) d\lambda,$$

$$z^{n+1} = \frac{q}{\Gamma(p)} \int_0^t \lambda^{q-1}(t_{n+1}-\lambda)^{p-1} f_3(x,y,z,\lambda) d\lambda. \tag{10.43}$$

The approximations of the above integral lead to the following:

$$x^{n+1} = \frac{q}{\Gamma(p)} \sum_{j=0}^n \int_{t_j}^{t_{j+1}} \lambda^{q-1}(t_{n+1}-\lambda)^{p-1} f_1(x,y,z,\lambda) d\lambda,$$

$$y^{n+1} = \frac{q}{\Gamma(p)} \sum_{j=0}^n \int_{t_j}^{t_{j+1}} \lambda^{q-1}(t_{n+1}-\lambda)^{p-1} f_2(x,y,z,\lambda) d\lambda,$$

$$z^{n+1} = \frac{q}{\Gamma(p)} \sum_{j=0}^n \int_{t_j}^{t_{j+1}} \lambda^{q-1}(t_{n+1}-\lambda)^{p-1} f_3(x,y,z,\lambda) d\lambda. \tag{10.44}$$

Now approximating the function $\lambda^{q-1} f_i(x,y,z,\lambda)$ for $i=1,2,3$ in expression (10.44) in the interval $[t_j, t_{j+1}]$ through the Lagrangian piece-wise interpolation, we get

$$P_j(\lambda) = \frac{\lambda - t_{j-1}}{t_j - t_{j-1}} t_j^{q-1} f_1(x^j, y^j, z^j, t_j) - \frac{\lambda - t_j}{t_j - t_{j-1}} t_{j-1}^{q-1} f_1(x^{j-1}, y^{j-1}, z^{j-1}, t_{j-1}),$$

$$Q_j(\lambda) = \frac{\lambda - t_{j-1}}{t_j - t_{j-1}} t_j^{q-1} f_2(x^j, y^j, z^j, t_j) - \frac{\lambda - t_j}{t_j - t_{j-1}} t_{j-1}^{q-1} f_2(x^{j-1}, y^{j-1}, z^{j-1}, t_{j-1}),$$

$$R_j(\lambda) = \frac{\lambda - t_{j-1}}{t_j - t_{j-1}} t_j^{q-1} f_3(x^j, y^j, z^j, t_j)$$
$$- \frac{\lambda - t_j}{t_j - t_{j-1}} t_{j-1}^{q-1} f_3(x^{j-1}, y^{j-1}, z^{j-1}, t_{j-1}). \tag{10.45}$$

The following result is obtained:

$$x^{n+1} = \frac{q}{\Gamma(p)} \sum_{j=0}^n \int_{t_j}^{t_{j+1}} \lambda^{q-1}(t_{n+1}-\lambda)^{p-1} P_j(\lambda) d\lambda,$$

$$y^{n+1} = \frac{q}{\Gamma(p)} \sum_{j=0}^n \int_{t_j}^{t_{j+1}} \lambda^{\kappa-1}(t_{n+1}-\lambda)^{p-1} Q_j(\lambda) d\lambda,$$

$$z^{n+1} = \frac{q}{\Gamma(p)} \sum_{j=0}^n \int_{t_j}^{t_{j+1}} \lambda^{\kappa-1}(t_{n+1}-\lambda)^{p-1} R_j(\lambda) d\lambda, \tag{10.46}$$

Solving (10.46), we get the result below

$$\begin{aligned}
x^{n+1} &= \frac{qh^p}{\Gamma(p+2)} \sum_{j=1}^{n} \Big[t_j^{q-1} f_1(x^j, y^j, z^j, t_j) \\
&\quad \times \Big((n+1-j)^p(n-j+2+p) - (n-j)^p(n-j+2+2p) \Big) \\
&\quad - t_{j-1}^{q-1} f_1(x^{j-1}, y^{j-1}, z^{j-1}, t_{j-1}) \\
&\quad \times \Big((n-j+1)^{p+1} - (n-j)^p(n-j+1+p) \Big) \Big], \\
y^{n+1} &= \frac{qh^p}{\Gamma(p+2)} \sum_{j=1}^{n} \Big[t_j^{q-1} f_2(x^j, y^j, z^j, t_j) \\
&\quad \times \Big((n+1-j)^p(n-j+2+p) - (n-j)^p(n-j+2+2p) \Big) \\
&\quad - t_{j-1}^{q-1} f_2(x^{j-1}, y^{j-1}, z^{j-1}, t_{j-1}) \\
&\quad \times \Big((n-j+1)^{p+1} - (n-j)^p(n-j+1+p) \Big) \Big], \\
z^{n+1} &= \frac{qh^p}{\Gamma(p+2)} \sum_{j=1}^{n} \Big[t_j^{q-1} f_3(x^j, y^j, z^j, t_j) \\
&\quad \times \Big((n+1-j)^p(n-j+2+p) - (n-j)^p(n-j+2+2p) \Big) \\
&\quad - t_{j-1}^{q-1} f_3(x^{j-1}, y^{j-1}, z^{j-1}, t_{j-1}) \\
&\quad \times \Big((n-j+1)^{p+1} - (n-j)^p(n-j+1+p) \Big) \Big]. \quad (10.47)
\end{aligned}$$

Next, we give the numerical scheme for the fractal-fractional Caputo-Fabrizio operator in the following subsection.

10.5.2 NUMERICAL PROCEDURE FOR FRACTAL-FRACTIONAL CAPUTO-FABRIZIO OPERATOR

The aim of this subsection is to present a numerical procedure for the solution of the given model in fractal-fractional Caputo-Fabrizio sense. In order to do this, we first write the model in the following form:

$$\begin{aligned}
{}^{CF}D_{0,t}^{p}\big(x(t)\big) &= qt^{q-1} f_1(x, y, z, t), \\
{}^{CF}D_{0,t}^{p}\big(y(t)\big) &= qt^{q-1} f_2(x, y, z, t), \\
{}^{CF}D_{0,t}^{p}\big(z(t)\big) &= qt^{q-1} f_3(x, y, z, t). \quad (10.48)
\end{aligned}$$

The application of the Caputo-Fabrizio integral yields to the following:

$$x(t) = x^0 + \frac{qt^{q-1}(1-p)}{M(p)}f_1(x,y,z,t) + \frac{pq}{M(p)}\int_0^t \lambda^{q-1}f_1(x,y,z,\lambda)d\lambda,$$

$$y(t) = y^0 + \frac{qt^{q-1}(1-p)}{M(p)}f_2(x,y,z,t) + \frac{pq}{M(p)}\int_0^t \lambda^{q-1}f_2(x,y,z,\lambda)d\lambda,$$

$$z(t) = z^0 + \frac{qt^{q-1}(1-p)}{M(p)}f_3(x,y,z,t)$$

$$+ \frac{pq}{M(p)}\int_0^t \lambda^{q-1}f_3(x,y,z,\lambda)d\lambda, \qquad (10.49)$$

For t_{n+1}, we have the following:

$$x^{n+1}(t) = x^n + \frac{qt^{q-1}(1-p)}{M(p)}f_1(x^n,y^n,z^n,t_n) + \frac{pq}{M(p)}\int_0^{t_{n+1}} \lambda^{q-1}f_1(x,y,z,\lambda)d\lambda,$$

$$y^{n+1}(t) = y^n + \frac{qt^{q-1}(1-p)}{M(p)}f_2(x^n,y^n,z^n,t_n) + \frac{pq}{M(p)}\int_0^{t_{n+1}} \lambda^{q-1}f_2(x,y,z,\lambda)d\lambda,$$

$$z^{n+1}(t) = z^n + \frac{qt^{q-1}(1-p)}{M(p)}f_3(x^n,y^n,z^n,t_n)$$

$$+ \frac{pq}{M(p)}\int_0^{t_{n+1}} \lambda^{q-1}f_3(x,y,z,\lambda)d\lambda. \qquad (10.50)$$

Further, we get

$$x^{n+1}(t) = x^n + \frac{qt^{q-1}(1-p)}{M(p)}$$

$$\times f_1(x^n,y^n,z^n,t_n) - \frac{qt_{n-1}^{q-1}(1-p)}{M(p)}f_1(x^{n-1},y^{n-1},z^{n-1},t_{n-1})$$

$$+ \frac{pq}{M(p)}\int_{t_n}^{t_{n+1}} \lambda^{q-1}f_1(x,y,z,\lambda)d\lambda,$$

$$y^{n+1}(t) = y^n + \frac{qt^{q-1}(1-p)}{M(p)}$$

$$\times f_2(x^n,y^n,z^n,t_n) - \frac{qt_{n-1}^{q-1}(1-p)}{M(p)}f_2(x^{n-1},y^{n-1},z^{n-1},t_{n-1})$$

$$+ \frac{pq}{M(p)}\int_{t_n}^{t_{n+1}} \lambda^{q-1}f_2(x,y,z,\lambda)d\lambda,$$

$$z^{n+1}(t) = z^n + \frac{qt^{q-1}(1-p)}{M(p)}$$

$$\times f_3(x^n,y^n,z^n,t_n) - \frac{qt_{n-1}^{q-1}(1-p)}{M(p)}f_3(x^{n-1},y^{n-1},z^{n-1},t_{n-1})$$

$$+ \frac{pq}{M(p)}\int_{t_n}^{t_{n+1}} \lambda^{q-1}f_3(x,y,z,\lambda)d\lambda. \qquad (10.51)$$

Fractal-Fractional Modified Chua Chaotic Attractor

The application of the Lagrange polynomial piece-wise interpolation and integrating, we arrive at

$$
\begin{aligned}
x^{n+1}(t) &= x^n + \frac{qt_n^{q-1}(1-p)}{M(p)} \\
&\times f_1(x^n, y^n, z^n, t_n) - \frac{qt_{n-1}^{q-1}(1-p)}{M(p)} f_1(x^{n-1}, y^{n-1}, z^{n-1}, t_{n-1}) \\
&+ \frac{pq}{M(p)} \left[\frac{3h}{2} t_n^{q-1} f_1(x^n, y^n, z_2^n, t_n) - \frac{h}{2} t_{n-1}^{q-1} f_1(x^{n-1}, y^{n-1}, z^{n-1}, t_{n-1}) \right], \\
y^{n+1}(t) &= y^n + \frac{qt_n^{q-1}(1-p)}{M(p)} f_2(x^n, y^n, z^n, t_n) - \frac{qt_{n-1}^{q-1}(1-p)}{M(p)} \\
&\times f_2(x^{n-1}, y^{n-1}, z^{n-1}, t_{n-1}) \\
&+ \frac{pq}{M(p)} \left[\frac{3h}{2} t_n^{q-1} f_1(x^n, y^n, z_2^n, t_n) - \frac{h}{2} t_{n-1}^{q-1} f_2(x^{n-1}, y^{n-1}, z^{n-1}, t_{n-1}) \right], \\
z^{n+1}(t) &= z^n + \frac{qt_n^{q-1}(1-p)}{M(p)} \\
&\times f_3(x^n, y^n, z^n, t_n) - \frac{qt_{n-1}^{q-1}(1-p)}{M(p)} f_1(x^{n-1}, y^{n-1}, z^{n-1}, t_{n-1}) \\
&+ \frac{pq}{M(p)} \left[\frac{3h}{2} t_n^{q-1} f_3(x^n, y^n, z_2^n, t_n) \right. \\
&\left. - \frac{h}{2} t_{n-1}^{q-1} f_3(x^{n-1}, y^{n-1}, z^{n-1}, t_{n-1}) \right]. \quad (10.52)
\end{aligned}
$$

Further, we achieve the following result:

$$
\begin{aligned}
x^{n+1} &= x^n + qt_n^{q-1}\left(\frac{1-p}{M(p)} + \frac{3ph}{2M(p)}\right) f_1(x^n, y^n, z^n, t_n) \\
&- qt_{n-1}^{q-1}\left(\frac{1-p}{M(p)} + \frac{ph}{2M(p)}\right) f_1(x^{n-1}, y^{n-1}, z^{n-1}, t_{n-1}), \\
y^{n+1} &= y^n + qt_n^{q-1}\left(\frac{1-p}{M(p)} + \frac{3ph}{2M(p)}\right) f_2(x^n, y^n, z^n, t_n) \\
&- qt_{n-1}^{q-1}\left(\frac{1-p}{M(p)} + \frac{ph}{2M(p)}\right) f_2(x^{n-1}, y^{n-1}, z^{n-1}, t_{n-1}), \\
z^{n+1} &= z^n + qt_n^{q-1}\left(\frac{1-p}{M(p)} + \frac{3ph}{2M(p)}\right) f_3(x^n, y^n, z^n, t_n) \\
&- qt_{n-1}^{q-1}\left(\frac{1-p}{M(p)} + \frac{ph}{2M(p)}\right) f_3(x^{n-1}, y^{n-1}, z^{n-1}, t_{n-1}). \quad (10.53)
\end{aligned}
$$

Next, a numerical scheme for the fractal-fractional system in the sense of Atangana-Baleanu derivative is shown the subsection below.

10.5.3 NUMERICAL PROCEDURE FOR FRACTAL-FRACTIONAL ATANGANA-BALEANU OPERATOR

In order to have a numerical procedure for the solution of fractal-fractional model in the sense of Atangana-Baleanu operator, we express first the model (10.2) in the following form:

$$^{ABR}D_{0,t}^{\alpha}\left(x(t)\right) = qt^{q-1}f_1(x,y,z,t),$$

$$^{ABR}D_{0,t}^{\alpha}\left(y(t)\right) = qt^{q-1}f_2(x,y,z,t),$$

$$^{ABR}D_{0,t}^{\alpha}\left(z(t)\right) = qt^{q-1}f_3(x,y,z,t). \quad (10.54)$$

The following is presented on the basis of Atangana-Baleanu integral:

$$x(t) = \frac{qt^{q-1}(1-p)}{AB(p)}f_1(x,y,z,t)$$
$$+ \frac{pq}{AB(p)\Gamma(p)}\int_0^t \lambda^{q-1}(t-\lambda)^{p-1}f_1(x,y,z,\lambda)d\lambda,$$

$$y(t) = \frac{qt^{q-1}(1-p)}{AB(p)}f_2(x,y,z,t)$$
$$+ \frac{pq}{AB(p)\Gamma(p)}\int_0^t \lambda^{q-1}(t-\lambda)^{p-1}f_2(x,y,z,\lambda)d\lambda,$$

$$z(t) = \frac{qt^{q-1}(1-p)}{AB(p)}f_3(x,y,z,t)$$
$$+ \frac{pq}{AB(p)\Gamma(p)}\int_0^t \lambda^{q-1}(t-\lambda)^{p-1}f_3(x,y,z,\lambda)d\lambda. \quad (10.55)$$

Further, at t_{n+1}, we get

$$x^{n+1} = \frac{qt_n^{q-1}(1-p)}{AB(\alpha)}f_1(x^n,y^n,z^n,t_n)$$
$$+ \frac{pq}{AB(p)\Gamma(p)}\int_0^{t_{n+1}} \lambda^{q-1}(t_{n+1}-\lambda)^{p-1}f_1(x,y,z,\lambda)d\lambda,$$

$$y^{n+1} = \frac{qt_n^{q-1}(1-p)}{AB(\alpha)}f_2(x^n,y^n,z^n,t_n)$$
$$+ \frac{pq}{AB(p)\Gamma(p)}\int_0^{t_{n+1}} \lambda^{q-1}(t_{n+1}-\lambda)^{p-1}f_2(x,y,z,\lambda)d\lambda,$$

$$z^{n+1} = \frac{qt_n^{q-1}(1-p)}{AB(\alpha)}f_3(x^n,y^n,z^n,t_n)$$
$$+ \frac{pq}{AB(p)\Gamma(p)}\int_0^{t_{n+1}} \lambda^{q-1}(t_{n+1}-\lambda)^{p-1}f_3(x,y,z,\lambda)d\lambda. \quad (10.56)$$

Fractal-Fractional Modified Chua Chaotic Attractor

We obtain the following after simplification of the integral in the above equations:

$$x^{n+1} = \frac{qt_n^{q-1}(1-p)}{AB(p)} f_1(x^n, y^n, z^n, t_n)$$
$$+ \frac{pq}{AB(p)\Gamma(p)} \sum_{j=0}^{n} \int_{t_j}^{t_{j+1}} \lambda^{q-1}(t_{n+1}-\lambda)^{p-1} f_1(x,y,z,\lambda) d\lambda,$$

$$y^{n+1} = \frac{qt_n^{q-1}(1-p)}{AB(p)} f_2(x^n, y^n, z^n, t_n)$$
$$+ \frac{pq}{AB(p)\Gamma(p)} \sum_{j=0}^{n} \int_{t_j}^{t_{j+1}} \lambda^{q-1}(t_{n+1}-\lambda)^{p-1} f_2(x,y,z,\lambda) d\lambda,$$

$$z^{n+1} = \frac{qt_n^{q-1}(1-p)}{AB(p)} f_3(x^n, y^n, z^n, t_n)$$
$$+ \frac{pq}{AB(p)\Gamma(p)} \sum_{j=0}^{n} \int_{t_j}^{t_{j+1}} \lambda^{q-1}(t_{n+1}-\lambda)^{p-1} f_3(x,y,z,\lambda) d\lambda. \quad (10.57)$$

Now by approximating the expressions in system above given by $\lambda^{q-1} f_1(x,y,z,\lambda)$, $\lambda^{q-1} f_2(x,y,z,\lambda)$, and $\lambda^{q-1} f_3(x,y,z,\lambda)$ in the given internal $[t_j, t_{j+1}]$, the following numerical scheme is presented

$$x^{n+1} = \frac{qt_n^{q-1}(1-p)}{AB(p)} f_1(x^n, y^n, z^n, t_n)$$
$$+ \frac{q(\Delta t)^p}{AB(p)\Gamma(p+2)} \sum_{j=1}^{n} \left[t_j^{q-1} f_1(x^j, y^j, z^j, t_j) \right.$$
$$\times \left((n+1-j)^p(n-j+2+\alpha) - (n-j)^p(n-j+2+2p) \right)$$
$$- t_{j-1}^{q-1} f_1(x^{j-1}, y^{j-1}, z^{j-1}, t_{j-1})$$
$$\left. \times \left((n-j+1)^{p+1} - (n-j)^p(n-j+1+p) \right) \right],$$

$$y^{n+1} = \frac{qt_n^{q-1}(1-p)}{AB(p)} f_2(x^n, y^n, z^n, t_n)$$
$$+ \frac{q(\Delta t)^p}{AB(p)\Gamma(p+2)} \sum_{j=1}^{n} \left[t_j^{q-1} f_2(x^j, y^j, z^j, t_j) \right.$$
$$\times \left((n+1-j)^p(n-j+2+\alpha) - (n-j)^p(n-j+2+2p) \right)$$
$$- t_{j-1}^{q-1} f_2(x^{j-1}, y^{j-1}, z^{j-1}, t_{j-1})$$
$$\left. \times \left((n-j+1)^{p+1} - (n-j)^p(n-j+1+p) \right) \right],$$

$$z^{n+1} = \frac{qt_n^{q-1}(1-p)}{AB(p)} f_3(x^n, y^n, z^n, t_n)$$

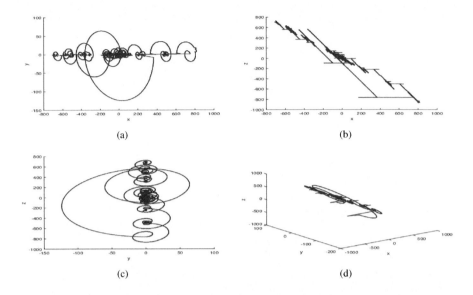

Figure 10.1 Model simulation of fractal-fractional Caputo model when $p = 1, q = 1$.

$$+ \frac{q(\Delta t)^p}{AB(p)\Gamma(p+2)} \sum_{j=1}^{n} \left[t_j^{q-1} f_3(x^j, y^j, z^j, t_j) \right.$$
$$\times \left((n+1-j)^p(n-j+2+\alpha) - (n-j)^p(n-j+2+2p) \right)$$
$$- t_{j-1}^{q-1} f_3(x^{j-1}, y^{j-1}, z^{j-1}, t_{j-1})$$
$$\left. \times \left((n-j+1)^{p+1} - (n-j)^p(n-j+1+p) \right) \right]. \tag{10.58}$$

10.6 NUMERICAL RESULTS

This section investigates the numerical solution of the fractal-fractional model in the sense of the three fractional operators that is the Caputo, Caputo-Fabrizio, and the Atangana-Baleanu. Figures 10.1–10.8 are the graphical results of the fractal-fractional model, in the sense of Caputo, CF, and AB derivatives. For the Caputo fractal-fractional model, we use different values of the fractal and fractional order parameters and presented the numerical results in the form of graphs, see Figures 10.1–10.3. The graphical results for the fractal-fractional model in the sense of Caputo-Fabrizio are shown in Figures 10.4–10.6 while the fractal-fractional model in the sense of Atangana-Baleanu operator is shown in Figures 10.7–10.8.

Fractal-Fractional Modified Chua Chaotic Attractor

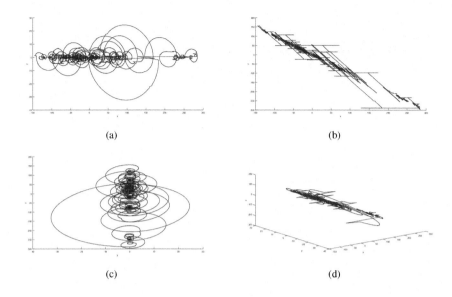

Figure 10.2 Model simulation of fractal-fractional Caputo model when $p = 0.99, q = 1$.

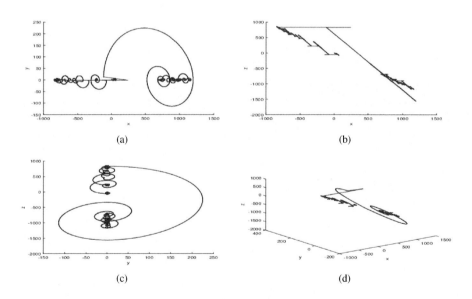

Figure 10.3 Model simulation of fractal-fractional Caputo model when $p = 1, q = 0.98$.

134   Numerical Methods for Fractal-Fractional Differential

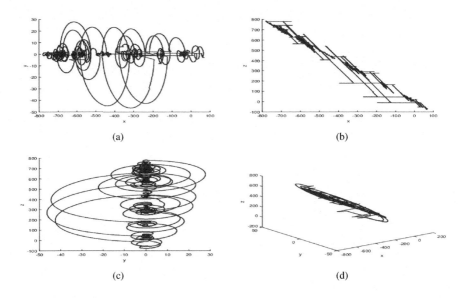

Figure 10.4 Model simulation of fractal-fractional Caputo-Fabrizio model when $p = 1, q = 1$.

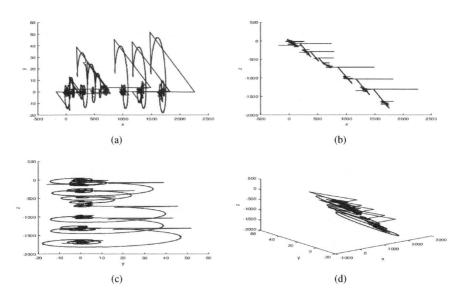

Figure 10.5 Model simulation of fractal-fractional Caputo-Fabrizio model when $p = 0.96, q = 1$.

Fractal-Fractional Modified Chua Chaotic Attractor 135

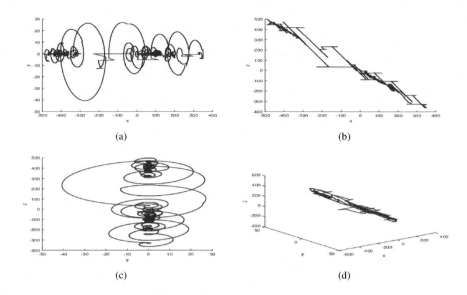

Figure 10.6 Model simulation of fractal-fractional Caputo-Fabrizio model when $p = 1, q = 0.96$.

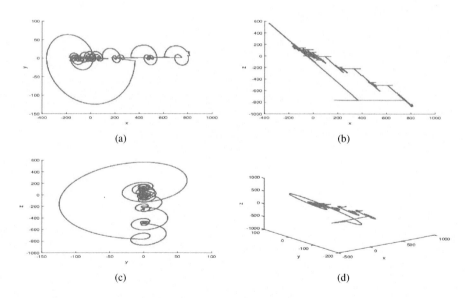

Figure 10.7 Model simulation of fractal-fractional Atangana-Baleanu model when $p = 1, q = 1$.

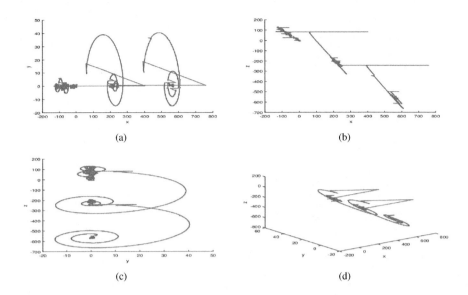

Figure 10.8 Model simulation of fractal-fractional Atangana-Baleanu model when $p = 0.98, q = 0.99$.

10.7 CONCLUSION

We investigated the dynamics of a new chaotic model with fractal-fractional operators. Initially, we formulated a chaotic model and chose the suitable initial conditions and parameters for which the model becomes chaotic. Then, we presented some of the existing equilibrium points of the model and obtained their stability results. Moreover, we presented a novel numerical approach for each fractal-fractional model and obtained graphical results with suitable fractal and fractional orders. The results show that the fractal-fractional operators are best to capture different attractors.

11 Application of Fractal-Fractional Operators to Study a New Chaotic Model

11.1 INTRODUCTION

The dynamical systems that are most sensitive to the initial conditions are the chaotic system or models. Due to the sensitivity of the initial values of the model variables, this chaotic behaviour exhibits nearly random behaviour, as stated in Sambas 2013 Design. Numerous nonlinear systems are known to exist in many fields, including chemistry, physics, biology, economics, engineering, and sociology. Planetary climate prediction models, data compression, neural network models, turbulence, nonlinear dynamical economics, processing information, circuits, and devices are a few examples of nonlinear chaotic systems. The Lorenz and Rossler systems are regarded as examples of low-dimensional systems in chaos dynamical systems that are often investigated by scholars. Due to the limited phase space, it should be emphasised that most results are given in 3D models [33]. In [34], the chaotic system defined by one direction of exponential spreading is discussed. A chaotic map is an evolution function that displays some type of chaotic behaviour in mathematics. A discrete-time or continuous-time parameter can be used to parameterise maps. Iterated functions are the most common form for discrete maps. In the study of dynamical systems, chaotic maps frequently appear. Fractals are often produced by chaotic maps. Some fractals are explored as sets rather than in terms of the map that creates them, even though fractals can be created via an iterative process. This frequently occurs because different iterative processes can be used to build the same fractal. A nonlinear system is one in which the output change is not proportional to the input change. Because most systems are intrinsically nonlinear, nonlinear problems are of interest to engineers, biologists, physicists, mathematicians, and many other scientists. In contrast to considerably simpler linear systems, nonlinear dynamical systems, which describe changes in variables over time, may appear chaotic, unexpected, or illogical. A nonlinear system of equations, which is a collection of simultaneous equations in which the unknowns or, in the case of differential equations, the unknown functions, appear as variables of a polynomial of degree greater than one or in the argument of a function that is not a polynomial of degree one, typically describes the behaviour of a nonlinear system in mathematics. In other words, the equations that must be solved in a nonlinear system of equations cannot be expressed as a linear combination of the variables or functions that are unknown to them. Whether or not the equations contain known linear functions, systems can be characterised as nonlinear. It has been discovered that many straightforward nonlinear systems of ordinary differential equations can reproduce actions that resemble attractors seen in the actual world. This chapter will extend a straightforward system of nonlinear equations to the realm of fractal-fractional nonlinear systems and conduct a numerical analysis of the results. This chapter present the fractal-fractional dynamics of a new chaotic model. Initially, we formulate the model and suggest some values for the parameters and initial conditions to show the model become chaotic. Then we apply the fractal-fractional operators in the sense of Caputo, CF, and AB on the given model. For the fractal-fractional model in each operator, a novel numerical approach is presented. Some suitable values for the fractal and fractional orders are considered, and the graphical results are shown.

11.2 MODEL FRAMEWORK

The purpose of this section is to formulate a chaotic model, apply the fractal-fractional operators, and discuss their stability at the given equilibrium points. The chaotic model is given by the following system of ordinary differential equations:

$$\begin{cases} \frac{dx(t)}{dt} = z - Ax + Byz, \\ \frac{dy(t)}{dt} = Cy - xz + z, \\ \frac{dz(t)}{dt} = Dxy - Ez, \end{cases} \tag{11.1}$$

where $x(t)$, $x(t)$, and $y(t)$ are the state variables while A, B, C, D, and E are the parameters. This system is chaotic for the given values of the parameters $A = 3$, $B = 2.7$, $C = 1.7$, $D = 2$, and $E = 9$ with the initial conditions on state variables are $x(0) = 2$, $y(0) = 0.1$ and $z(0) = 2$. The fractal-fractional representation of the model (11.1) is given by:

$$\begin{cases} {}^{FF}D_{0,t}^{p,q}\bigl(x(t)\bigr) = z - Ax + Byz, \\ {}^{FF}D_{0,t}^{p,q}\bigl(y(t)\bigr) = Cy - xz + z, \\ {}^{FF}D_{0,t}^{p,q}\bigl(z(t)\bigr) = Dxy - Ez, \end{cases} \tag{11.2}$$

where p represents the fractional order while q is fractal order.

11.3 EXISTENCE AND UNIQUENESS

We aim to show that the system (11.2) has a unique solution under some conditions that be presented here. It is assumed that for all $t \in [0, T]$, the functions $x(t)$, $y(t)$ and $z(t)$ are bounded, $\|x\|_\infty \le M_x$, $\|y\|_\infty \le M_y$ and $\|z\|_\infty \le M_z$. Further, we write the model (11.2) in the form:

$$f_x(t, x, y, z) = z - Ax + Byz,$$

Application of Fractal-Fractional Operators to Study a New Chaotic Model

$$f_y(t,x,y,z) = Cy - xz + z,$$

$$f_z(t,x,y,z) = Dxy - Ez.$$

If x, y, and z are bounded, then the functions f_x, f_y, and f_z are bounded too. Let x, y, and z ar bounded, then there exists M_x, M_y, and M_z such that

$$\sup_{t \in D_x} |x(t)| = \|x\|_\infty \leq M_x,$$

$$\sup_{t \in D_y} |y(t)| = \|y\|_\infty \leq M_y,$$

$$\sup_{t \in D_z} |z(t)| = \|z\|_\infty \leq M_z.$$

We first show that the functions satisfy the linear growth property.

$$\begin{aligned}
|f_x(t,x,y,z)| &\leq |z| + A|x| + B|y||z|, \\
&\leq \sup_{t \in D_z} |z| + A \sup_{t \in D_x} |x| + B \sup_{t \in D_y} |y| \sup_{t \in D_z} |z|, \\
&\leq \|z\|_\infty + A\|x\|_\infty + B\|y\|_\infty \|z\|_\infty, \\
&\leq M_z + AM_x + BM_y M_z = M_{f_x}, \\
&< \infty
\end{aligned}$$

$$\begin{aligned}
|f_y(t,x,y,z)| &= |Cy - xz + z| \\
&\leq |C||y| + |x||z| + |z|, \\
&\leq |C| \sup_{t \in D_y} |y| + \sup_{t \in D_x} |x| \sup_{t \in D_z} |z| + \sup_{t \in D_z} |z|, \\
&\leq C\|y\|_\infty + \|x\|_\infty \|z\|_\infty + \|z\|_\infty, \\
&\leq CM_y + M_x M_z + M_z = M_{f_y}, \\
&< \infty.
\end{aligned}$$

$$\begin{aligned}
|f_z(t,x,y,z)| &= |Dxy - Ez|, \\
&\leq |D||x||y| + |E||z|, \\
&\leq |D| \sup_{t \in D_x} |x| \sup_{t \in D_y} |y| + |E| \sup_{t \in D_z} |z|, \\
&\leq D\|x\|_\infty \|y\|_\infty + E\|z\|_\infty,
\end{aligned}$$

$$\leq DM_xM_y + EM_z = M_{f_z},$$
$$< \infty.$$

But on the other hand, we have that

$$|f_x(t,x_1,y,z) - f_x(t,x_2,y,z)| \leq |A||x_1 - x_2|,$$
$$\leq \frac{3}{2}|A||x_1 - x_2|,$$
$$|f_y(t,x,y_1,z) - f_y(t,x,y_2,z)| \leq |C||y_1 - y_2|,$$
$$\leq \frac{3}{2}|C||y_1 - y_2|,$$
$$|f_z(t,x,y,z_1) - f_z(t,x,y,z_2)| \leq |E||z_1 - z_2|,$$
$$\leq \frac{3}{2}|E||z_1 - z_2|.$$

f_x, f_y, and f_z verify the Lipschitz condition; however, there are contraction if

$$\max\{|C|,|E|,|A|\} < 1.$$

Alternatively, we shall verify that f_x, f_y, and f_z satisfy the linear growth and Lipschitz conditions

$$|f_x(t,x,y,z)|^2 = |z - Ax + Byz|^2,$$
$$\leq 3|z|^2 + 3A^2|x|^2 + 3B^2|y|^2|z|^2,$$
$$\leq 3M_z^2 + 3A^2|x|^2 + 3B^2M_y^2M_z^2,$$
$$\leq 3(M_z^2 + B^2M_y^2M_z^2)\left(1 + \frac{A^2|x|^2}{M_z^2 + B^2M_y^2M_z^2}\right),$$
$$\leq 3(M_z^2 + B^2M_y^2M_z^2)(1 + |x|^2),$$

if $\frac{A^2}{M_z^2 + B^2M_y^2M_z^2} < 1$. Therefore,

$$|f_x(t,x,y,z)|^2 \leq K_x(1 + |x|^2),$$

where $K_x = 3(M_z^2 + B^2M_y^2M_z^2)$.

$$|f_y(t,x,y,z)|^2 = |Cy - xz + z|^2,$$
$$\leq 3C^2|y|^2 + 3|x|^2|z|^2 + 3|z|^2,$$

$$\leq 3M_x^2 M_z^2 + 3M^2 z + 3C^2 |y|^2,$$

$$\leq 3(M_x^2 M_z^2 + M_z^2)\left(1 + \frac{C^2 |y|^2}{M_x^2 M_z^2 + M_z^2}\right),$$

$$\leq K_y(1 + |y|^2),$$

if $K_y = \frac{C^2}{M_x^2 M_z^2 + M_z^2}$.

$$|f_z(t,x,y,z)|^2 = |Dxy - Ez|^2,$$

$$\leq 2D^2 |x|^2 |y|^2 + 2E^2 |z|^2,$$

$$\leq 2D^2 M_x^2 M_y^2 + 2E^2 |z|^2,$$

$$\leq 2D^2 M_x^2 M_y^2 \left(1 + \frac{E^2 |z|^2}{D^2 M_x^2 M_y^2}\right),$$

$$\leq K_z(1 + |z|^2),$$

if $\frac{E^2}{D^2 M_x^2 M_y^2} < 1$ and $K_z = 2D^2 M_x^2 M_y^2$. On the other hand,

$$|f_x(t,x_1,y,z) - f_x(t,x_2,y,z)|^2 \leq A^2 |x_1 - x_2|^2,$$

$$\leq \frac{3}{2} A^2 |x_1 - x_2|^2,$$

$$\leq \overline{K}_x |x_1 - x_2|^2,$$

$$|f_y(t,x,y_1,z) - f_y(t,x,y_2,z)|^2 \leq C^2 |y_1 - y_2|^2,$$

$$\leq \frac{3}{2} C^2 |y_1 - y_2|^2,$$

$$\leq \overline{K}_y |y_1 - y_2|^2,$$

$$|f_z(t,x,y,z_1) - f_z(t,x,y,z_2)|^2 \leq E^2 |z_1 - z_2|^2,$$

$$\leq \frac{3}{2} E^2 |z_1 - z_2|,$$

$$\leq \overline{K}_z |z_1 - z_2|,$$

where $\overline{K}_x = \frac{3}{2} A^2$, $\overline{K}_y = \frac{3}{2} C^2$, and $\overline{K}_z = \frac{3}{2} E^2$.

11.3.1 EQUILIBRIUM POINTS AND ITS ANALYSIS

The equilibrium points of the system (11.2) can be obtained by setting

$$\begin{cases} 0 = z - Ax + Byz, \\ 0 = Cy - xz + z, \\ 0 = Dxy - Ez, \end{cases} \qquad (11.3)$$

and consider the numerical values of the parameters A, B, C, D, and E above, we get the following equilibrium points: $E_1 = (0,0,0)$, $E_2 = (3.31069, -2.42891, -1.78697)$, $E_3 = (-2.31069, 2.05854, -1.05703)$, $E_4 = (-2.31069, -2.42891, 1.24721)$, and $E_5 = (3.31069, 2.05854, 1.51449)$. A Jacobian matrix for the given system is obtained as follows:

$$J = \begin{pmatrix} -A & Bz & By+1 \\ -z & C & 1-x \\ Dy & Dx & -E \end{pmatrix}$$

at E_1, we get $\lambda_1 = -A$, $\lambda_2 = C$, and $\lambda_3 = -E$. So, the system is unstable at E_1. The other equilibrium points for the suggested values of the model parameters are not containing negative real parts, and hence, it is unstable.

11.4 NUMERICAL PROCEDURE FOR THE CHAOTIC MODEL

The aim of the present section is to present the novel procedure for solution of the chaotic problem in different fractal-fractional operators. We consider three different newly defined operators known as fractal-fractional Caputo, fractal-fractional Caputo-Fabrizio, and the fractal-fractional Atangana-Baleanu.

11.4.1 NUMERICAL PROCEDURE IN THE SENSE OF FRACTAL-FRACTIONAL-CAPUTO OPERATOR

This subsection determines the numerical procedure for the solution of the fractal-fractional model given by (11.2) in the sense of fractal-fractional-Caputo operator. We convert the model (11.2) to volterra form as the fractional integral is differentiable, so in the Riemann-Liouville sense we have

$$^{FFP}D_{0,t}^{p,q} f(t) = \frac{1}{\Gamma(1-p)} \frac{d}{dt} \int_0^t (t-q)^{-p} f(q) dq \frac{1}{qt^{q-1}}, \qquad (11.4)$$

we have the following result

$$\begin{aligned} ^{RL}D_{0,t}^p \left(x(t)\right) &= qt^{q-1} \left[z - Ax + Byz\right], \\ ^{RL}D_{0,t}^p \left(y(t)\right) &= qt^{q-1} \left[Cy - xz + z\right], \\ ^{RL}D_{0,t}^p \left(z(t)\right) &= qt^{q-1} \left[Dxy - Ez\right]. \end{aligned} \qquad (11.5)$$

Consider now replacing the RL derivative to Caputo derivative in order to make the use of the integer-order initial conditions. Then, we apply the Riemann-Liouville fractional integral on both sides to have the following:

$$x(t) = x(0) + \frac{q}{\Gamma(p)} \int_0^t \lambda^{q-1}(t-\lambda)^{p-1} f_1(x,y,z,\lambda) d\lambda,$$

$$y(t) = y(0) + \frac{q}{\Gamma(p)} \int_0^t \lambda^{q-1}(t-\lambda)^{p-1} f_2(x,y,z,\lambda) d\lambda,$$

$$z(t) = z(0) + \frac{q}{\Gamma(p)} \int_0^t \lambda^{q-1}(t-\lambda)^{p-1} f_3(x,y,z,\lambda) d\lambda, \qquad (11.6)$$

where

$$f_1(x,y,z,\lambda) = z - Ax + Byz,$$

$$f_2(x,y,z,\lambda) = Cy - xz + z,$$

$$f_3(x,y,z,\lambda) = Dxy - Ez.$$

We now describe here a novel procedure for the model given above using the approach at t_{n+1}. Then our model becomes

$$x^{n+1} = x^0 + \frac{q}{\Gamma(p)} \int_0^t \lambda^{q-1}(t_{n+1}-\lambda)^{p-1} f_1(x,y,z,\lambda) d\lambda,$$

$$y^{n+1} = y^0 + \frac{q}{\Gamma(p)} \int_0^t \lambda^{q-1}(t_{n+1}-\lambda)^{p-1} f_2(x,y,z,\lambda) d\lambda,$$

$$z^{n+1} = z^0 + \frac{q}{\Gamma(p)} \int_0^t \lambda^{q-1}(t_{n+1}-\lambda)^{p-1} f_3(x,y,z,\lambda) d\lambda. \qquad (11.7)$$

The approximations of the above integral lead to the following:

$$x^{n+1} = x^0 + \frac{q}{\Gamma(p)} \sum_{j=0}^{n} \int_{t_j}^{t_{j+1}} \lambda^{q-1}(t_{n+1}-\lambda)^{p-1} f_1(x,y,z,\lambda) d\lambda,$$

$$y^{n+1} = y^0 + \frac{q}{\Gamma(p)} \sum_{j=0}^{n} \int_{t_j}^{t_{j+1}} \lambda^{q-1}(t_{n+1}-\lambda)^{p-1} f_2(x,y,z,\lambda) d\lambda,$$

$$z^{n+1} = z^0 + \frac{q}{\Gamma(p)} \sum_{j=0}^{n} \int_{t_j}^{t_{j+1}} \lambda^{q-1}(t_{n+1}-\lambda)^{p-1} f_3(x,y,z,\lambda) d\lambda. \qquad (11.8)$$

Now approximating the function $\lambda^{q-1} f_i(x,y,z,\lambda)$ for $i = 1,2,3$ in expression (11.8) in the interval $[t_j, t_{j+1}]$ through the Lagrangian piece-wise interpolation, we get

$$P_j(\lambda) = \frac{\lambda - t_{j-1}}{t_j - t_{j-1}} t_j^{q-1} f_1(x^j, y^j, z^j, t_j) - \frac{\lambda - t_j}{t_j - t_{j-1}} t_{j-1}^{q-1} f_1(x^{j-1}, y^{j-1}, z^{j-1}, t_{j-1}),$$

$$Q_j(\lambda) = \frac{\lambda - t_{j-1}}{t_j - t_{j-1}} t_j^{q-1} f_2(x^j, y^j, z^j, t_j) - \frac{\lambda - t_j}{t_j - t_{j-1}} t_{j-1}^{q-1} f_2(x^{j-1}, y^{j-1}, z^{j-1}, t_{j-1}),$$

$$R_j(\lambda) = \frac{\lambda - t_{j-1}}{t_j - t_{j-1}} t_j^{q-1} f_3(x^j, y^j, z^j, t_j)$$
$$- \frac{\lambda - t_j}{t_j - t_{j-1}} t_{j-1}^{q-1} f_3(x^{j-1}, y^{j-1}, z^{j-1}, t_{j-1}). \tag{11.9}$$

We get the following:

$$x^{n+1} = x^0 + \frac{q}{\Gamma(p)} \sum_{j=0}^{n} \int_{t_j}^{t_{j+1}} \lambda^{q-1}(t_{n+1} - \lambda)^{p-1} P_j(\lambda) d\lambda,$$
$$y^{n+1} = y^0 + \frac{q}{\Gamma(p)} \sum_{j=0}^{n} \int_{t_j}^{t_{j+1}} \lambda^{\kappa-1}(t_{n+1} - \lambda)^{p-1} Q_j(\lambda) d\lambda,$$
$$z^{n+1} = z^0 + \frac{q}{\Gamma(p)} \sum_{j=0}^{n} \int_{t_j}^{t_{j+1}} \lambda^{\kappa-1}(t_{n+1} - \lambda)^{p-1} R_j(\lambda) d\lambda, \tag{11.10}$$

We have then by solving equation (11.10),

$$x^{n+1} = x^0 + \frac{q h^p}{\Gamma(p+2)} \sum_{j=1}^{n} \Big[t_j^{q-1} f_1(x^j, y^j, z^j, t_j)$$
$$\times \Big((n+1-j)^p(n-j+2+p) - (n-j)^p(n-j+2+2p)\Big)$$
$$- t_{j-1}^{q-1} f_1(x^{j-1}, y^{j-1}, z^{j-1}, t_{j-1})$$
$$\times \Big((n-j+1)^{p+1} - (n-j)^p(n-j+1+p)\Big)\Big],$$

$$y^{n+1} = y^0 + \frac{q h^p}{\Gamma(p+2)} \sum_{j=1}^{n} \Big[t_j^{q-1} f_2(x^j, y^j, z^j, t_j)$$
$$\times \Big((n+1-j)^p(n-j+2+p) - (n-j)^p(n-j+2+2p)\Big)$$
$$- t_{j-1}^{q-1} f_2(x^{j-1}, y^{j-1}, z^{j-1}, t_{j-1})$$
$$\times \Big((n-j+1)^{p+1} - (n-j)^p(n-j+1+p)\Big)\Big],$$

$$z^{n+1} = z^0 + \frac{q h^p}{\Gamma(p+2)} \sum_{j=1}^{n} \Big[t_j^{q-1} f_3(x^j, y^j, z^j, t_j)$$
$$\times \Big((n+1-j)^p(n-j+2+p) - (n-j)^p(n-j+2+2p)\Big)$$
$$- t_{j-1}^{q-1} f_3(x^{j-1}, y^{j-1}, z^{j-1}, t_{j-1})$$
$$\times \Big((n-j+1)^{p+1} - (n-j)^p(n-j+1+p)\Big)\Big]. \tag{11.11}$$

Next, in the following subsection, we present the numerical procedure for the fractal-fractional model in the sense of Caputo-Fabrizio operator.

11.4.2 NUMERICAL PROCEDURE FOR FRACTAL-FRACTIONAL CAPUTO-FABRIZIO OPERATOR

The aim of this subsection is to present a numerical procedure for the solution of the given model in fractal-fractional Caputo-Fabrizio sense. In order to do this, we first write the model in the following form:

$$\begin{aligned} {}^{CF}D_{0,t}^{p}\left(x(t)\right) &= qt^{q-1}f_1(x,y,z,t), \\ {}^{CF}D_{0,t}^{p}\left(y(t)\right) &= qt^{q-1}f_2(x,y,z,t), \\ {}^{CF}D_{0,t}^{p}\left(z(t)\right) &= qt^{q-1}f_3(x,y,z,t). \end{aligned} \qquad (11.12)$$

We obtain the following by using the Caputo-Fabrizio integral:

$$\begin{aligned} x(t) &= x^0 + \frac{qt^{q-1}(1-p)}{M(p)}f_1(x,y,z,t) \\ &\quad + \frac{pq}{M(p)}\int_0^t \lambda^{q-1}f_1(x,y,z,\lambda)d\lambda, \\ y(t) &= y^0 + \frac{qt^{q-1}(1-p)}{M(p)}f_2(x,y,z,t) \\ &\quad + \frac{pq}{M(p)}\int_0^t \lambda^{q-1}f_2(x,y,z,\lambda)d\lambda, \\ z(t) &= z^0 + \frac{qt^{q-1}(1-p)}{M(p)}f_3(x,y,z,t) \\ &\quad + \frac{pq}{M(p)}\int_0^t \lambda^{q-1}f_3(x,y,z,\lambda)d\lambda, \end{aligned} \qquad (11.13)$$

The following is presented at t_{n+1}:

$$\begin{aligned} x^{n+1}(t) &= x^n + \frac{qt^{q-1}(1-p)}{M(p)}f_1(x^n,y^n,z^n,t_n) \\ &\quad + \frac{pq}{M(p)}\int_0^{t_{n+1}} \lambda^{q-1}f_1(x,y,z,\lambda)d\lambda, \\ y^{n+1}(t) &= y^n + \frac{qt^{q-1}(1-p)}{M(p)}f_2(x^n,y^n,z^n,t_n) \\ &\quad + \frac{pq}{M(p)}\int_0^{t_{n+1}} \lambda^{q-1}f_2(x,y,z,\lambda)d\lambda, \\ z^{n+1}(t) &= z^n + \frac{qt^{q-1}(1-p)}{M(p)}f_3(x^n,y^n,z^n,t_n) \\ &\quad + \frac{pq}{M(p)}\int_0^{t_{n+1}} \lambda^{q-1}f_3(x,y,z,\lambda)d\lambda. \end{aligned} \qquad (11.14)$$

Further, we have the following:

$$\begin{aligned}
x^{n+1}(t) &= x^n + \frac{qt^{q-1}(1-p)}{M(p)} f_1(x^n, y^n, z^n, t_n) \\
&\quad - \frac{qt_{n-1}^{q-1}(1-p)}{M(p)} f_1(x^{n-1}, y^{n-1}, z^{n-1}, t_{n-1}) \\
&\quad + \frac{pq}{M(p)} \int_{t_n}^{t_{n+1}} \lambda^{q-1} f_1(x, y, z, \lambda) d\lambda, \\
y^{n+1}(t) &= y^n + \frac{qt^{q-1}(1-p)}{M(p)} f_2(x^n, y^n, z^n, t_n) \\
&\quad - \frac{qt_{n-1}^{q-1}(1-p)}{M(p)} f_2(x^{n-1}, y^{n-1}, z^{n-1}, t_{n-1}) \\
&\quad + \frac{pq}{M(p)} \int_{t_n}^{t_{n+1}} \lambda^{q-1} f_2(x, y, z, \lambda) d\lambda, \\
z^{n+1}(t) &= z^n + \frac{qt^{q-1}(1-p)}{M(p)} f_3(x^n, y^n, z^n, t_n) \\
&\quad - \frac{qt_{n-1}^{q-1}(1-p)}{M(p)} f_3(x^{n-1}, y^{n-1}, z^{n-1}, t_{n-1}) \\
&\quad + \frac{pq}{M(p)} \int_{t_n}^{t_{n+1}} \lambda^{q-1} f_3(x, y, z, \lambda) d\lambda. \quad (11.15)
\end{aligned}$$

Now using the Lagrange polynomial piece-wise interpolation and integrating, we obtain

$$\begin{aligned}
x^{n+1}(t) &= x^n + \frac{qt_n^{q-1}(1-p)}{M(p)} \\
&\quad \times f_1(x^n, y^n, z^n, t_n) - \frac{qt_{n-1}^{q-1}(1-p)}{M(p)} f_1(x^{n-1}, y^{n-1}, z^{n-1}, t_{n-1}) \\
&\quad + \frac{pq}{M(p)} \left[\frac{3h}{2} t_n^{q-1} f_1(x^n, y^n, z_2^n, t_n) \right.\\
&\quad \left. - \frac{h}{2} t_{n-1}^{q-1} f_1(x^{n-1}, y^{n-1}, z^{n-1}, t_{n-1}) \right], \\
y^{n+1}(t) &= y^n + \frac{qt_n^{q-1}(1-p)}{M(p)} \\
&\quad \times f_2(x^n, y^n, z^n, t_n) - \frac{qt_{n-1}^{q-1}(1-p)}{M(p)} f_2(x^{n-1}, y^{n-1}, z^{n-1}, t_{n-1}) \\
&\quad + \frac{pq}{M(p)} \left[\frac{3h}{2} t_n^{q-1} f_1(x^n, y^n, z_2^n, t_n) \right.\\
&\quad \left. - \frac{h}{2} t_{n-1}^{q-1} f_2(x^{n-1}, y^{n-1}, z^{n-1}, t_{n-1}) \right],
\end{aligned}$$

Application of Fractal-Fractional Operators to Study a New Chaotic Model

$$\begin{aligned}z^{n+1}(t) &= z^n + \frac{qt_n^{q-1}(1-p)}{M(p)} \\ &\quad f_3(x^n,y^n,z^n,t_n) - \frac{qt_{n-1}^{q-1}(1-p)}{M(p)} f_1(x^{n-1},y^{n-1},z^{n-1},t_{n-1}) \\ &\quad + \frac{pq}{M(p)}\left[\frac{3h}{2}t_n^{q-1}f_3(x^n,y^n,z_2^n,t_n)\right. \\ &\quad \left. - \frac{h}{2}t_{n-1}^{q-1}f_3(x^{n-1},y^{n-1},z^{n-1},t_{n-1})\right]. \end{aligned} \quad (11.16)$$

Further, we obtain the following:

$$\begin{aligned} x^{n+1} &= x^n + qt_n^{q-1}\left(\frac{1-p}{M(p)} + \frac{3ph}{2M(p)}\right)f_1(x^n,y^n,z^n,t_n) \\ &\quad - qt_{n-1}^{q-1}\left(\frac{1-p}{M(p)} + \frac{ph}{2M(p)}\right)f_1(x^{n-1},y^{n-1},z^{n-1},t_{n-1}), \\ y^{n+1} &= y^n + qt_n^{q-1}\left(\frac{1-p}{M(p)} + \frac{3ph}{2M(p)}\right)f_2(x^n,y^n,z^n,t_n) \\ &\quad - qt_{n-1}^{q-1}\left(\frac{1-p}{M(p)} + \frac{ph}{2M(p)}\right)f_2(x^{n-1},y^{n-1},z^{n-1},t_{n-1}), \\ z^{n+1} &= z^n + qt_n^{q-1}\left(\frac{1-p}{M(p)} + \frac{3ph}{2M(p)}\right)f_3(x^n,y^n,z^n,t_n) \\ &\quad - qt_{n-1}^{q-1}\left(\frac{1-p}{M(p)} + \frac{ph}{2M(p)}\right)f_3(x^{n-1},y^{n-1},z^{n-1},t_{n-1}). \quad (11.17)\end{aligned}$$

Next, we present the numerical procedure for the fractal-fractional in the sense of Atangana-Baleanu derivative in the following subsection:

11.4.3 NUMERICAL PROCEDURE FOR FRACTAL-FRACTIONAL ATANGANA-BALEANU OPERATOR

In order to have a numerical procedure for the solution of fractal-fractional model in the sense of Atangana-Baleanu operator, we express first the model (11.2) in the following form:

$$\begin{aligned} {}^{ABR}D_{0,t}^{\alpha}\left(x(t)\right) &= qt^{q-1}f_1(x,y,z,t), \\ {}^{ABR}D_{0,t}^{\alpha}\left(y(t)\right) &= qt^{q-1}f_2(x,y,z,t), \\ {}^{ABR}D_{0,t}^{\alpha}\left(z(t)\right) &= qt^{q-1}f_3(x,y,z,t). \end{aligned} \quad (11.18)$$

The following is presented on the basis of Atangana-Baleanu integral:

$$x(t) = \frac{qt^{q-1}(1-p)}{AB(p)}f_1(x,y,z,t)$$

$$+ \frac{pq}{AB(p)\Gamma(p)} \int_0^t \lambda^{q-1}(t-\lambda)^{p-1} f_1(x,y,z,\lambda) d\lambda,$$

$$y(t) = \frac{qt^{q-1}(1-p)}{AB(p)} f_2(x,y,z,t)$$

$$+ \frac{pq}{AB(p)\Gamma(p)} \int_0^t \lambda^{q-1}(t-\lambda)^{p-1} f_2(x,y,z,\lambda) d\lambda,$$

$$z(t) = \frac{qt^{q-1}(1-p)}{AB(p)} f_3(x,y,z,t)$$

$$+ \frac{pq}{AB(p)\Gamma(p)} \int_0^t \lambda^{q-1}(t-\lambda)^{p-1} f_3(x,y,z,\lambda) d\lambda. \quad (11.19)$$

Further, at t_{n+1}, we get,

$$x^{n+1} = \frac{qt_n^{q-1}(1-p)}{AB(\alpha)} f_1(x^n,y^n,z^n,t_n)$$

$$+ \frac{pq}{AB(p)\Gamma(p)} \int_0^{t_{n+1}} \lambda^{q-1}(t_{n+1}-\lambda)^{p-1} f_1(x,y,z,\lambda) d\lambda,$$

$$y^{n+1} = \frac{qt_n^{q-1}(1-p)}{AB(\alpha)} f_2(x^n,y^n,z^n,t_n)$$

$$+ \frac{pq}{AB(p)\Gamma(p)} \int_0^{t_{n+1}} \lambda^{q-1}(t_{n+1}-\lambda)^{p-1} f_2(x,y,z,\lambda) d\lambda,$$

$$z^{n+1} = \frac{qt_n^{q-1}(1-p)}{AB(\alpha)} f_3(x^n,y^n,z^n,t_n)$$

$$+ \frac{pq}{AB(p)\Gamma(p)} \int_0^{t_{n+1}} \lambda^{q-1}(t_{n+1}-\lambda)^{p-1} f_3(x,y,z,\lambda) d\lambda. \quad (11.20)$$

We obtain the following after simplification the integral in the above equations:

$$x^{n+1} = \frac{qt_n^{q-1}(1-p)}{AB(p)} f_1(x^n,y^n,z^n,t_n)$$

$$+ \frac{pq}{AB(p)\Gamma(p)} \sum_{j=0}^n \int_{t_j}^{t_{j+1}} \lambda^{q-1}(t_{n+1}-\lambda)^{p-1} f_1(x,y,z,\lambda) d\lambda,$$

$$y^{n+1} = \frac{qt_n^{q-1}(1-p)}{AB(p)} f_2(x^n,y^n,z^n,t_n)$$

$$+ \frac{pq}{AB(p)\Gamma(p)} \sum_{j=0}^n \int_{t_j}^{t_{j+1}} \lambda^{q-1}(t_{n+1}-\lambda)^{p-1} f_2(x,y,z,\lambda) d\lambda,$$

$$z^{n+1} = \frac{qt_n^{q-1}(1-p)}{AB(p)} f_3(x^n,y^n,z^n,t_n)$$

$$+ \frac{pq}{AB(p)\Gamma(p)} \sum_{j=0}^n \int_{t_j}^{t_{j+1}} \lambda^{q-1}(t_{n+1}-\lambda)^{p-1} f_3(x,y,z,\lambda) d\lambda. \quad (11.21)$$

Now approximating the expressions in (11.21) given by $\lambda^{q-1} f_1(x,y,z,\lambda)$, $\lambda^{q-1} f_2(x,y,z,\lambda)$, and $\lambda^{q-1} f_3(x,y,z,\lambda)$ in the given internal $[t_j, t_{j+1}]$, the following

Application of Fractal-Fractional Operators to Study a New Chaotic Model   **149**

numerical scheme is presented:

$$\begin{aligned}
x^{n+1} &= \frac{qt_n^{q-1}(1-p)}{AB(p)} f_1(x^n, y^n, z^n, t_n) \\
&+ \frac{q(\Delta t)^p}{AB(p)\Gamma(p+2)} \sum_{j=1}^{n} \left[t_j^{q-1} f_1(x^j, y^j, z^j, t_j) \right. \\
&\times \left((n+1-j)^p(n-j+2+\alpha) - (n-j)^p(n-j+2+2p) \right) \\
&- t_{j-1}^{q-1} f_1(x^{j-1}, y^{j-1}, z^{j-1}, t_{j-1}) \\
&\left. \times \left((n-j+1)^{p+1} - (n-j)^p(n-j+1+p) \right) \right], \\
y^{n+1} &= \frac{qt_n^{q-1}(1-p)}{AB(p)} f_2(x^n, y^n, z^n, t_n) \\
&+ \frac{q(\Delta t)^p}{AB(p)\Gamma(p+2)} \sum_{j=1}^{n} \left[t_j^{q-1} f_2(x^j, y^j, z^j, t_j) \right. \\
&\times \left((n+1-j)^p(n-j+2+\alpha) - (n-j)^p(n-j+2+2p) \right) \\
&- t_{j-1}^{q-1} f_2(x^{j-1}, y^{j-1}, z^{j-1}, t_{j-1}) \\
&\left. \times \left((n-j+1)^{p+1} - (n-j)^p(n-j+1+p) \right) \right], \\
z^{n+1} &= \frac{qt_n^{q-1}(1-p)}{AB(p)} f_3(x^n, y^n, z^n, t_n) \\
&+ \frac{q(\Delta t)^p}{AB(p)\Gamma(p+2)} \sum_{j=1}^{n} \left[t_j^{q-1} f_3(x^j, y^j, z^j, t_j) \right. \\
&\times \left((n+1-j)^p(n-j+2+\alpha) - (n-j)^p(n-j+2+2p) \right) \\
&- t_{j-1}^{q-1} f_3(x^{j-1}, y^{j-1}, z^{j-1}, t_{j-1}) \\
&\left. \times \left((n-j+1)^{p+1} - (n-j)^p(n-j+1+p) \right) \right]. \quad (11.22)
\end{aligned}$$

11.5 NUMERICAL RESULTS

The numerical results for the chaotic model are presented by considering specific values of the fractal and fractal order parameters and obtained the graphical results shown in Figures 11.1–11.6. The fractal-fractional model in the sense of Caputo operator we have the graphical results 11.1 – 11.2, the Caputo-Fabrizio model see Figures (11.3-11.4) and for the Atangana-Baleanu model (11.5-11.6).

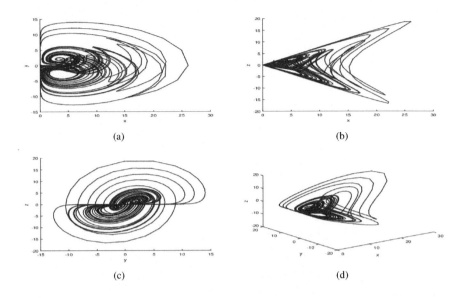

Figure 11.1 Model simulation of fractal-fractional Caputo model when $p = 1, q = 1$.

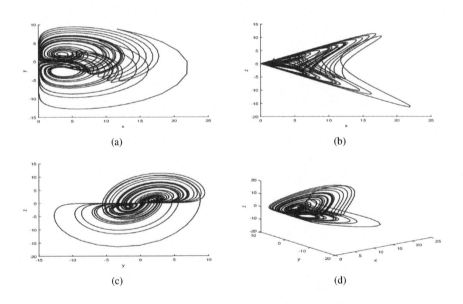

Figure 11.2 Model simulation of fractal-fractional Caputo model when $p = 0.99, q = 0.98$.

Application of Fractal-Fractional Operators to Study a New Chaotic Model

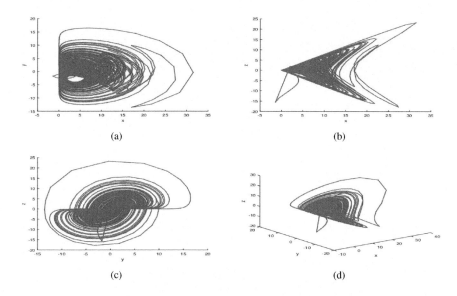

Figure 11.3 Model simulation of fractal-fractional Caputo-Fabrizio model when $p = 1$, $q = 1$.

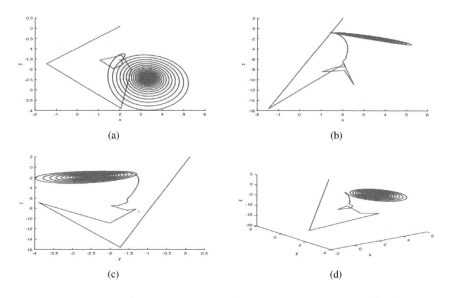

Figure 11.4 Model simulation of fractal-fractional Caputo-Fabrizio model when $p = 0.96, q = 1$.

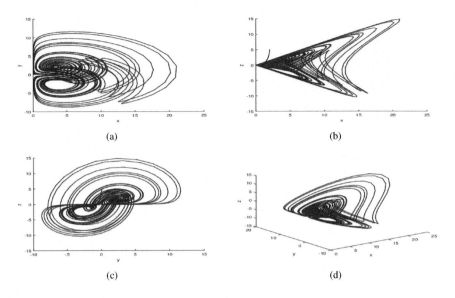

Figure 11.5 Model simulation of fractal-fractional Atangana-Baleanu model when $p = 1, q = 1$.

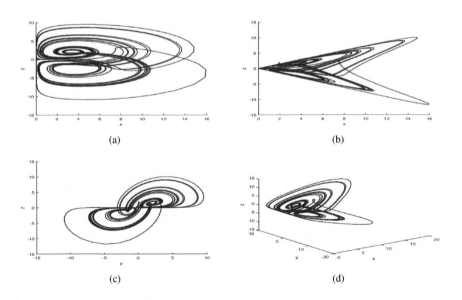

Figure 11.6 Model simulation of fractal-fractional Atangana-Baleanu model when $p = 0.99, q = 0.98$.

11.6 CONCLUSION

We presented in this chapter the dynamics of a new chaotic model in the sense of a newly defined fractal-fractional operators in the sense of Caputo, CF, and Atangana-Baleanu. The results indicate that each value of the fractal and fractional order parameters provides reasonable chaotic graphs which show the importance of the fractal-fractional operators for a dynamic system.

12 Fractal-Fractional Operators and Their Application to a Chaotic System with Sinusoidal Component

12.1 INTRODUCTION

The trigonometric functions in mathematics are real functions that connect the angle of a right-angled triangle to the ratios of its two side lengths. They are also known as circle functions, angle functions, or goniometric functions. All geosciences, including navigation, solid mechanics, celestial mechanics, geodesy, and many more, utilise them extensively. They are some of the most straightforward periodic functions, and as a result, Fourier analysis is frequently employed to examine periodic phenomena. The sine, cosine, and tangent are the trigonometric functions that are employed in modern mathematics the most frequently. Their corresponding reciprocals, which are less common, are the cosecant, secant, and cotangent. Each of these six trigonometric functions has an equivalent inverse function as well as a hyperbolic function that serves as an analogue. The trigonometric functions are crucial in physics as well. Simple harmonic motion, which simulates a variety of natural events, such as the movement of a mass linked to spring and, at small angles, the pendular motion of a mass hanging by a string, is described, for instance, using the sine and cosine functions. One-dimensional projections of uniform circular motion are the sine and cosine functions. The study of general periodic functions also benefits from the usage of trigonometric functions. Modelling recurring phenomena like sound or light waves is made easier by the distinctive wave patterns of periodic functions. The current mathematical trend is to create geometry from calculus rather than the other way around. Trigonometric functions are therefore defined using calculus techniques unless they are exceedingly simple. At every place where they are defined, that is, for the sine, cosine, and tangent, trigonometric functions are differentiable and analytical. Differential equations with trigonometric functions have been discovered to accurately reproduce several contemporary issues that have emerged in the previous several decades across all branches of science, technology, and engineering. It has been discovered that these equations can recreate some chaotic phenomena. Thus, a set of ordinary differential equations with trigonometric functions is examined in this chapter. In this chapter, we study the dynamics of a chaos model in different fractal-fractional operators. We apply the newly introduced fractal-fractional operator in the sense of Caputo, Caputo-Fabrizio, and Atangana-Baleanu. We present a novel numerical approach for the solution of fractal-fractional model. Various graphical results show that the fractal-fractional approach to study chaos model provides rich dynamics.

12.2 MODEL DESCRIPTIONS

We consider here a model that describes chaotic behaviour from literature. The chaotic model we consider here is given by the following equations:

$$\begin{cases} \frac{dx(t)}{dt} = -Ax + sin(y), \\ \frac{dy(t)}{dt} = -Ay + sin(z), \\ \frac{dz(t)}{dt} = -Az + sin(x), \end{cases} \quad (12.1)$$

where $x(t)$, $y(t)$, and $z(t)$ are the state variables while A is the only parameter of the given model. This system is chaotic for the given values of the parameters $A = 0.19$ with the initial conditions on state variables are $x(0) = -12$, $y(0) = 22$, and $z(0) = 0$. The fractal-fractional representation of the model (12.1) is given by:

$$\begin{cases} {}^{FF}D_{0,t}^{p,q}\left(x(t)\right) = -Ax + sin(y)), \\ {}^{FF}D_{0,t}^{p,q}\left(y(t)\right) = -Ay + sin(z), \\ {}^{FF}D_{0,t}^{p,q}\left(z(t)\right) = -Az + sin(x), \end{cases} \quad (12.2)$$

where p represents the fractional order while q is fractal order.

12.3 EXISTENCE AND UNIQUENESS

We will present the conditions under which the model under investigation (12.2) admits a unique solution. Further, we write the model (12.2) in the form:

$$\begin{aligned} f_x(t,x,y,z) &= -Ax + sin(y), \\ f_y(t,x,y,z) &= -Ay + sin(z), \\ f_z(t,x,y,z) &= -Az + sin(x). \end{aligned}$$

It is assumed that for all $t \in [0,T]$, the functions $x(t)$, $y(t)$, and $z(t)$ are bounded, $\|x\|_\infty \leq M_x$, $\|y\|_\infty \leq M_y$ and $\|z\|_\infty \leq M_z$. Thus,

$$\begin{aligned} |f_x(t,x,y,z)| &= |-Ax + sin(y)|, \\ &\leq |A||x| + |siny|, \end{aligned}$$

Fractal-Fractional Operators and Their Application to a Chaotic System

$$\leq |A||x|+1,$$

$$\leq |A|\sup_{t\in D_x}|x(t)|+1,$$

$$\leq |A|M_x+1,$$

$$< \infty.$$

$$|f_y(t,x,y,z)| = |-Ay+sin(z)|,$$

$$\leq A|y|+|sinz|,$$

$$\leq |A||y|+1,$$

$$\leq |A|\sup_{t\in D_y}|y|+1,$$

$$\leq |A|M_y+1,$$

$$< \infty.$$

$$|f_z(t,x,y,z)| = |-Az+sin(x)| \leq |A||z|+|sinx|,$$

$$\leq |A||z|+1,$$

$$\leq A\sup_{t\in D_z}|z|+1,$$

$$\leq |A|M_z+1,$$

$$< \infty.$$

$$|f_x(t,x,y,z)|^2 = |-Ax+sin(y)|^2,$$

$$\leq 2A^2|x|^2+2|siny|^2,$$

we note that $siny \approx y-\frac{y^3}{6},...$, then,

$$|f_x(t,x,y,z)|^2 \leq 2A^2|x|^2+2|y-\frac{y^3}{6}|^2,$$

$$\leq 2A^2|x|^2+4|y|^2+\frac{4}{36}|y|^6,$$

$$\leq 2A^2|x|^2+4M_y^2+\frac{1}{9}M_y^6,$$

$$\leq 2(M_y^2 + \frac{M_y^6}{18})\left(1 + \frac{A^2|x|^2}{M_y^2 + \frac{M_y^6}{18}}\right),$$

$$\leq 2(M_y^2 + \frac{M_y^6}{18})(1 + |x|^2),$$

if $\frac{A^2}{M_y^2 + \frac{M_y^6}{18}} < 1$, then

$$|f_x(t,x,y,z)|^2 \leq K_x(1 + |x|^2),$$

where $K_x = 2(M_y^2 + \frac{M_y^6}{18})$. In the same way,

$$|f_y(t,x,y,z)|^2 \leq K_y(1 + |y|^2),$$

where $K_y = 2(M_z^2 + \frac{M_z^6}{18})$, and $\frac{A^2}{M_z^2 + \frac{M_z^6}{18}} < 1$.

$$|f_z(t,x,y,z)|^2 \leq K_z(1 + |z|^2),$$

where $K_z = 2(M_x^2 + \frac{M_x^6}{18})$, and $\frac{A^2}{M_x^2 + \frac{M_x^6}{18}} < 1$. Thus, the system satisfies the linear growth property if

$$\max\left\{\frac{A^2}{M_x^2 + \frac{M_x^6}{18}}, \frac{A^2}{M_z^2 + \frac{M_z^6}{18}}, \frac{A^2}{M_y^2 + \frac{M_y^6}{18}}\right\} < 1.$$

We now show the Lipschitz conditions:

$$|f_x(t,x_1,y,z) - f_x(t,x_2,y,z)|^2 = A^2|x_1 - x_2|^2,$$
$$\leq \frac{3}{2}A^2|x_1 - x_2|^2,$$
$$\leq \overline{K}_x|x_1 - x_2|^2,$$
$$|f_y(t,x,y_1,z) - f_y(t,x,y_2,z)|^2 = A^2|y_1 - y_2|^2,$$
$$= \frac{3}{2}A^2|y_1 - y_2|^2,$$
$$\leq \overline{K}_y|y_1 - y_2|^2,$$
$$|f_z(t,x,y,z_1) - f_z(t,x,y,z_2)|^2 = A^2|z_1 - z_2|^2,$$
$$\leq \overline{K}_z|z_1 - z_2|^2.$$

So, the system satisfies the linear growth property and the Lipschitz conditions. Hence, the system admits a unique system of solution.

12.4 EQUILIBRIUM POINTS

It can be observed that the model has the only trivial equilibrium upon solving the equation (12.2), and we have $E_0 = (0,0,0)$. We obtain the Jacobian of the chaotic system given by

$$J = \begin{pmatrix} -A & \cos(y(t)) & 0 \\ 0 & -A & \cos(z(t)) \\ \cos(x(t)) & 0 & -A \end{pmatrix} \quad (12.3)$$

at E_0, we obtain the eigenvalues $\lambda_1 = 1 - A$, $\lambda_2 = -\frac{1}{2}i\left(-2iA - i + \sqrt{3}\right)$, $\lambda_3 = \frac{1}{2}i\left(2iA + i + \sqrt{3}\right)$. It is clear that the eigenvalues given by $\lambda_{2,3}$ have negative real parts under certain conditions while λ_1 can be negative if $A > 1$ and in our simulation, we use $A = 0.19$, so the system is unstable.

12.5 NUMERICAL PROCEDURE FOR THE CHAOTIC MODEL

The aim of the present section is to present the novel procedure for solution of the chaotic problem in different fractal-fractional operators. We consider three different newly defined operators known as fractal-fractional Caputo, fractal-fractional Caputo-Fabrizio, and the fractal-fractional Atangana-Baleanu.

12.5.1 NUMERICAL PROCEDURE IN THE SENSE OF FRACTAL-FRACTIONAL-CAPUTO OPERATOR

This subsection determines the numerical procedure for the solution of the fractal-fractional model given by (12.2) in the sense of fractal-fractional-Caputo operator. We convert the model (12.2) to volterra form as the fractional integral is differentiable, so in the Riemann-Liouville sense, we have,

$$^{FFP}D_{0,t}^{p,q}f(t) = \frac{1}{\Gamma(1-p)} \frac{d}{dt} \int_0^t (t-q)^{-p} f(q) dv \frac{1}{qt^{q-1}}, \quad (12.4)$$

we have the following result:

$$\begin{aligned} ^{RL}D_{0,t}^{p}\left(x(t)\right) &= qt^{q-1}\left[-Ax + \sin(y)\right], \\ ^{RL}D_{0,t}^{p}\left(y(t)\right) &= qt^{q-1}\left[-Ay + \sin(z)\right], \\ ^{RL}D_{0,t}^{p}\left(z(t)\right) &= qt^{q-1}\left[-Az + \sin(x)\right]. \end{aligned} \quad (12.5)$$

Replacing the RL derivative to Caputo derivative where we can get the appropriate initial conditions for the integer order. Then, applying the Riemann-Liouville fractional integral on both sides, we get the following:

$$x(t) = \frac{q}{\Gamma(p)} \int_0^t \lambda^{q-1}(t-\lambda)^{p-1} f_1(x,y,z,\lambda) d\lambda,$$

$$y(t) = \frac{q}{\Gamma(p)} \int_0^t \lambda^{q-1}(t-\lambda)^{p-1} f_2(x,y,z,\lambda) d\lambda,$$

$$z(t) = \frac{q}{\Gamma(p)} \int_0^t \lambda^{q-1}(t-\lambda)^{p-1} f_3(x,y,z,\lambda) d\lambda, \quad (12.6)$$

where

$$f_1(x,y,z,\lambda) = -Ax + \sin(y),$$

$$f_2(x,y,z,\lambda) = -Ay + \sin(z),$$

$$f_3(x,y,z,\lambda) = -Az + \sin(x).$$

We now describe here a novel procedure for the model given above using the approach at t_{n+1}. Then, our model becomes

$$x^{n+1} = \frac{q}{\Gamma(p)} \int_0^t \lambda^{q-1}(t_{n+1}-\lambda)^{p-1} f_1(x,y,z,\lambda) d\lambda,$$

$$y^{n+1} = \frac{q}{\Gamma(p)} \int_0^t \lambda^{q-1}(t_{n+1}-\lambda)^{p-1} f_2(x,y,z,\lambda) d\lambda,$$

$$z^{n+1} = \frac{q}{\Gamma(p)} \int_0^t \lambda^{q-1}(t_{n+1}-\lambda)^{p-1} f_3(x,y,z,\lambda) d\lambda. \quad (12.7)$$

The approximations of the above integral lead to the following:

$$x^{n+1} = \frac{q}{\Gamma(p)} \sum_{j=0}^n \int_{t_j}^{t_{j+1}} \lambda^{q-1}(t_{n+1}-\lambda)^{p-1} f_1(x,y,z,\lambda) d\lambda,$$

$$y^{n+1} = \frac{q}{\Gamma(p)} \sum_{j=0}^n \int_{t_j}^{t_{j+1}} \lambda^{q-1}(t_{n+1}-\lambda)^{p-1} f_2(x,y,z,\lambda) d\lambda,$$

$$z^{n+1} = \frac{q}{\Gamma(p)} \sum_{j=0}^n \int_{t_j}^{t_{j+1}} \lambda^{q-1}(t_{n+1}-\lambda)^{p-1} f_3(x,y,z,\lambda) d\lambda. \quad (12.8)$$

Now approximating the function $\lambda^{q-1} f_i(x,y,z,\lambda)$ for $i = 1, 2, 3$ in expression (12.8) in the interval $[t_j, t_{j+1}]$ through the Lagrangian piece-wise interpolation, we get

$$P_j(\lambda) = \frac{\lambda - t_{j-1}}{t_j - t_{j-1}} t_j^{q-1} f_1(x^j, y^j, z^j, t_j) - \frac{\lambda - t_j}{t_j - t_{j-1}} t_{j-1}^{q-1} f_1(x^{j-1}, y^{j-1}, z^{j-1}, t_{j-1}),$$

$$Q_j(\lambda) = \frac{\lambda - t_{j-1}}{t_j - t_{j-1}} t_j^{q-1} f_2(x^j, y^j, z^j, t_j) - \frac{\lambda - t_j}{t_j - t_{j-1}} t_{j-1}^{q-1} f_2(x^{j-1}, y^{j-1}, z^{j-1}, t_{j-1}),$$

$$R_j(\lambda) = \frac{\lambda - t_{j-1}}{t_j - t_{j-1}} t_j^{q-1} f_3(x^j, y^j, z^j, t_j)$$

Fractal-Fractional Operators and Their Application to a Chaotic System

$$-\frac{\lambda - t_j}{t_j - t_{j-1}} t_{j-1}^{q-1} f_3(x^{j-1}, y^{j-1}, z^{j-1}, t_{j-1}). \tag{12.9}$$

We get the following:

$$\begin{aligned}
x^{n+1} &= \frac{q}{\Gamma(p)} \sum_{j=0}^{n} \int_{t_j}^{t_{j+1}} \lambda^{q-1}(t_{n+1} - \lambda)^{p-1} P_j(\lambda) d\lambda, \\
y^{n+1} &= \frac{q}{\Gamma(p)} \sum_{j=0}^{n} \int_{t_j}^{t_{j+1}} \lambda^{\kappa-1}(t_{n+1} - \lambda)^{p-1} Q_j(\lambda) d\lambda, \\
z^{n+1} &= \frac{q}{\Gamma(p)} \sum_{j=0}^{n} \int_{t_j}^{t_{j+1}} \lambda^{\kappa-1}(t_{n+1} - \lambda)^{p-1} R_j(\lambda) d\lambda,
\end{aligned} \tag{12.10}$$

The solutions of (12.10) lead to the scheme below

$$\begin{aligned}
x^{n+1} &= \frac{qh^p}{\Gamma(p+2)} \sum_{j=1}^{n} \Big[t_j^{q-1} f_1(x^j, y^j, z^j, t_j) \\
&\quad \times \Big((n+1-j)^p(n-j+2+p) - (n-j)^p(n-j+2+2p) \Big) \\
&\quad - t_{j-1}^{q-1} f_1(x^{j-1}, y^{j-1}, z^{j-1}, t_{j-1}) \\
&\quad \times \Big((n-j+1)^{p+1} - (n-j)^p(n-j+1+p) \Big) \Big], \\
y^{n+1} &= \frac{qh^p}{\Gamma(p+2)} \sum_{j=1}^{n} \Big[t_j^{q-1} f_2(x^j, y^j, z^j, t_j) \\
&\quad \times \Big((n+1-j)^p(n-j+2+p) - (n-j)^p(n-j+2+2p) \Big) \\
&\quad - t_{j-1}^{q-1} f_2(x^{j-1}, y^{j-1}, z^{j-1}, t_{j-1}) \\
&\quad \times \Big((n-j+1)^{p+1} - (n-j)^p(n-j+1+p) \Big) \Big], \\
z^{n+1} &= \frac{qh^p}{\Gamma(p+2)} \sum_{j=1}^{n} \Big[t_j^{q-1} f_3(x^j, y^j, z^j, t_j) \\
&\quad \times \Big((n+1-j)^p(n-j+2+p) - (n-j)^p(n-j+2+2p) \Big) \\
&\quad - t_{j-1}^{q-1} f_3(x^{j-1}, y^{j-1}, z^{j-1}, t_{j-1}) \\
&\quad \times \Big((n-j+1)^{p+1} - (n-j)^p(n-j+1+p) \Big) \Big].
\end{aligned} \tag{12.11}$$

Next, in the following subsection, we present the numerical procedure for the fractal-fractional model in the sense of Caputo-Fabrizio operator.

12.5.2 NUMERICAL PROCEDURE FOR FRACTAL-FRACTIONAL CAPUTO-FABRIZIO OPERATOR

The aim of this subsection is to present a numerical procedure for the solution of the given model in fractal-fractional Caputo-Fabrizio sense. In order to do this, we first write the model in the following form:

$$\begin{aligned}
{}^{CF}D_{0,t}^{p}\left(x(t)\right) &= qt^{q-1}f_1(x,y,z,t), \\
{}^{CF}D_{0,t}^{p}\left(y(t)\right) &= qt^{q-1}f_2(x,y,z,t), \\
{}^{CF}D_{0,t}^{p}\left(z(t)\right) &= qt^{q-1}f_3(x,y,z,t).
\end{aligned} \quad (12.12)$$

We obtain the following by using the Caputo-Fabrizio integral:

$$\begin{aligned}
x(t) &= x^0 + \frac{qt^{q-1}(1-p)}{M(p)}f_1(x,y,z,t) + \frac{pq}{M(p)}\int_0^t \lambda^{q-1}f_1(x,y,z,\lambda)d\lambda, \\
y(t) &= y^0 + \frac{qt^{q-1}(1-p)}{M(p)}f_2(x,y,z,t) + \frac{pq}{M(p)}\int_0^t \lambda^{q-1}f_2(x,y,z,\lambda)d\lambda, \\
z(t) &= z^0 + \frac{qt^{q-1}(1-p)}{M(p)}f_3(x,y,z,t) \\
&\quad + \frac{pq}{M(p)}\int_0^t \lambda^{q-1}f_3(x,y,z,\lambda)d\lambda,
\end{aligned} \quad (12.13)$$

The following is presented at t_{n+1}:

$$\begin{aligned}
x^{n+1}(t) &= x^n + \frac{qt^{q-1}(1-p)}{M(p)}f_1(x^n,y^n,z^n,t_n) + \frac{pq}{M(p)}\int_0^{t_{n+1}} \lambda^{q-1}f_1(x,y,z,\lambda)d\lambda, \\
y^{n+1}(t) &= y^n + \frac{qt^{q-1}(1-p)}{M(p)}f_2(x^n,y^n,z^n,t_n) + \frac{pq}{M(p)}\int_0^{t_{n+1}} \lambda^{q-1}f_2(x,y,z,\lambda)d\lambda, \\
z^{n+1}(t) &= z^n + \frac{qt^{q-1}(1-p)}{M(p)}f_3(x^n,y^n,z^n,t_n) \\
&\quad + \frac{pq}{M(p)}\int_0^{t_{n+1}} \lambda^{q-1}f_3(x,y,z,\lambda)d\lambda.
\end{aligned} \quad (12.14)$$

Further, we have the following:

$$\begin{aligned}
x^{n+1}(t) &= x^n + \frac{qt^{q-1}(1-p)}{M(p)} \\
&\quad \times f_1(x^n,y^n,z^n,t_n) - \frac{qt_{n-1}^{q-1}(1-p)}{M(p)}f_1(x^{n-1},y^{n-1},z^{n-1},t_{n-1}) \\
&\quad + \frac{pq}{M(p)}\int_{t_n}^{t_{n+1}} \lambda^{q-1}f_1(x,y,z,\lambda)d\lambda,
\end{aligned}$$

Fractal-Fractional Operators and Their Application to a Chaotic System

$$y^{n+1}(t) = y^n + \frac{qt^{q-1}(1-p)}{M(p)}$$
$$\times f_2(x^n, y^n, z^n, t_n) - \frac{qt_{n-1}^{q-1}(1-p)}{M(p)} f_2(x^{n-1}, y^{n-1}, z^{n-1}, t_{n-1})$$
$$+ \frac{pq}{M(p)} \int_{t_n}^{t_{n+1}} \lambda^{q-1} f_2(x, y, z, \lambda) d\lambda,$$

$$z^{n+1}(t) = z^n + \frac{qt^{q-1}(1-p)}{M(p)}$$
$$\times f_3(x^n, y^n, z^n, t_n) - \frac{qt_{n-1}^{q-1}(1-p)}{M(p)} f_3(x^{n-1}, y^{n-1}, z^{n-1}, t_{n-1})$$
$$+ \frac{pq}{M(p)} \int_{t_n}^{t_{n+1}} \lambda^{q-1} f_3(x, y, z, \lambda) d\lambda. \tag{12.15}$$

Now using the Lagrange polynomial piece-wise interpolation and integrating, we obtain

$$x^{n+1}(t) = x^n + \frac{qt_n^{q-1}(1-p)}{M(p)}$$
$$\times f_1(x^n, y^n, z^n, t_n) - \frac{qt_{n-1}^{q-1}(1-p)}{M(p)} f_1(x^{n-1}, y^{n-1}, z^{n-1}, t_{n-1})$$
$$+ \frac{pq}{M(p)} \left[\frac{3h}{2} t_n^{q-1} f_1(x^n, y^n, z_2^n, t_n) \right.$$
$$\left. - \frac{h}{2} t_{n-1}^{q-1} f_1(x^{n-1}, y^{n-1}, z^{n-1}, t_{n-1}) \right],$$

$$y^{n+1}(t) = y^n + \frac{qt_n^{q-1}(1-p)}{M(p)}$$
$$\times f_2(x^n, y^n, z^n, t_n) - \frac{qt_{n-1}^{q-1}(1-p)}{M(p)} f_2(x^{n-1}, y^{n-1}, z^{n-1}, t_{n-1})$$
$$+ \frac{pq}{M(p)} \left[\frac{3h}{2} t_n^{q-1} f_1(x^n, y^n, z_2^n, t_n) \right.$$
$$\left. - \frac{h}{2} t_{n-1}^{q-1} f_2(x^{n-1}, y^{n-1}, z^{n-1}, t_{n-1}) \right],$$

$$z^{n+1}(t) = z^n + \frac{qt_n^{q-1}(1-p)}{M(p)}$$
$$\times f_3(x^n, y^n, z^n, t_n) - \frac{qt_{n-1}^{q-1}(1-p)}{M(p)} f_1(x^{n-1}, y^{n-1}, z^{n-1}, t_{n-1})$$
$$+ \frac{pq}{M(p)} \left[\frac{3h}{2} t_n^{q-1} f_3(x^n, y^n, z_2^n, t_n) \right.$$
$$\left. - \frac{h}{2} t_{n-1}^{q-1} f_3(x^{n-1}, y^{n-1}, z^{n-1}, t_{n-1}) \right]. \tag{12.16}$$

Further, we obtain the following:

$$\begin{aligned}
x^{n+1} &= x^n + qt_n^{q-1}\left(\frac{1-p}{M(p)} + \frac{3ph}{2M(p)}\right)f_1(x^n,y^n,z^n,t_n) \\
&\quad - qt_{n-1}^{q-1}\left(\frac{1-p}{M(p)} + \frac{ph}{2M(p)}\right)f_1(x^{n-1},y^{n-1},z^{n-1},t_{n-1}), \\
y^{n+1} &= y^n + qt_n^{q-1}\left(\frac{1-p}{M(p)} + \frac{3ph}{2M(p)}\right)f_2(x^n,y^n,z^n,t_n) \\
&\quad - qt_{n-1}^{q-1}\left(\frac{1-p}{M(p)} + \frac{ph}{2M(p)}\right)f_2(x^{n-1},y^{n-1},z^{n-1},t_{n-1}), \\
z^{n+1} &= z^n + qt_n^{q-1}\left(\frac{1-p}{M(p)} + \frac{3ph}{2M(p)}\right)f_3(x^n,y^n,z^n,t_n) \\
&\quad - qt_{n-1}^{q-1}\left(\frac{1-p}{M(p)} + \frac{ph}{2M(p)}\right)f_3(x^{n-1},y^{n-1},z^{n-1},t_{n-1}). \quad (12.17)
\end{aligned}$$

Next, we present the numerical procedure for the fractal-fractional in the sense of Atangana-Baleanu derivative in the following subsection.

12.5.3 NUMERICAL PROCEDURE FOR FRACTAL-FRACTIONAL ATANGANA-BALEANU OPERATOR

In order to have a numerical procedure for the solution of fractal-fractional model in the sense of Atangana-Baleanu operator, we express first the model (12.2) in the following form:

$$\begin{aligned}
{}^{ABR}D_{0,t}^{\alpha}\left(x(t)\right) &= qt^{q-1}f_1(x,y,z,t), \\
{}^{ABR}D_{0,t}^{\alpha}\left(y(t)\right) &= qt^{q-1}f_2(x,y,z,t), \\
{}^{ABR}D_{0,t}^{\alpha}\left(z(t)\right) &= qt^{q-1}f_3(x,y,z,t). \quad (12.18)
\end{aligned}$$

The following is presented on the basis of Atangana-Baleanu integral:

$$\begin{aligned}
x(t) &= \frac{qt^{q-1}(1-p)}{AB(p)}f_1(x,y,z,t) \\
&\quad + \frac{pq}{AB(p)\Gamma(p)}\int_0^t \lambda^{q-1}(t-\lambda)^{p-1}f_1(x,y,z,\lambda)d\lambda, \\
y(t) &= \frac{qt^{q-1}(1-p)}{AB(p)}f_2(x,y,z,t) \\
&\quad + \frac{pq}{AB(p)\Gamma(p)}\int_0^t \lambda^{q-1}(t-\lambda)^{p-1}f_2(x,y,z,\lambda)d\lambda, \\
z(t) &= \frac{qt^{q-1}(1-p)}{AB(p)}f_3(x,y,z,t)
\end{aligned}$$

Fractal-Fractional Operators and Their Application to a Chaotic System

$$+\frac{pq}{AB(p)\Gamma(p)}\int_0^t \lambda^{q-1}(t-\lambda)^{p-1}f_3(x,y,z,\lambda)d\lambda. \quad (12.19)$$

Further, at t_{n+1}, we get,

$$x^{n+1} = \frac{qt_n^{q-1}(1-p)}{AB(\alpha)}f_1(x^n,y^n,z^n,t_n)$$

$$+\frac{pq}{AB(p)\Gamma(p)}\int_0^{t_{n+1}} \lambda^{q-1}(t_{n+1}-\lambda)^{p-1}f_1(x,y,z,\lambda)d\lambda,$$

$$y^{n+1} = \frac{qt_n^{q-1}(1-p)}{AB(\alpha)}f_2(x^n,y^n,z^n,t_n)$$

$$+\frac{pq}{AB(p)\Gamma(p)}\int_0^{t_{n+1}} \lambda^{q-1}(t_{n+1}-\lambda)^{p-1}f_2(x,y,z,\lambda)d\lambda,$$

$$z^{n+1} = \frac{qt_n^{q-1}(1-p)}{AB(\alpha)}f_3(x^n,y^n,z^n,t_n)$$

$$+\frac{pq}{AB(p)\Gamma(p)}\int_0^{t_{n+1}} \lambda^{q-1}(t_{n+1}-\lambda)^{p-1}f_3(x,y,z,\lambda)d\lambda. \quad (12.20)$$

We obtain the following after simplification the integral in the above equations:

$$x^{n+1} = \frac{qt_n^{q-1}(1-p)}{AB(p)}f_1(x^n,y^n,z^n,t_n)$$

$$+\frac{pq}{AB(p)\Gamma(p)}\sum_{j=0}^n \int_{t_j}^{t_{j+1}} \lambda^{q-1}(t_{n+1}-\lambda)^{p-1}f_1(x,y,z,\lambda)d\lambda,$$

$$y^{n+1} = \frac{qt_n^{q-1}(1-p)}{AB(p)}f_2(x^n,y^n,z^n,t_n)$$

$$+\frac{pq}{AB(p)\Gamma(p)}\sum_{j=0}^n \int_{t_j}^{t_{j+1}} \lambda^{q-1}(t_{n+1}-\lambda)^{p-1}f_2(x,y,z,\lambda)d\lambda,$$

$$z^{n+1} = \frac{qt_n^{q-1}(1-p)}{AB(p)}f_3(x^n,y^n,z^n,t_n)$$

$$+\frac{pq}{AB(p)\Gamma(p)}\sum_{j=0}^n \int_{t_j}^{t_{j+1}} \lambda^{q-1}(t_{n+1}-\lambda)^{p-1}$$

$$\times f_3(x,y,z,\lambda)d\lambda. \quad (12.21)$$

Now approximating the expressions in (12.21) given by $\lambda^{q-1}f_1(x,y,z,\lambda)$, $\lambda^{q-1}f_2(x,y,z,\lambda)$; and $\lambda^{q-1}f_3(x,y,z,\lambda)$ in the given internal $[t_j,t_{j+1}]$, the following numerical scheme is presented:

$$x^{n+1} = \frac{qt_n^{q-1}(1-p)}{AB(p)}f_1(x^n,y^n,z^n,t_n)$$

$$+\frac{q(\Delta t)^p}{AB(p)\Gamma(p+2)}\sum_{j=1}^n \left[t_j^{q-1}f_1(x^j,y^j,z^j,t_j)\right.$$

$$\times \left((n+1-j)^p(n-j+2+\alpha) - (n-j)^p(n-j+2+2p)\right)$$
$$-t_{j-1}^{q-1} f_1(x^{j-1}, y^{j-1}, z^{j-1}, t_{j-1})$$
$$\times \left((n-j+1)^{p+1} - (n-j)^p(n-j+1+p)\right)\Bigg],$$

$$y^{n+1} = \frac{qt_n^{q-1}(1-p)}{AB(p)} f_2(x^n, y^n, z^n, t_n)$$
$$+ \frac{q(\Delta t)^p}{AB(p)\Gamma(p+2)} \sum_{j=1}^{n} \left[t_j^{q-1} f_2(x^j, y^j, z^j, t_j) \right.$$
$$\times \left((n+1-j)^p(n-j+2+\alpha) - (n-j)^p(n-j+2+2p)\right)$$
$$-t_{j-1}^{q-1} f_2(x^{j-1}, y^{j-1}, z^{j-1}, t_{j-1})$$
$$\times \left((n-j+1)^{p+1} - (n-j)^p(n-j+1+p)\right)\Bigg],$$

$$z^{n+1} = \frac{qt_n^{q-1}(1-p)}{AB(p)} f_3(x^n, y^n, z^n, t_n)$$
$$+ \frac{q(\Delta t)^p}{AB(p)\Gamma(p+2)} \sum_{j=1}^{n} \left[t_j^{q-1} f_3(x^j, y^j, z^j, t_j) \right.$$
$$\times \left((n+1-j)^p(n-j+2+\alpha) - (n-j)^p(n-j+2+2p)\right)$$
$$-t_{j-1}^{q-1} f_3(x^{j-1}, y^{j-1}, z^{j-1}, t_{j-1})$$
$$\times \left((n-j+1)^{p+1} - (n-j)^p(n-j+1+p)\right)\Bigg]. \quad (12.22)$$

12.6 NUMERICAL RESULTS

For the proposed fractal-fractional model (12.2); we present here the numerical results by considering the numerical procedure above. The numerical results given in Figures 12.1–12.6 represent the behaviour of the fractal-fractional model in the sense of the Caputo, CF, and AB operators. The model with each fractal-fractional operator behaves differently and provides different chaotic results.

12.7 CONCLUSION

A mathematical model considered in this chapter is chaotic for the specific values of the parameters and the initial conditions. The model is analysed for their equilibrium points and then different fractal-fractional operators applied and obtain a suitable and efficient numerical schemes for their solution. Some selective graphs are presented for each operators in order to study the chaotic behaviour of the model.

Fractal-Fractional Operators and Their Application to a Chaotic System

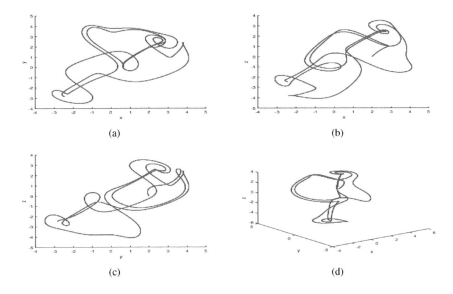

Figure 12.1 Model simulation of fractal-fractional Caputo model when $p = 1, q = 1$.

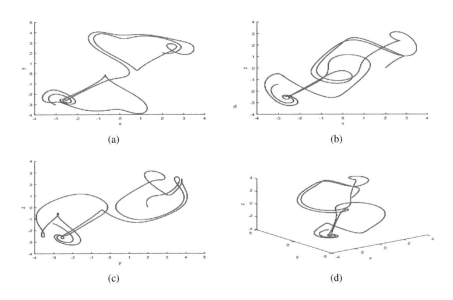

Figure 12.2 Model simulation of fractal-fractional Caputo model when $p = 0.99, q = 0.99$.

168 Numerical Methods for Fractal-Fractional Differential

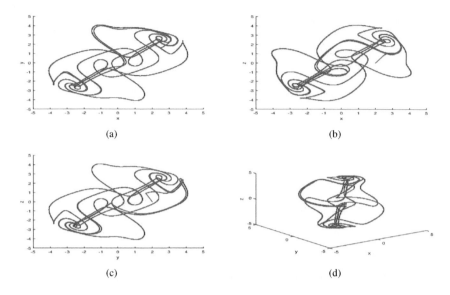

Figure 12.3 Model simulation of fractal-fractional Caputo-Fabrizio model when $p = 1$, $q = 1$.

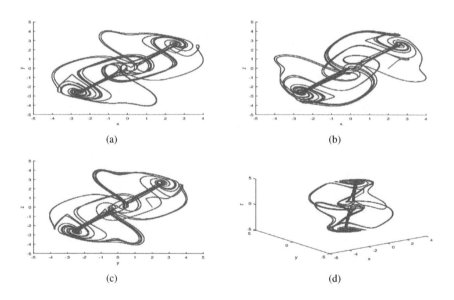

Figure 12.4 Model simulation of fractal-fractional Caputo-Fabrizio model when $p = 0.99, q = 0.99$.

Fractal-Fractional Operators and Their Application to a Chaotic System

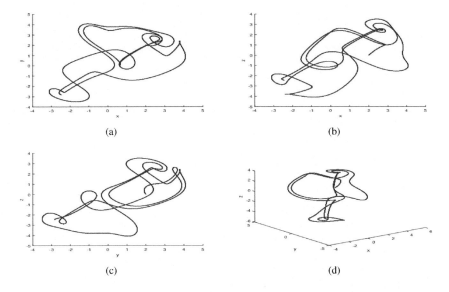

Figure 12.5 Model simulation of fractal-fractional Atangana-Baleanu model when $p = 1, q = 1$.

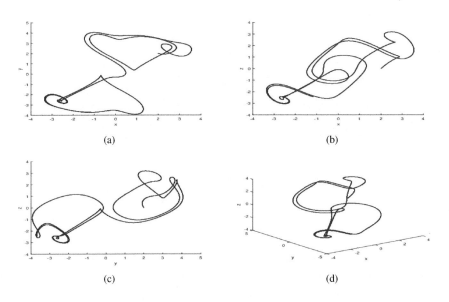

Figure 12.6 Model simulation of fractal-fractional Atangana-Baleanu model when $p = 0.99, q = 0.99$.

13 Application of Fractal-Fractional Operators to Four-Scroll Chaotic System

13.1 INTRODUCTION

A papyrus, parchment, or paper roll with writing on it is referred to as a scroll. Pages, which might occasionally be distinct papyrus or parchment sheets adhered together at the edges, are typically used to divide up a scroll. A continuous roll of writing material may be divided into scrolls. The scroll is often unrolled so that one page can be seen at a time for writing or reading, with the other pages being rolled and stored to the left and right of the viewable page. The (double, four, multi)-scroll attractor is a peculiar attractor found in the mathematics of dynamical systems and is derived from a practical electrical chaotic circuit with a single nonlinear resistor. An equation with three segments and a system of three nonlinear ordinary differential equations are frequently used to describe the double-scroll system. Due to Chua's circuits' straightforward architecture, the system can be easily simulated numerically and physically. The X, Y, and Z output signals of a Chua's circuit are used to observe this shape on an oscilloscope. Given that it resembles two Saturn-like rings joined by whirling lines in three dimensions, this chaotic attractor is also known as the double scroll. The (double, four, or multiple)-scroll attractor has an endless number of fractal-like layers, as demonstrated by numerical analysis of the attractor's geometrical structure. At all scales, it seems like every cross section is a fractal. In the recent past, it has also been revealed that the double scroll contains secret attractors. A number of academicians with diverse backgrounds have shown an interest in these systems of nonlinear ordinary differential equations. An attractor with four scrolls will be looked at in this chapter. In various field of engineering research use chaos theory in their work especially the widely used use of chaos in oscillators, see [83, 190, 249, 260], epilepsy [183], random bit generators [115], neural networks [40, 246], in population ecology, we refer the readers to see [162, 248], text encryption [57], image encryption [240], neurons [35, 77, 247], voice encryption [250], robotics [251], weather systems [70], economy [229], chemical reactions [247] and secure communication systems [208] etc. The study of multi-scroll chaotic attractors has gained popularity and piqued academics' curiosity. The systems of Lorenz, Lu, and Tigan are regarded as classic instances of two-scroll chaotic attractors [45, 113, 150]. The two-scroll chaotic attractors that considered the

DOI: 10.1201/9781003359258-13

recent examples are the Lien system [236],Vaidyanathan systems [143], Zheng system [283], etc. The examples new three-scroll system include Dadras system [56], Pan system [182], Vaidyanathan 3-scroll system [245], etc. The examples of four-scroll chaotic system are also explored in the literature, which include, Wang system [193], Lu 4-scroll system [153], Liu-Chen 4-scroll system [148], Sampath 4-scroll system [209], Akgul system [5], etc. The four-scroll generated chaotic system, Yang and Li discussed In 2003 with feedback control, for details see in [275]. In the present chapter, we aim to study the dynamics of a new chaos four-scroll chaotic system in light if different fractal-fractional operators with novel schemes based on linear interpolations. We first examine the existence and uniqueness of the system and then study their equilibrium points and its analysis. After that, we consider the model by applying the three different fractal-fractional operators and obtain their numerical simulations for the values of the parameters that the system become chaotic. For each operator, we have a numerical scheme and provide a set for figures for each case when using different fractal and fractional orders.

13.2 MODEL DESCRIPTIONS

In the present section, we consider the four-scroll chaotic system from literature [210], given by the following system of differential equations:

$$\begin{cases} \frac{dx(t)}{dt} = a(y-x) + byz, \\ \frac{dy(t)}{dt} = -y - 10y|y| + 4xz, \\ \frac{dz(t)}{dt} = cz - xy, \end{cases} \tag{13.1}$$

where $x(t)$, $y(t)$, and $z(t)$ are the state variables while a, b, and c describe the parameters of the model. This system is chaotic for the given values of the parameters $a = 3, b = 12, c = 14$ with the initial conditions on state variables are $x(0) = 0.1$, $y(0) = 0.1$, and $z(0) = 0.1$. The fractal-fractional representation of the model (13.1) is given by:

$$\begin{cases} {}^{FF}D_{0,t}^{p,\,q}\left(x(t)\right) = a(y-x) + byz, \\ {}^{FF}D_{0,t}^{p,\,q}\left(y(t)\right) = -y - 10y|y| + 4xz, \\ {}^{FF}D_{0,t}^{p,\,q}\left(z(t)\right) = cz - xy, \end{cases} \tag{13.2}$$

where p represents the fractional order while q is fractal order.

13.3 EXISTENCE AND UNIQUENESS

In this section, we study the existence and puniness of the model (13.2). It is assumed that for all $t \in [0,T]$, the functions $x(t)$, $y(t)$, and $z(t)$ are bounded, $\|x\|_\infty \leq M_x$, $\|y\|_\infty \leq M_y$, and $\|z\|_\infty \leq M_z$. Further, we write the model (13.2) in the form:

$$f_x(t,x,y,z) = a(y-x) + byz,$$

$$f_y(t,x,y,z) = -y - 10y|y| + 4xz,$$

$$f_z(t,x,y,z) = cz - xy.$$

We prove that f_x, f_y, and f_z satisfy the linear growth and Lipschitz conditions. We first show that the functions satisfy the linear growth property.

$$\begin{aligned}
\|f_x(t,x,y,z)\| &= \|a(y-x) + byz\|, \\
&\leq a\|y\| + a\|x\| + b\|y\|\|z\|, \\
&\leq a\|y\|_\infty + a\|x\|_\infty + b\|y\|_\infty\|z\|_\infty, \\
&\leq aM_y + aM_x + bM_yM_z, \\
&< \infty.
\end{aligned}$$

$$\begin{aligned}
\|f_y(t,x,y,z)\| &= \|-y - 10y|y| + 4xz\|, \\
&\leq \|y\| + 10\|y\|\|y\| + 4\|x\|\|z\|, \\
&\leq a\|y\|_\infty + 10\|y\|_\infty\|y\|_\infty + 4\|x\|_\infty\|z\|_\infty, \\
&\leq aM_y + 10M_y^2 + 4M_xM_z, \\
&< \infty.
\end{aligned}$$

$$\begin{aligned}
\|f_z(t,x,y,z)\| &= \|cz - xy\|, \\
&\leq c\|z\| + \|x\|\|y\|, \\
&\leq c\|z\|_\infty + \|x\|_\infty\|y\|_\infty, \\
&\leq cM_z + M_xM_y, \\
&< \infty.
\end{aligned}$$

We now show the Lipschitz conditions:

$$|f_x(t,x_1,y,z) - f_x(t,x_2,y,z)| \leq a|x_1 - x_2|.$$

$$|f_y(t,x,y_1,z) - f_y(t,x,y_2,z)| \leq |y_1 - y_2 - 10(y_1 - y_2)|(y_1 - y_2)\|,$$

$$\leq |y_1 - y_2| + 10|y_1 - y_2||y_1 - y_2|,$$
$$\leq |y_1 - y_2|\left(1 + 10|y_1 - y_2|\right),$$
$$\leq |y_1 - y_2|(1 + 10(|y_1| + |y_2|)),$$
$$\leq |y_1 - y_2|(1 + 10(M_{y_1} + M_{y_2})),$$
$$\leq K|y_1 - y_2|,$$

where $K = (1 + 10(M_{y_1} + M_{y_2}))$.

$$|f_z(t,x,y,z_1) - f_z(t,x,y,z_2)| \leq c|z_1 - z_2|.$$

Alternatively, we have the following:

$$|f_x(t,x,y,z)|^2 = |ay - ax + byz|^2,$$
$$\leq 3a^2|y|^2 + 3a^2|x|^2 + 3b^2|y|^2|z|^2,$$
$$\leq 3a^2 M_y^2 + 3a^2|x|^2 + 3b^2 M_y^2 M_z^2,$$
$$\leq 3(a^2 M_y^2 + b^2 M_y^2 M_z^2)\left(1 + \frac{a^2|x|^2}{(a^2 M_y^2 + b^2 M_y^2 M_z^2)}\right),$$

if $\frac{a^2}{(a^2 M_y^2 + b^2 M_y^2 M_z^2)} < 1$, then

$$|f_x(t,x,y,z)|^2 \leq K_x(1 + |x|^2),$$

where $K_x = 3(a^2 M_y^2 + b^2 M_y^2 M_z^2)$.

$$|f_y(t,x,y,z)|^2 = |-y - 10y|y| + 4xz|^2,$$
$$\leq 3|y|^2 + 300|y|^4 + 48|x|^2|z|^2,$$
$$\leq 3|y|^2(1 + 100M_y^2) + 48M_x^2 M_z^2,$$
$$\leq 3M_x^2 M_z^2\left(1 + \frac{1 + 100M_y^2}{16M_x^2 M_z^2}\right),$$

if $\frac{1 + 100M_y^2}{16M_x^2 M_z^2} < 1$, then,

$$|f_y(t,x,y,z)|^2 \leq K_y(1 + |y|^2),$$

$K_y = 3M_x^2 M_z^2$.

$$|f_z(t,x,y,z)|^2 = |cz - xy|^2,$$

$$\leq 2c^2|z|^2 + 2|x|^2|y|^2,$$
$$\leq 2c^2|z|^2 + 2M_x^2 M_y^2,$$
$$\leq 2M_x^2 M_y^2 \left(1 + \frac{c^2|z|^2}{M_x^2 M_y^2}\right),$$

if $\frac{c^2}{M_x^2 M_y^2} < 1$, then

$$|f_z(t,x,y,z)|^2 \leq K_z(1+|z|^2),$$

where $K_z = 2M_x^2 M_y^2$. Also,

$$|f_x(t,x_1,y,z) - f_x(t,x_2,y,z)|^2 = a^2|x_1 - x_2|^2,$$
$$\leq \frac{3}{2}a^2|x_1 - x_2|^2,$$
$$|f_z(t,x,y,z_1) - f_z(t,x,y,z_2)|^2 = c^2|z_1 - z_2|^2,$$
$$\leq \frac{3}{2}c^2|z_1 - z_2|^2,$$
$$|f_y(t,x,y_1,z) - f_y(t,x,y_2,z)|^2 = |(y_1 - y_2) - 10(y_1 - y_2)|y_1 - y_2||^2,$$
$$\leq 2|y_1 - y_2|^2 + 20|y_1 - y_2|^4,$$
$$\leq 2|y_1 - y_2|^2 (1 + 20M_{y_1}^2 + 20M_{y_2}^2),$$
$$\leq \overline{K}_y |y_1 - y_2|^2.$$

So, the system satisfies the linear growth property and the Lipschitz conditions. Hence, the system admits a unique system of solution.

13.4 EQUILIBRIUM POINTS

The equilibrium points of the model (13.2) are obtained by setting

$$\begin{cases} 0 = a(y-x) + byz, \\ 0 = -y - 10y|y| + 4xz, \\ 0 = cz - xy, \end{cases}$$

we have $E_0 = (0,0,0)$, $E_1 = (3.0837, 0.8509, 0.6560)$, and $E_2 = (-3.0837, -0.8509, 0.6560)$ by using the parameter values described above. Among these equilibrium points, E_0 is unstable equilibrium point, whereas the remaining two equilibrium points are saddle-foci equilibrium points. Thus, all the three equilibrium points are unstable which shows that the system is a self-excited attractor.

13.5 NUMERICAL PROCEDURE FOR THE CHAOTIC MODEL

This section discusses the new algorithms for the solution of fractal-fractional differential equations using the concept of linear interpolations. We consider three different kernels, such as the power law kernel, exponential law kernel, and the Mittag-Leffler. We provide for each case a numerical scheme in the following:

13.5.1 NUMERICAL SCHEME FOR POWER LAW KERNEL USING LINEAR INTERPOLATION

In this case, we consider the following fractal-fractional Cauchy problem,

$$\begin{aligned} {}_0^{FFP}D_t^{p,q}\mathbf{g}(t) &= f(t,\mathbf{g}(t)), \text{ if } t>0, \\ \mathbf{g}(0) &= \mathbf{g}_0, \text{ if } t=0. \end{aligned} \quad (13.3)$$

It is assumed that $f(t,\mathbf{g}(t))$ is twice differentiable and that $\mathbf{g}(t)$ is at least continuous

$$\begin{aligned} {}_0^{RL}D_t^p\mathbf{g}(t) &= qt^{q-1}f(t,\mathbf{g}(t)), \text{ if } t>0, \\ \mathbf{g}(0) &= \mathbf{g}_0, \text{ if } t=0. \end{aligned} \quad (13.4)$$

We write further

$$\begin{aligned} \mathbf{g}(t) &= \frac{q}{\Gamma(p)}\int_0^t \tau^{q-1}f(\tau,\mathbf{g}(\tau))(t-\tau)^{p-1}d\tau, \text{ if } t>0, \\ \mathbf{g}(0) &= \mathbf{g}_0, \text{ if } t=0, \end{aligned} \quad (13.5)$$

at $t = t_{n+1}$, we have

$$\mathbf{g}(t_{n+1}) = \frac{q}{\Gamma(p)}\int_0^{t_{n+1}} \tau^{q-1}f(\tau,\mathbf{g}(\tau))(t_{n+1}-\tau)^{p-1}d\tau. \quad (13.6)$$

Also, we write

$$\mathbf{g}(t_{n+1}) = \frac{q}{\Gamma(p)}\sum_{j=0}^n \int_{t_j}^{t_{j+1}} \tau^{q-1}f(\tau,\mathbf{g}(\tau))(t_{n+1}-\tau)^{p-1}d\tau. \quad (13.7)$$

Thus, within $[t_j, t_{j+1}]$, we approximate the function

$$f(\tau,\mathbf{g}(\tau)) \approx P_j(\tau) = f(t_j,\mathbf{g}_j) + (t-t_j)\frac{f(t_{j+1},\mathbf{g}_{j+1}) - f(t_j,\mathbf{g}_j)}{\Delta t}, \quad (13.8)$$

replacing yields

$$\mathbf{g}(t_{n+1}) \approx \mathbf{g}_{n+1} = \frac{q}{\Gamma(p)}\sum_{j=0}^n \int_{t_j}^{t_{j+1}} \tau^{q-1}(t_{n+1}-\tau)^{p-1}$$

$$\times \left(f(t_j, \mathbf{g}_j) + (t - t_j) \frac{f(t_{j+1}, \mathbf{g}_{j+1}) - f(t_j, \mathbf{g}_j)}{\Delta t} \right) d\tau,$$

$$= \frac{q}{\Gamma(p)} \sum_{j=0}^{n} f(t_j, \mathbf{g}_j) \int_{t_j}^{t_{j+1}} \tau^{q-1} (t_{n+1} - \tau)^{p-1} d\tau$$

$$+ \frac{q}{\Gamma(p)} \sum_{j=0}^{n} \frac{f(t_{j+1}, \mathbf{g}_{j+1}) - f(t_j, \mathbf{g}_j)}{\Delta t}$$

$$\times \int_{t_j}^{t_{j+1}} (t_{n+1} - \tau)^{p-1} (t - t_j) d\tau. \tag{13.9}$$

Noting that

$$\int_{t_j}^{t_{j+1}} \tau^{q-1} (t_{n+1} - \tau)^{p-1} d\tau = \int_{0}^{t_{j+1}} \tau^{q-1} (t_{n+1} - \tau)^{p-1} d\tau$$
$$- \int_{0}^{t_j} \tau^{q-1} (t_{n+1} - \tau)^{p-1} d\tau, \tag{13.10}$$

where

$$\int_{0}^{t_{j+1}} \tau^{q-1} (t_{n+1} - \tau)^{p-1} d\tau = t_{n+1}^{p+q-1} \int_{0}^{\frac{t_{j+1}}{t_{n+1}}} \mathbf{g}^{q-1} (1 - \mathbf{g})^p \mathbf{g}^{-1} d\mathbf{g},$$
$$= t_{n+1}^{p+q-1} B\left(\frac{t_{j+1}}{t_{n+1}}, q, p \right). \tag{13.11}$$

$$\int_{0}^{t_j} \tau^{q-1} (t_{n+1} - \tau)^{p-1} d\tau = t_{n+1}^{p+q-1} B\left(\frac{t_j}{t_{n+1}}, q, p \right). \tag{13.12}$$

Therefore,

$$\int_{t_j}^{t_{j+1}} \tau^{q-1} (t_{n+1} - \tau)^{p-1} (\tau - t_j) d\tau = \int_{t_j}^{t_{j+1}} \tau^q (t_{n+1} - \tau)^{p-1} d\tau$$
$$- t_j \int_{t_j}^{t_{j+1}} \tau^{q-1} (t_{n+1} - \tau)^{p-1} d\tau,$$
$$= t_{n+1}^{p+q} \widehat{B}_{q+1} - t_j \widehat{B}_q t_{n+1}^{p+q-1},$$

where

$$\widehat{B}_{q+1} = \left\{ B\left(\frac{t_{j+1}}{t_{n+1}}, q+1, p \right) - B\left(\frac{t_j}{t_{n+1}}, q+1, p \right) \right\},$$
$$\widehat{B}_q = \left\{ B\left(\frac{t_{j+1}}{t_{n+1}}, q, p \right) - B\left(\frac{t_j}{t_{n+1}}, q, p \right) \right\}.$$

Replacing yields, and further simplifications give

$$\mathbf{g}_{n+1} = \frac{q}{\Gamma(p)} (\Delta t)^{p+q-1} \sum_{j=0}^{n} f(t_j, \mathbf{g}_j)(n+1)^{p+q-1}$$

$$\times \left\{ B\left(\frac{t_{j+1}}{t_{n+1}},q,p\right) - B\left(\frac{t_j}{t_{n+1}},q,p\right) \right\}$$

$$+ \frac{q}{\Gamma(p)}(\Delta t)^{p+q-1} \sum_{j=0}^{n-1} \left[f(t_{j+1},\mathbf{g}_{j+1}) - f(t_j,\mathbf{g}_j) \right]$$

$$\times \left\{ (n+1)^{p+q} B\left(\frac{t_{j+1}}{t_{n+1}},q+1,p\right) - n^{p+q} B\left(\frac{t_j}{t_{n+1}},q+1,p\right) \right.$$

$$\left. - j \left[B\left(\frac{t_{j+1}}{t_{n+1}},q,p\right) - B\left(\frac{t_j}{t_{n+1}},q,p\right) \right](n+1)^{p+q-1} \right\}$$

$$+ \frac{q(\Delta t)^{p+q-1}}{\Gamma(p)} \left[f(t_n+h,\mathbf{g}^u_{n+1}) - f(t_n,\mathbf{g}_n) \right]$$

$$\times \left\{ (n+1)^{p+q}\left(B(1,q+1,p) - B\left(\frac{t_n}{t_{n+1}},q+1,p\right)\right) \right.$$

$$\left. - n\left(B(1,q,p) - B\left(\frac{t_n}{t_{n+1}},q,p\right)\right)(n+1)^{p+q-1} \right\} \quad (13.13)$$

where

$$\mathbf{g}^u_{n+1} = \mathbf{g}_0 + \frac{q}{\Gamma(p+1)}(\Delta t)^{p+q-1} \sum_{j=0}^{n} f(t_j,\mathbf{g}_j)(n+1)^{p+q-1}$$

$$\times \left\{ B\left(\frac{t_{j+1}}{t_{n+1}},q,p\right) - B\left(\frac{t_j}{t_{n+1}},q,p\right) \right\}$$

For our proposed system, the scheme follows:

$$x_{n+1} = \frac{q}{\Gamma(p)}(\Delta t)^{p+q-1} \sum_{j=0}^{n} f_1(t_j,x_j,y_j,z_j)(n+1)^{p+q-1}$$

$$\times \left\{ B\left(\frac{t_{j+1}}{t_{n+1}},q,p\right) - B\left(\frac{t_j}{t_{n+1}},q,p\right) \right\} + \frac{q}{\Gamma(p)}(\Delta t)^{p+q-1}$$

$$\times \sum_{j=0}^{n-1} \left[f_1(t_{j+1},x_{j+1},y_{j+1},z_{j+1}) \right.$$

$$\left. - f_1(t_j,x_j,y_j,z_j) \right]$$

$$\times \left\{ (n+1)^{p+q} B\left(\frac{t_{j+1}}{t_{n+1}},q+1,p\right) - n^{p+q} B\left(\frac{t_j}{t_{n+1}},q+1,p\right) \right.$$

$$\left. - j \left[B\left(\frac{t_{j+1}}{t_{n+1}},q,p\right) - B\left(\frac{t_j}{t_{n+1}},q,p\right) \right](n+1)^{p+q-1} \right\}$$

$$+ \frac{q(\Delta t)^{p+q-1}}{\Gamma(p)} \left[f_1(t_n+h,x^u_{n+1},y^u_{n+1},z^u_{n+1}) \right.$$

$$\left. - f_1(t_n,x_n,y_n,z_n) \right]$$

$$\times \left\{ (n+1)^{p+q}\left(B(1,q+1,p) - B\left(\frac{t_n}{t_{n+1}},q+1,p\right)\right) \right.$$

$$-n\left(B(1,q,p) - B\left(\frac{t_n}{t_{n+1}},q,p\right)\right)(n+1)^{p+q-1}\bigg\},$$

$$y_{n+1} = \frac{q}{\Gamma(p)}(\Delta t)^{p+q-1} \sum_{j=0}^{n} f_2(t_j,x_j,y_j,z_j)(n+1)^{p+q-1}$$

$$\times \left\{B\left(\frac{t_{j+1}}{t_{n+1}},q,p\right) - B\left(\frac{t_j}{t_{n+1}},q,p\right)\right\} + \frac{q}{\Gamma(p)}(\Delta t)^{p+q-1}$$

$$\times \sum_{j=0}^{n-1} \left[f_2(t_{j+1},x_{j+1},y_{j+1},z_{j+1})\right.$$

$$\left. - f_2(t_j,x_j,y_j,z_j)\right]$$

$$\times \left\{(n+1)^{p+q}B\left(\frac{t_{j+1}}{t_{n+1}},q+1,p\right) - n^{p+q}B\left(\frac{t_j}{t_{n+1}},q+1,p\right)\right.$$

$$\left. - j\left[B\left(\frac{t_{j+1}}{t_{n+1}},q,p\right) - B\left(\frac{t_j}{t_{n+1}},q,p\right)\right](n+1)^{p+q-1}\right\}$$

$$+ \frac{q(\Delta t)^{p+q-1}}{\Gamma(p)}\left[f_2(t_n+h,x_{n+1}^u,y_{n+1}^u,z_{n+1}^u)\right.$$

$$\left. - f_2(t_n,x_n,y_n,z_n)\right]$$

$$\times \left\{(n+1)^{p+q}\left(B(1,q+1,p) - B\left(\frac{t_n}{t_{n+1}},q+1,p\right)\right)\right.$$

$$\left. - n\left(B(1,q,p) - B\left(\frac{t_n}{t_{n+1}},q,p\right)\right)(n+1)^{p+q-1}\right\},$$

$$z_{n+1} = \frac{q}{\Gamma(p)}(\Delta t)^{p+q-1} \sum_{j=0}^{n} f_3(t_j,x_j,y_j,z_j)(n+1)^{p+q-1}$$

$$\times \left\{B\left(\frac{t_{j+1}}{t_{n+1}},q,p\right) - B\left(\frac{t_j}{t_{n+1}},q,p\right)\right\} + \frac{q}{\Gamma(p)}(\Delta t)^{p+q-1}$$

$$\times \sum_{j=0}^{n-1} \left[f_3(t_{j+1},x_{j+1},y_{j+1},z_{j+1})\right.$$

$$\left. - f_3(t_j,x_j,y_j,z_j)\right]$$

$$\times \left\{(n+1)^{p+q}B\left(\frac{t_{j+1}}{t_{n+1}},q+1,p\right) - n^{p+q}B\left(\frac{t_j}{t_{n+1}},q+1,p\right)\right.$$

$$\left. - j\left[B\left(\frac{t_{j+1}}{t_{n+1}},q,p\right) - B\left(\frac{t_j}{t_{n+1}},q,p\right)\right](n+1)^{p+q-1}\right\}$$

$$+ \frac{q(\Delta t)^{p+q-1}}{\Gamma(p)}\left[f_3(t_n+h,x_{n+1}^u,y_{n+1}^u,z_{n+1}^u)\right.$$

$$\left. - f_3(t_n,x_n,y_n,z_n)\right]$$

$$\times \left\{(n+1)^{p+q}\left(B(1,q+1,p) - B\left(\frac{t_n}{t_{n+1}},q+1,p\right)\right)\right.$$

$$-n\left(B(1,q,p) - B\left(\frac{t_n}{t_{n+1}},q,p\right)\right)(n+1)^{p+q-1}\right\}, \quad (13.14)$$

where

$$x_{n+1}^u = \frac{q}{\Gamma(p+1)}(\Delta t)^{p+q-1} \sum_{j=0}^n f_1(t_j,x_j,y_j,z_j)$$
$$\times (n+1)^{p+q-1}\left\{B\left(\frac{t_{j+1}}{t_{n+1}},q,p\right) - B\left(\frac{t_j}{t_{n+1}},q,p\right)\right\},$$

$$y_{n+1}^u = \frac{q}{\Gamma(p+1)}(\Delta t)^{p+q-1} \sum_{j=0}^n f_2(t_j,x_j,y_j,z_j)$$
$$\times (n+1)^{p+q-1}\left\{B\left(\frac{t_{j+1}}{t_{n+1}},q,p\right) - B\left(\frac{t_j}{t_{n+1}},q,p\right)\right\},$$

$$z_{n+1}^u = \frac{q}{\Gamma(p+1)}(\Delta t)^{p+q-1} \sum_{j=0}^n f_3(t_j,x_j,y_j,z_j)$$
$$\times (n+1)^{p+q-1}\left\{B\left(\frac{t_{j+1}}{t_{n+1}},q,p\right) - B\left(\frac{t_j}{t_{n+1}},q,p\right)\right\}.$$

13.5.2 NUMERICAL SCHEME FOR EXPONENTIAL DECAY KERNEL USING LINEAR INTERPOLATIONS

In this subsection, we consider the fractal-fractional Cauchy problem with exponential kernel, given by,

$$_0^{FFE}D_t^{p,q}\mathbf{g}(t) = f(t,\mathbf{g}(t)), \text{ if } t > 0,$$

$$\mathbf{g}(0) = \mathbf{g}_0, \text{ if } t = 0.$$

where $\mathbf{g}(t) = (x(t), y(t), z(t))$ are the equations of the model proposed.

$$_0^{CF}D_t^p\mathbf{g}(t) = qt^{q-1}f(t,\mathbf{g}(t)), \text{ if } t > 0,$$

$$\mathbf{g}(0) = \mathbf{g}_0, \text{ if } t = 0.$$

$$\mathbf{g}(t) = (1-p)qt^{q-1}f(t,\mathbf{g}(t)) + pq\int_0^t \tau^{q-1}f(\tau,\mathbf{g}(\tau))d\tau, \ y(0) = y_0,$$

at $t = t_{n+1}$, we have

$$\mathbf{g}(t_{n+1}) = (1-p)qt_{n+1}^{q-1}f(t_{n+1},\mathbf{g}_{n+1}) + pq\int_0^{t_{n+1}} \tau^{q-1}f(\tau,\mathbf{g}(\tau))d\tau,$$

at $t = t_n$, we have

$$\mathbf{g}(t_n) = (1-p)qt_n^{q-1}f(t_n,\mathbf{g}_n) + pq\int_0^{t_n} \tau^{q-1}f(\tau,\mathbf{g}(\tau))d\tau.$$

Application of Fractal-Fractional Operators to Four-Scroll Chaotic System

Therefore,

$$\mathbf{g}(t_{n+1}) = \mathbf{g}(t_n) + (1-p)q\left[t_{n+1}^{q-1}f(t_{n+1},\mathbf{g}_{n+1}) - t_n^{q-1}f(t_n,\mathbf{g}(t_n))\right]$$
$$+pq\int_{t_n}^{t_{n+1}} \tau^{q-1}f(\tau,\mathbf{g}(\tau))d\tau.$$

Using the polynomial approximations of $f(\tau,\mathbf{g}(\tau))$ within $[t_n,t_{n+1}]$ yields,

$$\mathbf{g}_{n+1} = \mathbf{g}_n + (1-p)q\left[t_{n+1}^{q-1}f(t_{n+1},\mathbf{g}_{n+1}^u) - t_n^{q-1}f(t_n,\mathbf{g}_n)\right]$$
$$+ p\left[f(t_n,\mathbf{g}_n)(t_{n+1}^q - t_n^q) + \frac{f(t_{n+1},\mathbf{g}_{n+1}^u) - f(t_n,\mathbf{g}_n)}{\Delta t}\right.$$
$$\left.\left\{\frac{q}{q+1}(t_{n+1}^{q+1} - t_n^{q+1}) - t_n(t_{n+1}^q - t_n^q)\right\}\right]. \tag{13.15}$$

Further, we have

$$\mathbf{g}_{n+1} = \mathbf{g}_n + (1-p)q(\Delta t)^{q-1}$$
$$\left\{(n+1)^{q-1}f(t_n+h,\mathbf{g}_{n+1}^u) - n^{q-1}f(t_n,\mathbf{g}_n)\right\}$$
$$+p(\Delta t)^q\left\{f(t_n,\mathbf{g}_n)((n+1)^q - n^q)\right.$$
$$+(f(t_n+h,\mathbf{g}_{n+1}^u) - f(t_n,\mathbf{g}_n))\Big\}$$
$$\times\left\{\frac{q}{q+1}[(n+1)^{q+1} - n^{q+1}] - n\Big((n+1)^q - n^q\Big)\right\},$$

where

$$\mathbf{g}_{n+1}^u = \mathbf{g}_0 + (1-p)q(\Delta t)^{q-1}n^{q-1}f(t_n,\mathbf{g}_n)$$
$$+p\sum_{j=0}^{n} f(t_j,\mathbf{g}_j)\left[\frac{q}{q+1}[t_{j+1}^{q+1} - t_j^{q+1}] - t_j(t_{j+1}^q - t_j^q)\right].$$

For our system, the scheme can be written as

$$x_{n+1} = x_n + (1-p)q(\Delta t)^{q-1}$$
$$\left\{(n+1)^{q-1}f_1(t_n+h,x_{n+1}^u,y_{n+1}^u,z_{n+1}^u)\right.$$
$$\left. - n^{q-1}f_1(t_n,x_n,y_n,z_n)\right\}$$
$$+p(\Delta t)^q\Big\{f_1(t_n,x_n,y_n,z_n)((n+1)^q - n^q)$$
$$+(f_1(t_n+h,x_{n+1}^u,y_{n+1}^u,z_{n+1}^u) - f_1(t_n,x_n,y_n,z_n))\Big\}$$

$$\times \left\{ \frac{q}{q+1}[(n+1)^{q+1} - n^{q+1}] - n\left((n+1)^q - n^q\right) \right\},$$

$$y_{n+1} = y_n + (1-p)q(\Delta t)^{q-1}$$
$$\left\{ (n+1)^{q-1} f_2(t_n + h, x^u_{n+1}, y^u_{n+1}, z^u_{n+1}) \right.$$
$$\left. - n^{q-1} f_2(t_n, x_n, y_n, z_n) \right\}$$
$$+ p(\Delta t)^q \left\{ f_2(t_n, x_n, y_n, z_n)((n+1)^q - n^q) \right.$$
$$\left. + (f_2(t_n + h, x^u_{n+1}, y^u_{n+1}, z^u_{n+1}) - f_2(t_n, x_n, y_n, z_n)) \right\}$$
$$\times \left\{ \frac{q}{q+1}[(n+1)^{q+1} - n^{q+1}] - n\left((n+1)^q - n^q\right) \right\},$$

$$z_{n+1} = z_n + (1-p)q(\Delta t)^{q-1}$$
$$\left\{ (n+1)^{q-1} f_3(t_n + h, x^u_{n+1}, y^u_{n+1}, z^u_{n+1}) \right.$$
$$\left. - n^{q-1} f_3(t_n, x_n, y_n, z_n) \right\}$$
$$+ p(\Delta t)^q \left\{ f_3(t_n, x_n, y_n, z_n)((n+1)^q - n^q) \right.$$
$$\left. + (f_3(t_n + h, x^u_{n+1}, y^u_{n+1}, z^u_{n+1}) - f_3(t_n, x_n, y_n, z_n)) \right\}$$
$$\times \left\{ \frac{q}{q+1}[(n+1)^{q+1} - n^{q+1}] - n\left((n+1)^q - n^q\right) \right\},$$

where

$$x^u_{n+1} = x_0 + (1-p)q(\Delta t)^{q-1} n^{q-1} f_1(t_n, x_n, y_n, z_n)$$
$$+ p \sum_{j=0}^{n} f_1(t_j, x_j, y_j, z_j) \left[\frac{q}{q+1}[t^{q+1}_{j+1} - t^{q+1}_j] - t_j(t^q_{j+1} - t^q_j) \right],$$

$$y^u_{n+1} = y_0 + (1-p)q(\Delta t)^{q-1} n^{q-1} f_2(t_n, x_n, y_n, z_n)$$
$$+ p \sum_{j=0}^{n} f_2(t_j, x_j, y_j, z_j) \left[\frac{q}{q+1}[t^{q+1}_{j+1} - t^{q+1}_j] - t_j(t^q_{j+1} - t^q_j) \right],$$

$$z^u_{n+1} = z_0 + (1-p)q(\Delta t)^{q-1} n^{q-1} f_3(t_n, x_n, y_n, z_n)$$
$$+ p \sum_{j=0}^{n} f_3(t_j, x_j, y_j, z_j) \left[\frac{q}{q+1}[t^{q+1}_{j+1} - t^{q+1}_j] - t_j(t^q_{j+1} - t^q_j) \right].$$

13.5.3 NUMERICAL SCHEME FOR GENERALISED MITTAG-LEFFLER KERNEL USING LINEAR INTERPOLATIONS

We consider in this section a fractal-fractional Cauchy problem with the generalised Mittag-Leffler Kernel:

$$_0^{FFM}D_t^{p,q}\mathbf{g}(t) = f(t,\mathbf{g}(t)), \text{ if } t > 0,$$

$$\mathbf{g}(0) = \mathbf{g}_0, \text{ if } t = 0.$$

We have for AB-RL case

$$_0^{ABR}D_t^p\mathbf{g}(t) = qt^{q-1}f(t,\mathbf{g}(t)), \text{ if } t > 0,$$

$$\mathbf{g}(0) = \mathbf{g}_0, \text{ if } t = 0.$$

$$\mathbf{g}(t) = (1-p)qt^{q-1}f(t,\mathbf{g}(t))$$

$$+\frac{pq}{\Gamma(p)}\int_0^t \tau^{q-1}(t-\tau)^{p-1}f(\tau,\mathbf{g}(\tau))d\tau, \ \mathbf{g}(0) = \mathbf{g}_0.$$

At $t = t_{n+1}$, we have

$$\mathbf{g}(t_{n+1}) = (1-p)qt_{n+1}^{q-1}f(t_{n+1},\mathbf{g}(t_{n+1}))$$

$$+\frac{pq}{\Gamma(p)}\int_0^{t_{n+1}} \tau^{q-1}(t_{n+1}-\tau)^{p-1}f(\tau,\mathbf{g}(\tau))d\tau.$$

$$\mathbf{g}(t_{n+1}) = (1-p)qt_{n+1}^{q-1}f(t_{n+1},\mathbf{g}(t_{n+1}))$$

$$+\frac{pq}{\Gamma(p)}\sum_{j=0}^{n}\int_{t_j}^{t_{j+1}}(t_{n+1}-\tau)^{p-1}\tau^{q-1}f(\tau,\mathbf{g}(\tau))d\tau.$$

Approximating the function $f(\tau,\mathbf{g}(\tau))$ by polynomial approximations, we have

$$\mathbf{g}_{n+1} = qt_{n+1}^q(1-p)f(t_{n+1},\mathbf{g}_{n+1}^u)$$

$$+\frac{pq}{\Gamma(p)}\sum_{j=0}^{n}f(t_j,\mathbf{g}_j)\left\{t_{n+1}^{p+q-1}\left[B\left(\frac{t_{j+1}}{t_{n+1}},q,p\right)\right.\right.$$

$$\left.\left.-B\left(\frac{t_j}{t_{n+1}},q,p\right)\right\}\right]$$

$$+\frac{pq}{\Gamma(p)}\sum_{j=0}^{n-1}\left(\frac{f(t_{j+1},\mathbf{g}_{j+1})-f(t_j,\mathbf{g}_j)}{\Delta t}\right)$$

$$\left\{t_{n+1}^{p+q}\left[B\left(\frac{t_{j+1}}{t_{n+1}},q+1,p\right)-B\left(\frac{t_j}{t_{n+1}},q+1,p\right)\right]\right\}$$

$$-t_j t_{n+1}^{p+q-1}\left[B\left(\frac{t_{j+1}}{t_{n+1}},q,p\right)-B\left(\frac{t_j}{t_{n+1}},q,p\right)\right]\Big\}$$
$$+\frac{pq}{\Gamma(p)}\left(\frac{f(t_n+h,\mathbf{g}_{n+1}^u)-f(t_n,\mathbf{g}_n)}{\Delta t}\right)$$
$$\times(\Delta t)^{p+q}\Big\{(n+1)^{p+q}\{B(1,q+1,p)-B(\frac{t_n}{t_{n+1}},q+1,p)\}$$
$$-n(n+1)^{p+q-1}\{B(1,q,p)-B(\frac{t_n}{t_{n+1}},q,p)\}\Big\} \tag{13.16}$$

where

$$\begin{aligned}\mathbf{g}_{n+1}^u &= (1-p)qf(t_n,\mathbf{g}_n)\\ &+\frac{pq}{\Gamma(p)}\sum_{j=0}^{n}f(t_j,\mathbf{g}_j)(n+1)^{p+q-1}\\ &\times\left[B\left(\frac{t_{j+1}}{t_{n+1}},q,p\right)-B\left(\frac{t_j}{t_{n+1}},q,p\right)\right]. \end{aligned}\tag{13.17}$$

The scheme for the case of Mittag-Leffler kernel can be applied to our system of equations as follows:

$$\begin{aligned}x_{n+1} &= qt_{n+1}^q(1-p)f_1(t_{n+1},x_{n+1}^u,y_{n+1}^u,z_{n+1}^u)\\ &+\frac{pq}{\Gamma(p)}\sum_{j=0}^{n}f_1(t_j,x_j,y_j,z_j)\Big\{t_{n+1}^{p+q-1}\left[B\left(\frac{t_{j+1}}{t_{n+1}},q,p\right)\right.\\ &\left.-B\left(\frac{t_j}{t_{n+1}},q,p\right)\right\}\Big]\\ &+\frac{pq}{\Gamma(p)}\sum_{j=0}^{n-1}\left(\frac{f_1(t_{j+1},x_{j+1},y_{j+1},z_{j+1})-f_1(t_j,x_j,y_j,z_j)}{\Delta t}\right)\\ &\Big\{t_{n+1}^{p+q}\left[B\left(\frac{t_{j+1}}{t_{n+1}},q+1,p\right)-B\left(\frac{t_j}{t_{n+1}},q+1,p\right)\right]\\ &-t_j t_{n+1}^{p+q-1}\left[B\left(\frac{t_{j+1}}{t_{n+1}},q,p\right)-B\left(\frac{t_j}{t_{n+1}},q,p\right)\right]\Big\}\\ &+\frac{pq}{\Gamma(p)}\left(\frac{f_1(t_n+h,x_{n+1}^u,y_{n+1}^u,z_{n+1}^u)-f_1(t_n,x_n,y_n,z_n)}{\Delta t}\right)\\ &\times(\Delta t)^{p+q}\Big\{(n+1)^{p+q}\{B(1,q+1,p)-B(\frac{t_n}{t_{n+1}},q+1,p)\}\\ &-n(n+1)^{p+q-1}\{B(1,q,p)-B(\frac{t_n}{t_{n+1}},q,p)\}\Big\},\\ y_{n+1} &= qt_{n+1}^q(1-p)f_2(t_{n+1},x_{n+1}^u,y_{n+1}^u,z_{n+1}^u)\\ &+\frac{pq}{\Gamma(p)}\sum_{j=0}^{n}f_2(t_j,x_j,y_j,z_j)\Big\{t_{n+1}^{p+q-1}\left[B\left(\frac{t_{j+1}}{t_{n+1}},q,p\right)\right.\end{aligned}$$

$$-B\left(\frac{t_j}{t_{n+1}},q,p\right)\Big\}\Big]$$
$$+\frac{pq}{\Gamma(p)}\sum_{j=0}^{n-1}\left(\frac{f_2(t_{j+1},x_{j+1},y_{j+1},z_{j+1})-f_2(t_j,x_j,y_J,z_j)}{\Delta t}\right)$$
$$\left\{t_{n+1}^{p+q}\left[B\left(\frac{t_{j+1}}{t_{n+1}},q+1,p\right)-B\left(\frac{t_j}{t_{n+1}},q+1,p\right)\right]\right.$$
$$\left.-t_j t_{n+1}^{p+q-1}\left[B\left(\frac{t_{j+1}}{t_{n+1}},q,p\right)-B\left(\frac{t_j}{t_{n+1}},q,p\right)\right]\right\}$$
$$+\frac{pq}{\Gamma(p)}\left(\frac{f_2(t_n+h,x_{n+1}^u,y_{n+1}^u,z_{n+1}^u)-f_2(t_n,x_n,y_n,z_n)}{\Delta t}\right)$$
$$\times(\Delta t)^{p+q}\left\{(n+1)^{p+q}\{B(1,q+1,p)-B(\frac{t_n}{t_{n+1}},q+1,p)\}\right.$$
$$\left.-n(n+1)^{p+q-1}\{B(1,q,p)-B(\frac{t_n}{t_{n+1}},q,p)\}\right\},$$

$$z_{n+1} = qt_{n+1}^q(1-p)f_3(t_{n+1},x_{n+1}^u,y_{n+1}^u,z_{n+1}^u)$$
$$+\frac{pq}{\Gamma(p)}\sum_{j=0}^{n}f_3(t_j,x_j,y_j,z_j)\left\{t_{n+1}^{p+q-1}\left[B\left(\frac{t_{j+1}}{t_{n+1}},q,p\right)\right.\right.$$
$$\left.\left.-B\left(\frac{t_j}{t_{n+1}},q,p\right)\right\}\right]$$
$$+\frac{pq}{\Gamma(p)}\sum_{j=0}^{n-1}\left(\frac{f_3(t_{j+1},x_{j+1},y_{j+1},z_{j+1})-f_2(t_j,x_j,y_j,z_j)}{\Delta t}\right)$$
$$\left\{t_{n+1}^{p+q}\left[B\left(\frac{t_{j+1}}{t_{n+1}},q+1,p\right)-B\left(\frac{t_j}{t_{n+1}},q+1,p\right)\right]\right.$$
$$\left.-t_j t_{n+1}^{p+q-1}\left[B\left(\frac{t_{j+1}}{t_{n+1}},q,p\right)-B\left(\frac{t_j}{t_{n+1}},q,p\right)\right]\right\}$$
$$+\frac{pq}{\Gamma(p)}\left(\frac{f_3(t_n+h,x_{n+1}^u,y_{n+1}^u,z_{n+1}^u)-f_3(t_n,x_n,y_n,z_n)}{\Delta t}\right)$$
$$\times(\Delta t)^{p+q}\left\{(n+1)^{p+q}\{B(1,q+1,p)-B(\frac{t_n}{t_{n+1}},q+1,p)\}\right.$$
$$\left.-n(n+1)^{p+q-1}\{B(1,q,p)-B(\frac{t_n}{t_{n+1}},q,p)\}\right\},$$

where

$$x_{n+1}^u = (1-p)qf_1(t_n,x_n,y_n,z_n)$$
$$+\frac{pq}{\Gamma(p)}\sum_{j=0}^{n}f_1(t_j,x_j,y_j,z_j)(n+1)^{p+q-1}$$
$$\times\left[B\left(\frac{t_{j+1}}{t_{n+1}},q,p\right)-B\left(\frac{t_j}{t_{n+1}},q,p\right)\right],$$

$$y^u_{n+1} = (1-p)qf_2(t_n,x_n,y_n,z_n)$$
$$+ \frac{pq}{\Gamma(p)} \sum_{j=0}^{n} f_2(t_j,x_j,y_j,z_j)(n+1)^{p+q-1}$$
$$\times \left[B\left(\frac{t_{j+1}}{t_{n+1}},q,p\right) - B\left(\frac{t_j}{t_{n+1}},q,p\right) \right],$$
$$z^u_{n+1} = (1-p)qf_3(t_n,x_n,y_n,z_n)$$
$$+ \frac{pq}{\Gamma(p)} \sum_{j=0}^{n} f_3(t_j,x_j,y_j,z_j)(n+1)^{p+q-1}$$
$$\times \left[B\left(\frac{t_{j+1}}{t_{n+1}},q,p\right) - B\left(\frac{t_j}{t_{n+1}},q,p\right) \right]. \quad (13.18)$$

13.6 NUMERICAL RESULTS

We use the schemes presented above for the fractal-fractional model in the sense of well-known three operators that is the Caputo, Caputo-Fabrizio, and the Atangana-Baleanu. Using initial conditions and parameter values described above, we observe that the model becomes chaotic and then we choose a set of suitable and some selected graphs for the fractal and fractional order parameters, see for Caputo Figures 13.1–13.2, for CF see Figures 13.3–13.4, and for Atangana-Baleanu, see

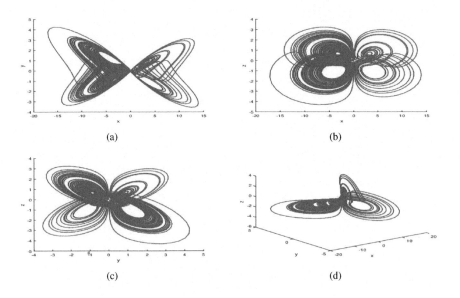

Figure 13.1 Model simulation of fractal-fractional Caputo model when $p=1, q=1$.

Application of Fractal-Fractional Operators to Four-Scroll Chaotic System

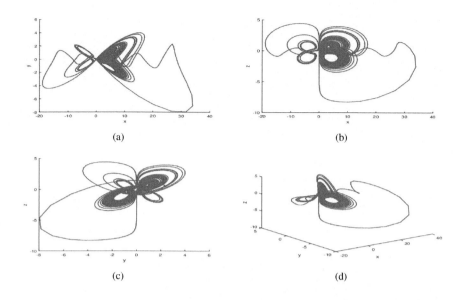

Figure 13.2 Model simulation of fractal-fractional Caputo model when $p = 0.99, q = 0.96$.

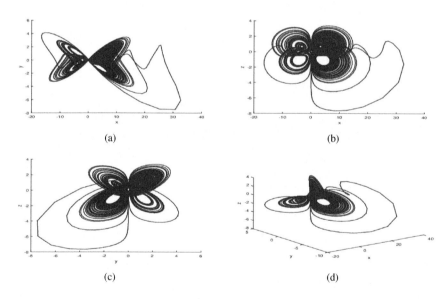

Figure 13.3 Model simulation of fractal-fractional Caputo-Fabrizio model when $p = 1$, $q = 1$.

Figures 13.5–13.7. The results of the chaotic model with these new fractal-fractional operators provide efficient results.

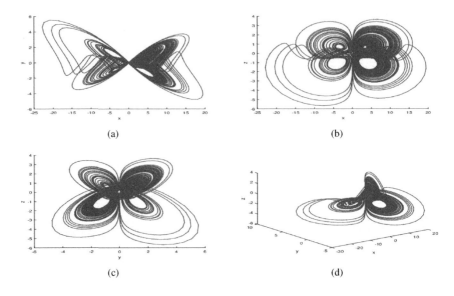

Figure 13.4 Model simulation of fractal-fractional Caputo-Fabrizio model when $p = 1$, $q = 0.98$.

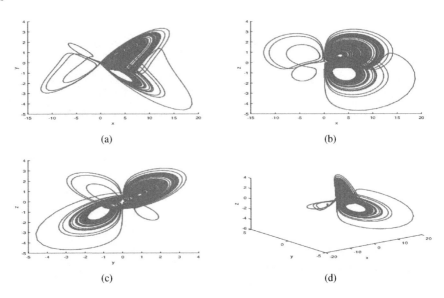

Figure 13.5 Model simulation of fractal-fractional Atangana-Baleanu model when $p = 1, q = 1$.

13.7 CONCLUSION

A new chaotic model with fractal-fractional operators using linear interpolations has been examined. We study the existence and uniqueness of the fractal-fractional

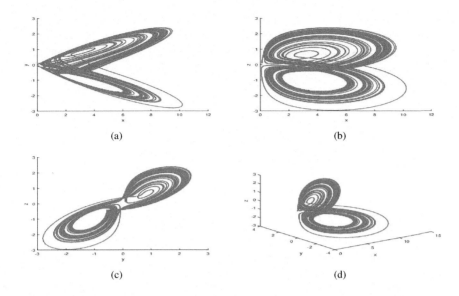

Figure 13.6 Model simulation of fractal-fractional Atangana-Baleanu model when $p = 0.99, q = 0.99$.

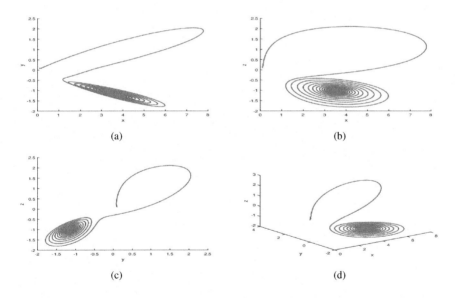

Figure 13.7 Model simulation of fractal-fractional Atangana-Baleanu model when $p = 0.98, q = 0.94$.

system and found that the system solutions exist and bounded. Further, we found some suitable values for the model parameters for which the system behaved chaotically. Then, we constructed the model in each fractal-fractional operators using the new schemes based on linear interpolations and provided the numerical results graphically. For some suitable set of fractal-fractional orders, we obtained graphically results.

14 Application of Fractal-Fractional Operators to a Novel Chaotic Model

14.1 INTRODUCTION

Diabetes mellitus, also referred to as diabetes, is a collection of metabolic illnesses characterised by persistently elevated blood sugar levels (hyperglyceamia). Frequent urination, increased thirst, and increased appetite are common symptoms. Diabetes can lead to a wide range of health issues if neglected. Supersaturated hyperglycaemia, low blood sugar, and even mortality are examples of acute complications. Cardiovascular disease, stroke, chronic kidney disease, foot ulcers, eye damage, nerve damage, and cognitive impairment are examples of serious long-term consequences [217]. Type 1 diabetes is brought on by the pancreas' inability to produce enough insulin as a result of beta cell loss. Previously, his condition was known as insulin-dependent diabetes mellitus. Beta cell loss is brought on by an autoimmune reaction. This autoimmune reaction's origin is uncertain [166,217]. While type 1 diabetes typically first manifests in childhood or adolescence, it can also strike adults. Insulin resistance, a disease in which cells do not react to insulin as it should, is the precursor to type 2 diabetes. A shortage of insulin may also develop as the condition worsens. Previously, this kind of diabetes was known as non-insulin-dependent diabetes mellitus (DM) [166, 200, 217]. Older persons are more likely to have type 2 diabetes. Globally, 425 million individuals with diabetes in 2017, up from 108 million in 1980 and an estimated 382 million in 2013. With the changing age makeup of the world's population taken into account, the prevalence of diabetes among adults has increased from 4.7% in 1980 to 8.8% now [200]. About 90% of cases are type 2 instances [200]. Although some data suggest that rates are roughly equal for men and women, male excess in diabetes has been observed in many populations with higher type 2 incidence. This may be because of sex-related differences in insulin sensitivity, the effects of obesity and regional body fat deposition, and other risk factors like high blood pressure, tobacco use, and alcohol consumption. Numerous scholars from various backgrounds have focused their attention on understanding the biological behaviours of this disease, its impacts on the human body, and other aspects because of the disease's terrible impact throughout the world. To better understand the biological process of this in a human body, mathematicians have, on the other hand, proposed several mathematical models. The numerical simulations produced by these models occasionally exhibit certain chaotic tendencies. These mathematical simulations of natural phenomena have made it simpler to comprehend their dynamics. The mathematical model of the hypothetical interaction between glucose and insulin is studied in [159] naturally, the researchers' ongoing efforts led them to develop a mathematical model for it. Diabetes, sometimes referred to as DM, is a condition that involves the most typical metabolic diseases, according to [215]. The patients with diabetes have elevated blood sugar levels. Because it is challenging to control their blood sugar levels [215]. The literature provided evidence of the rise in diabetes patients worldwide, according to [215]. The truth is the death rate from diabetes is shown by citations [3,276]. From 2012 to 2015, there will be between 1.5 and 5.0 million individuals [3,276]. Another finding in the study [276] that examined diabetes in 415 million people roughly, of which 8.3% were adults [276]. The comprehension of this ailment can be thoroughly researched by scientists, and they can offer a considerate procedure. The peptide hormone insulin regulates blood sugar levels. In diabetes, either no insulin is secreted or the body cells do not take insulin into account. For more information on the many types of diabetes and how they are classified, see, for instance [84,204,233]. Regarding the mathematical modelling of this illness, we point readers to the following sources for additional helpful information: [4,53,67,69,79,117,122,144,167]. We consequently extend an existing mathematical model with classical differential operators to the domain of fractal-fractional derivative in this chapter in order to incorporate the effect of nonlocal behaviours into the model. We also offer some analysis. We consider a fractal-fractional chaos model and present their analysis. We study its dynamics under different kernels using the new numerical approaches. The new numerical approaches are based on linear interpolations. We give first the model and then present its existence and uniqueness and further, and we study their equilibrium points. We construct numerical scheme based on linear interpolations for each kernel and then present the numerical results graphically for each case.

14.2 MODEL DESCRIPTIONS

The aim of the present section is to present the novel procedure for solution of the chaotic problem in different fractal-fractional operators. We consider three different newly defined operators known as fractal-fractional Caputo, fractal-fractional Caputo-Fabrizio, and the fractal-fractional Atangana-Baleanu. The chaotic model we consider here is given by the following equations:

$$\begin{cases} \frac{dx(t)}{dt} = z - Ax + Byz, \\ \frac{dy(t)}{dt} = Cy - xz + z, \\ \frac{dz(t)}{dt} = -Dy, \end{cases} \quad (14.1)$$

where $x(t)$, $y(t)$, and $z(t)$ are the state variables while A, B, C, and D are parameters. This system is chaotic for the given values of the parameters $A = 3$, $B = 2.7$, $C = 1.7$ and $D = 12.57$, with the initial conditions on state variables are $x(0) = 2$, $y(0) = 3$, and $z(0) = 0.01$. The model equations in (14.1) will take the following shape upon

Application of Fractal-Fractional Operators to a Novel Chaotic Model

applying the concept of fractal-fractional operators:

$$\begin{cases} {}^{FF}D_{0,t}^{p,q}\Big(x(t)\Big) = z - Ax + Byz, \\ {}^{FF}D_{0,t}^{p,q}\Big(y(t)\Big) = Cy - xz + z, \\ {}^{FF}D_{0,t}^{p,q}\Big(z(t)\Big) = -Dy, \end{cases} \quad (14.2)$$

where p represents the fractional order while q is fractal order.

14.3 EXISTENCE AND UNIQUENESS

In the present section, we aim to study the existence and uniqueness of the model (14.2). It is assumed that for all $t \in [0,T]$, the functions $x(t)$, $y(t)$, and $z(t)$ are bounded, $\|x\|_\infty \leq M_x$, $\|y\|_\infty \leq M_y$, and $\|z\|_\infty \leq M_z$. Further, we write the model (14.2) in the form:

$$\begin{aligned} f_x(t,x,y,z) &= z - Ax + Byz, \\ f_y(t,x,y,z) &= Cy - xz + z, \\ f_z(t,x,y,z) &= -Dy. \end{aligned}$$

where the parameters A, B, C, and D are assumed to be positive. We prove that f_x, f_y, and f_z satisfy the linear growth and Lipschitz conditions. We first show that the functions satisfy the linear growth property.

$$\begin{aligned} |f_x(t,x,y,z)| &\leq |z| + A|x| + B|y||z|, \\ &\leq \sup_{t \in D_z}|z(t)| + A\sup_{t \in D_x}|x(t)| + B\sup_{t \in D_y}|y(t)|\sup_{t \in D_z}|z(t)|, \\ &\leq M_z + AM_x + BM_yM_z, \\ &< \infty \end{aligned}$$

$$\begin{aligned} |f_y(t,x,y,z)| &= |Cy - xz + z|, \\ &\leq C|y| + |x||z| + |z|, \\ &\leq C\sup_{t \in D_y}|y(t)| + \sup_{t \in D_x}|x(t)|\sup_{t \in D_z}|z(t)| + \sup_{t \in D_z}|z(t)|, \\ &\leq CM_y + M_xM_z + M_z, \\ &< \infty. \end{aligned}$$

$$|f_z(t,x,y,z)| = |-Dy| \leq D\sup_{t \in D_y}|y(t)|,$$

$$\leq DM_y,$$
$$< \infty.$$

$$\begin{aligned}
|f_x(t,x,y,z)|^2 &= |z - Ax + Byz|^2 \\
&\leq 3|z|^2 + 3A^2|x|^2 + 3B^2|y|^2|z|^2, \\
&\leq 3M_z^2 + 3A^2|x|^2 + 3B^2M_y^2M_z^2, \\
&\leq 3(M_z^2 + BM_y^2M_z^2)\left(1 + \frac{A^2}{M_z^2 + BM_y^2M_z^2}|x|^2\right), \\
&\leq K_x(1 + |x|^2),
\end{aligned}$$

if $\frac{A^2}{M_z^2 + BM_y^2M_z^2}$, $K_x = 3(M_z^2 + BM_y^2M_z^2)$.

$$\begin{aligned}
|f_y(t,x,y,z)|^2 &= |Cy + z - xz|^2, \\
&\leq 3C^2|y|^2 + 3|z|^2 + 3|x|^2|z|^2, \\
&\leq 3C^2|y|^2 + 3M_z^2 + 3M_x^2M_z^2, \\
&\leq 3(M_z^2 + M_x^2M_z^2)\left(1 + \frac{C^2}{M_z^2 + M_x^2M_z^2}|y|^2\right), \\
&\leq K_y(1 + |y|^2),
\end{aligned}$$

if $\frac{C^2}{M_z^2 + M_x^2M_z^2} < 1$, and $K_y = 3(M_z^2 + M_x^2M_z^2)$.

$$\begin{aligned}
|f_z(t,x,y,z)|^2 &\leq D^2|y|^2(1 + |z|^2), \\
&\leq D^2M_y^2(1 + |z|^2).
\end{aligned}$$

Therefore, if the following inequality is obtained,

$$\max\left\{\frac{C^2}{M_z^2 + M_x^2M_z^2}, \frac{A^2}{M_z^2 + BM_y^2M_z^2},\right\} < 1.$$

The system satisfies the linear growth condition. Now, we verify the Lipschitz condition

$$|f_x(t,x_1,y,z) - f_x(t,x_2,y,z)|^2 \leq \frac{3}{2}A^2|x_1 - x_2|^2,$$

$$|f_y(t,x,y_1,z) - f_y(t,x,y_2,z)| \leq \frac{3}{2}c^2|y_1 - y_2|^2,$$

$$|f_z(t,x,y,z_1) - f_z(t,x,y,z_2)| = 0 \leq |z_1 - z_2|^2.$$

Therefore, our system admits a unique solution if the conditions under the linear growth are obtained.

14.3.1 EQUILIBRIUM POINTS AND THEIR ANALYSIS

Using the system (14.2), we find the possible equilibrium points by equating the right-hand side of model equal to zero and obtain the two equilibrium points, which are $E_0 = (0,0,0)$ and $E_1 = (1,0,A)$. The stability analysis of the model at these equilibrium points shall be discussed in the following: The Jacobian matrix of the system (14.2) is given by

$$J(E_0) = \begin{pmatrix} -A & Bz^* & By^* + 1 \\ -z^* & C & 1-x^* \\ 0 & -D & 0 \end{pmatrix} \quad (14.3)$$

At the equilibrium point E_0, the eigenvalues of the Jacobian matrix are $\lambda_1 = -A$, $\lambda_2 = \frac{1}{2}\left(c - \sqrt{c^2 - 4d}\right)$, $\lambda_3 = \frac{1}{2}\left(\sqrt{c^2 - 4d} + c\right)$. It can observed that $\lambda_{2,3}$ does not have negative real parts, and hence, the equilibrium point E_0 is unstable. The equilibrium point E_1 is analysed and obtained the possible eigenvalues of the Jacobian matrix $\lambda_1 = -1.44299 + 4.65778I$, $\lambda_2 = -1.44299 - 4.65778I$, and $\lambda_3 = 1.58598$. The model at E_1 is unstable due to λ_3 is positive. Hence, the model is unstable at both the equilibrium points E_0 and E_1.

14.4 NUMERICAL SCHEMES BASED ON LINEAR INTERPOLATIONS

This section discusses the new concept using the linear interpolations to solve the fractal-fractional-based systems numerically. The details for each kernel are given in the following:

14.5 NUMERICAL SCHEME FOR POWER LAW KERNEL

In this case, we consider the following fractal-fractional Cauchy problem,

$$^{FFP}_0 D_t^{p,q} \mathbf{g}(t) = f(t, \mathbf{g}(t)), \text{ if } t > 0,$$

$$\mathbf{g}(0) = \mathbf{g}_0, \text{ if } t = 0. \quad (14.4)$$

It is assumed that $f(t, \mathbf{g}(t))$ is twice differentiable and that $\mathbf{g}(t)$ is at least continuous

$$^{RL}_0 D_t^p \mathbf{g}(t) = qt^{q-1} f(t, \mathbf{g}(t)), \text{ if } t > 0,$$

We write further

$$\mathbf{g}(t) = \frac{q}{\Gamma(p)} \int_0^t \tau^{q-1} f(\tau, \mathbf{g}(\tau))(t-\tau)^{p-1} d\tau, \text{ if } t > 0,$$
$$\mathbf{g}(0) = \mathbf{g}_0, \text{ if } t = 0, \quad (14.6)$$

at $t = t_{n+1}$, we have

$$\mathbf{g}(t_{n+1}) = \frac{q}{\Gamma(p)} \int_0^{t_{n+1}} \tau^{q-1} f(\tau, \mathbf{g}(\tau))(t_{n+1}-\tau)^{p-1} d\tau. \quad (14.7)$$

Also, we write

$$\mathbf{g}(t_{n+1}) = \frac{q}{\Gamma(p)} \sum_{j=0}^n \int_{t_j}^{t_{j+1}} \tau^{q-1} f(\tau, \mathbf{g}(\tau))(t_{n+1}-\tau)^{p-1} d\tau. \quad (14.8)$$

Thus, within $[t_j, t_{j+1}]$, we approximate the function

$$f(\tau, \mathbf{g}(\tau)) \approx P_j(\tau) = f(t_j, \mathbf{g}_j) + (t-t_j) \frac{f(t_{j+1}, \mathbf{g}_{j+1}) - f(t_j, \mathbf{g}_j)}{\Delta t}, \quad (14.9)$$

replacing yields

$$\mathbf{g}(t_{n+1}) \approx \mathbf{g}_{n+1} = \frac{q}{\Gamma(p)} \sum_{j=0}^n \int_{t_j}^{t_{j+1}} \tau^{q-1} (t_{n+1}-\tau)^{p-1}$$
$$\times \left(f(t_j, \mathbf{g}_j) + (t-t_j) \frac{f(t_{j+1}, \mathbf{g}_{j+1}) - f(t_j, \mathbf{g}_j)}{\Delta t} \right) d\tau,$$
$$= \frac{q}{\Gamma(p)} \sum_{j=0}^n f(t_j, \mathbf{g}_j) \int_{t_j}^{t_{j+1}} \tau^{q-1} (t_{n+1}-\tau)^{p-1} d\tau$$
$$+ \frac{q}{\Gamma(p)} \sum_{j=0}^n \frac{f(t_{j+1}, \mathbf{g}_{j+1}) - f(t_j, \mathbf{g}_j)}{\Delta t}$$
$$\times \int_{t_j}^{t_{j+1}} (t_{n+1}-\tau)^{p-1} (t-t_j) d\tau. \quad (14.10)$$

Noting that

$$\int_{t_j}^{t_{j+1}} \tau^{q-1} (t_{n+1}-\tau)^{p-1} d\tau = \int_0^{t_{j+1}} \tau^{q-1} (t_{n+1}-\tau)^{p-1} d\tau$$
$$- \int_0^{t_j} \tau^{q-1} (t_{n+1}-\tau)^{p-1} d\tau, \quad (14.11)$$

where

$$\int_0^{t_{j+1}} \tau^{q-1} (t_{n+1}-\tau)^{p-1} d\tau = t_{n+1}^{p+q-1} \int_0^{\frac{t_{j+1}}{t_{n+1}}} \mathbf{g}^{q-1} (1-\mathbf{g})^p \mathbf{g}^{-1} d\mathbf{g},$$

Application of Fractal-Fractional Operators to a Novel Chaotic Model

$$= t_{n+1}^{p+q-1} B\left(\frac{t_{j+1}}{t_{n+1}}, q, p\right). \tag{14.12}$$

$$\int_0^{t_j} \tau^{q-1}(t_{n+1} - \tau)^{p-1} d\tau = t_{n+1}^{p+q-1} B\left(\frac{t_j}{t_{n+1}}, q, p\right). \tag{14.13}$$

Therefore,

$$\int_{t_j}^{t_{j+1}} \tau^{q-1}(t_{n+1} - \tau)^{p-1}(\tau - t_j) d\tau = \int_{t_j}^{t_{j+1}} \tau^q (t_{n+1} - \tau)^{p-1} d\tau$$
$$- t_j \int_{t_j}^{t_{j+1}} \tau^{q-1}(t_{n+1} - \tau)^{p-1} d\tau,$$
$$= t_{n+1}^{p+q} \widehat{B}_{q+1} - t_j \widehat{B}_q t_{n+1}^{p+q-1},$$

where

$$\widehat{B}_{q+1} = \left\{B\left(\frac{t_{j+1}}{t_{n+1}}, q+1, p\right) - B\left(\frac{t_j}{t_{n+1}}, q+1, p\right)\right\},$$

$$\widehat{B}_q = \left\{B\left(\frac{t_{j+1}}{t_{n+1}}, q, p\right) - B\left(\frac{t_j}{t_{n+1}}, q, p\right)\right\}.$$

Replacing yields, and further simplifications give

$$\mathbf{g}_{n+1} = \frac{q}{\Gamma(p)}(\Delta t)^{p+q-1} \sum_{j=0}^{n} f(t_j, \mathbf{g}_j)(n+1)^{p+q-1}$$
$$\times \left\{B\left(\frac{t_{j+1}}{t_{n+1}, q, p}\right) - B\left(\frac{t_j}{t_{n+1}, q, p}\right)\right\}$$
$$+ \frac{q}{\Gamma(p)}(\Delta t)^{p+q-1} \sum_{j=0}^{n-1} \left[f(t_{j+1}, \mathbf{g}_{j+1}) - f(t_j, \mathbf{g}_j)\right]$$
$$\times \left\{(n+1)^{p+q} B\left(\frac{t_{j+1}}{t_{n+1}}, q+1, p\right) - n^{p+q} B\left(\frac{t_j}{t_{n+1}}, q+1, p\right)\right.$$
$$\left. -j\left[B\left(\frac{t_{j+1}}{t_{n+1}}, q, p\right) - B\left(\frac{t_j}{t_{n+1}}, q, p\right)\right](n+1)^{p+q-1}\right\}$$
$$+ \frac{q(\Delta t)^{p+q-1}}{\Gamma(p)} \left[f(t_n + h, \mathbf{g}_{n+1}^u) - f(t_n, \mathbf{g}_n)\right]$$
$$\times \left\{(n+1)^{p+q}\left(B(1, q+1, p) - B\left(\frac{t_n}{t_{n+1}}, q+1, p\right)\right)\right.$$
$$\left. -n\left(B(1, q, p) - B\left(\frac{t_n}{t_{n+1}}, q, p\right)\right)(n+1)^{p+q-1}\right\} \tag{14.14}$$

where

$$\mathbf{g}_{n+1}^u = \mathbf{g}_0 + \frac{q}{\Gamma(p+1)}(\Delta t)^{p+q-1} \sum_{j=0}^{n} f(t_j, \mathbf{g}_j)(n+1)^{p+q-1}$$

$$\times \left\{ B\left(\frac{t_{j+1}}{t_{n+1}}, q, p\right) - B\left(\frac{t_j}{t_{n+1}}, q, p\right) \right\}$$

For our proposed system, the scheme follows:

$$\begin{aligned}
x_{n+1} &= \frac{q}{\Gamma(p)}(\Delta t)^{p+q-1} \sum_{j=0}^{n} f_1(t_j, x_j, y_j, z_j)(n+1)^{p+q-1} \\
&\quad \times \left\{ B\left(\frac{t_{j+1}}{t_{n+1}}, q, p\right) - B\left(\frac{t_j}{t_{n+1}}, q, p\right) \right\} + \frac{q}{\Gamma(p)}(\Delta t)^{p+q-1} \\
&\quad \times \sum_{j=0}^{n-1} \Big[f_1(t_{j+1}, x_{j+1}, y_{j+1}, z_{j+1}) \\
&\quad - f_1(t_j, x_j, y_j, z_j) \Big] \\
&\quad \times \left\{ (n+1)^{p+q} B\left(\frac{t_{j+1}}{t_{n+1}}, q+1, p\right) - n^{p+q} B\left(\frac{t_j}{t_{n+1}}, q+1, p\right) \right. \\
&\quad \left. - j \left[B\left(\frac{t_{j+1}}{t_{n+1}}, q, p\right) - B\left(\frac{t_j}{t_{n+1}}, q, p\right) \right](n+1)^{p+q-1} \right\} \\
&\quad + \frac{q(\Delta t)^{p+q-1}}{\Gamma(p)} \Big[f_1(t_n + h, x^u_{n+1}, y^u_{n+1}, z^u_{n+1}) \\
&\quad - f_1(t_n, x_n, y_n, z_n) \Big] \\
&\quad \times \left\{ (n+1)^{p+q} \left(B(1, q+1, p) - B\left(\frac{t_n}{t_{n+1}}, q+1, p\right) \right) \right. \\
&\quad \left. - n \left(B(1, q, p) - B\left(\frac{t_n}{t_{n+1}}, q, p\right) \right)(n+1)^{p+q-1} \right\}, \\
y_{n+1} &= \frac{q}{\Gamma(p)}(\Delta t)^{p+q-1} \sum_{j=0}^{n} f_2(t_j, x_j, y_j, z_j)(n+1)^{p+q-1} \\
&\quad \times \left\{ B\left(\frac{t_{j+1}}{t_{n+1}}, q, p\right) - B\left(\frac{t_j}{t_{n+1}}, q, p\right) \right\} + \frac{q}{\Gamma(p)}(\Delta t)^{p+q-1} \\
&\quad \times \sum_{j=0}^{n-1} \Big[f_2(t_{j+1}, x_{j+1}, y_{j+1}, z_{j+1}) \\
&\quad - f_2(t_j, x_j, y_j, z_j) \Big] \\
&\quad \times \left\{ (n+1)^{p+q} B\left(\frac{t_{j+1}}{t_{n+1}}, q+1, p\right) - n^{p+q} B\left(\frac{t_j}{t_{n+1}}, q+1, p\right) \right. \\
&\quad \left. - j \left[B\left(\frac{t_{j+1}}{t_{n+1}}, q, p\right) - B\left(\frac{t_j}{t_{n+1}}, q, p\right) \right](n+1)^{p+q-1} \right\} \\
&\quad + \frac{q(\Delta t)^{p+q-1}}{\Gamma(p)} \Big[f_2(t_n + h, x^u_{n+1}, y^u_{n+1}, z^u_{n+1}) \\
&\quad - f_2(t_n, x_n, y_n, z_n) \Big]
\end{aligned}$$

Application of Fractal-Fractional Operators to a Novel Chaotic Model

$$\times \left\{ (n+1)^{p+q} \left(B(1,q+1,p) - B\left(\frac{t_n}{t_{n+1}}, q+1, p\right) \right) \right.$$
$$\left. - n \left(B(1,q,p) - B\left(\frac{t_n}{t_{n+1}}, q, p\right) \right) (n+1)^{p+q-1} \right\},$$

$$z_{n+1} = \frac{q}{\Gamma(p)} (\Delta t)^{p+q-1} \sum_{j=0}^{n} f_3(t_j, x_j, y_j, z_j)(n+1)^{p+q-1}$$
$$\times \left\{ B\left(\frac{t_{j+1}}{t_{n+1}}, q, p\right) - B\left(\frac{t_j}{t_{n+1}}, q, p\right) \right\} + \frac{q}{\Gamma(p)} (\Delta t)^{p+q-1}$$
$$\times \sum_{j=0}^{n-1} \left[f_3(t_{j+1}, x_{j+1}, y_{j+1}, z_{j+1}) \right.$$
$$\left. - f_3(t_j, x_j, y_j, z_j) \right]$$
$$\times \left\{ (n+1)^{p+q} B\left(\frac{t_{j+1}}{t_{n+1}}, q+1, p\right) - n^{p+q} B\left(\frac{t_j}{t_{n+1}}, q+1, p\right) \right.$$
$$\left. - j \left[B\left(\frac{t_{j+1}}{t_{n+1}}, q, p\right) - B\left(\frac{t_j}{t_{n+1}}, q, p\right) \right] (n+1)^{p+q-1} \right\}$$
$$+ \frac{q(\Delta t)^{p+q-1}}{\Gamma(p)} \left[f_3(t_n + h, x_{n+1}^u, y_{n+1}^u, z_{n+1}^u) \right.$$
$$\left. - f_3(t_n, x_n, y_n, z_n) \right]$$
$$\times \left\{ (n+1)^{p+q} \left(B(1,q+1,p) - B\left(\frac{t_n}{t_{n+1}}, q+1, p\right) \right) \right.$$
$$\left. - n \left(B(1,q,p) - B\left(\frac{t_n}{t_{n+1}}, q, p\right) \right) (n+1)^{p+q-1} \right\}, \tag{14.15}$$

where

$$x_{n+1}^u = \frac{q}{\Gamma(p+1)} (\Delta t)^{p+q-1} \sum_{j=0}^{n} f_1(t_j, x_j, y_j, z_j)$$
$$\times (n+1)^{p+q-1} \left\{ B\left(\frac{t_{j+1}}{t_{n+1}}, q, p\right) - B\left(\frac{t_j}{t_{n+1}}, q, p\right) \right\},$$

$$y_{n+1}^u = \frac{q}{\Gamma(p+1)} (\Delta t)^{p+q-1} \sum_{j=0}^{n} f_2(t_j, x_j, y_j, z_j)$$
$$\times (n+1)^{p+q-1} \left\{ B\left(\frac{t_{j+1}}{t_{n+1}}, q, p\right) - B\left(\frac{t_j}{t_{n+1}}, q, p\right) \right\},$$

$$z_{n+1}^u = \frac{q}{\Gamma(p+1)} (\Delta t)^{p+q-1} \sum_{j=0}^{n} f_3(t_j, x_j, y_j, z_j)$$
$$\times (n+1)^{p+q-1} \left\{ B\left(\frac{t_{j+1}}{t_{n+1}}, q, p\right) - B\left(\frac{t_j}{t_{n+1}}, q, p\right) \right\}.$$

The scheme above discussed is utilised here and obtain the graphical results for the fractal-fractional model in the sense of Caputo derivative, see Figures 14.1–14.2.

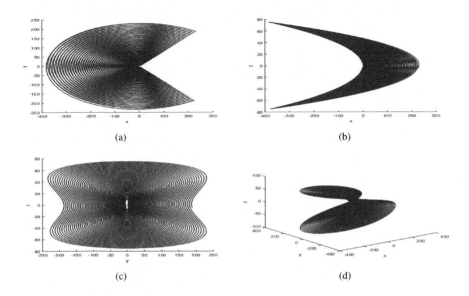

Figure 14.1 Model simulation of fractal-fractional Caputo model when $p = 1, q = 1$.

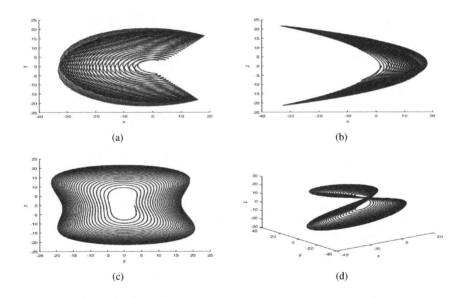

Figure 14.2 Model simulation of fractal-fractional Caputo model when $p = 0.99, q = 0.99$.

Next, in the following subsection, we present the numerical procedure for the fractal-fractional model in the sense of Caputo-Fabrizio operator.

14.5.1 NUMERICAL SCHEME FOR EXPONENTIAL DECAY KERNEL USING LINEAR INTERPOLATIONS

In this subsection, we consider the fractal-fractional Cauchy problem with exponential kernel, given by

$$_0^{FFE}D_t^{p,q}\mathbf{g}(t) = f(t,\mathbf{g}(t)), \text{ if } t > 0,$$

$$\mathbf{g}(0) = \mathbf{g}_0, \text{ if } t = 0.$$

where $\mathbf{g}(t) = (x(t), y(t), z(t))$ are the equations of the model proposed.

$$_0^{CF}D_t^p\mathbf{g}(t) = qt^{q-1}f(t,\mathbf{g}(t)), \text{ if } t > 0,$$

$$\mathbf{g}(0) = \mathbf{g}_0, \text{ if } t = 0.$$

$$\mathbf{g}(t) = (1-p)qt^{q-1}f(t,\mathbf{g}(t)) + pq\int_0^t \tau^{q-1}f(\tau,\mathbf{g}(\tau))d\tau, \ y(0) = y_0,$$

at $t = t_{n+1}$, we have

$$\mathbf{g}(t_{n+1}) = (1-p)qt_{n+1}^{q-1}f(t_{n+1},\mathbf{g}_{n+1}) + pq\int_0^{t_{n+1}} \tau^{q-1}f(\tau,\mathbf{g}(\tau))d\tau,$$

at $t = t_n$, we have

$$\mathbf{g}(t_n) = (1-p)qt_n^{q-1}f(t_n,\mathbf{g}_n) + pq\int_0^{t_n} \tau^{q-1}f(\tau,\mathbf{g}(\tau))d\tau.$$

Therefore,

$$\mathbf{g}(t_{n+1}) = \mathbf{g}(t_n) + (1-p)q\left[t_{n+1}^{q-1}f(t_{n+1},\mathbf{g}_{n+1}) - t_n^{q-1}f(t_n,\mathbf{g}(t_n))\right]$$

$$+ pq\int_{t_n}^{t_{n+1}} \tau^{q-1}f(\tau,\mathbf{g}(\tau))d\tau.$$

Using the polynomial approximations of $f(\tau,\mathbf{g}(\tau))$ within $[t_n,t_{n+1}]$ yields

$$\mathbf{g}_{n+1} = \mathbf{g}_n + (1-p)q\left[t_{n+1}^{q-1}f(t_{n+1},\mathbf{g}_{n+1}^u) - t_n^{q-1}f(t_n,\mathbf{g}_n)\right]$$

$$+ p\left[f(t_n,\mathbf{g}_n)(t_{n+1}^q - t_n^q) + \frac{f(t_{n+1},\mathbf{g}_{n+1}^u) - f(t_n,\mathbf{g}_n)}{\Delta t}\right.$$

$$\left.\left\{\frac{q}{q+1}(t_{n+1}^{q+1} - t_n^{q+1}) - t_n(t_{n+1}^q - t_n^q)\right\}\right]. \tag{14.16}$$

Further, we have

$$\mathbf{g}_{n+1} = \mathbf{g}_n + (1-p)q(\Delta t)^{q-1}$$

$$\left\{(n+1)^{q-1}f(t_n+h,\mathbf{g}_{n+1}^u) - n^{q-1}f(t_n,\mathbf{g}_n)\right\}$$
$$+p(\Delta t)^q\left\{f(t_n,\mathbf{g}_n)((n+1)^q - n^q)\right.$$
$$+(f(t_n+h,\mathbf{g}_{n+1}^u) - f(t_n,\mathbf{g}_n))\Big\}$$
$$\times\left\{\frac{q}{q+1}[(n+1)^{q+1} - n^{q+1}] - n\Big((n+1)^q - n^q\Big)\right\},$$

where

$$\mathbf{g}_{n+1}^u = \mathbf{g}_0 + (1-p)q(\Delta t)^{q-1}n^{q-1}f(t_n,\mathbf{g}_n)$$
$$+p\sum_{j=0}^{n}f(t_j,\mathbf{g}_j)\left[\frac{q}{q+1}[t_{j+1}^{q+1} - t_j^{q+1}] - t_j(t_{j+1}^q - t_j^q)\right].$$

For our system, the scheme can be written as

$$x_{n+1} = x_n + (1-p)q(\Delta t)^{q-1}$$
$$\left\{(n+1)^{q-1}f_1(t_n+h,x_{n+1}^u,y_{n+1}^u,z_{n+1}^u)\right.$$
$$-n^{q-1}f_1(t_n,x_n,y_n,z_n)\Big\}$$
$$+p(\Delta t)^q\left\{f_1(t_n,x_n,y_n,z_n)((n+1)^q - n^q)\right.$$
$$+(f_1(t_n+h,x_{n+1}^u,y_{n+1}^u,z_{n+1}^u) - f_1(t_n,x_n,y_n,z_n))\Big\}$$
$$\times\left\{\frac{q}{q+1}[(n+1)^{q+1} - n^{q+1}] - n\Big((n+1)^q - n^q\Big)\right\},$$

$$y_{n+1} = y_n + (1-p)q(\Delta t)^{q-1}$$
$$\left\{(n+1)^{q-1}f_2(t_n+h,x_{n+1}^u,y_{n+1}^u,z_{n+1}^u)\right.$$
$$-n^{q-1}f_2(t_n,x_n,y_n,z_n)\Big\}$$
$$+p(\Delta t)^q\left\{f_2(t_n,x_n,y_n,z_n)((n+1)^q - n^q)\right.$$
$$+(f_2(t_n+h,x_{n+1}^u,y_{n+1}^u,z_{n+1}^u) - f_2(t_n,x_n,y_n,z_n))\Big\}$$
$$\times\left\{\frac{q}{q+1}[(n+1)^{q+1} - n^{q+1}] - n\Big((n+1)^q - n^q\Big)\right\},$$

$$z_{n+1} = z_n + (1-p)q(\Delta t)^{q-1}$$
$$\left\{(n+1)^{q-1}f_3(t_n+h,x_{n+1}^u,y_{n+1}^u,z_{n+1}^u)\right.$$

$$-n^{q-1}f_3(t_n,x_n,y_n,z_n)\Big\}$$
$$+p(\Delta t)^q\Big\{f_3(t_n,x_n,y_n,z_n)((n+1)^q-n^q)$$
$$+(f_3(t_n+h,x^u_{n+1},y^u_{n+1},z^u_{n+1})-f_3(t_n,x_n,y_n,z_n))\Big\}$$
$$\times\Big\{\frac{q}{q+1}[(n+1)^{q+1}-n^{q+1}]-n\Big((n+1)^q-n^q\Big)\Big\},$$

where

$$x^u_{n+1} = x_0+(1-p)q(\Delta t)^{q-1}n^{q-1}f_1(t_n,x_n,y_n,z_n)$$
$$+p\sum_{j=0}^n f_1(t_j,x_j,y_j,z_j)\left[\frac{q}{q+1}[t^{q+1}_{j+1}-t^{q+1}_j]-t_j(t^q_{j+1}-t^q_j)\right],$$
$$y^u_{n+1} = y_0+(1-p)q(\Delta t)^{q-1}n^{q-1}f_2(t_n,x_n,y_n,z_n)$$
$$+p\sum_{j=0}^n f_2(t_j,x_j,y_j,z_j)\left[\frac{q}{q+1}[t^{q+1}_{j+1}-t^{q+1}_j]-t_j(t^q_{j+1}-t^q_j)\right],$$
$$z^u_{n+1} = z_0+(1-p)q(\Delta t)^{q-1}n^{q-1}f_3(t_n,x_n,y_n,z_n)$$
$$+p\sum_{j=0}^n f_3(t_j,x_j,y_j,z_j)\left[\frac{q}{q+1}[t^{q+1}_{j+1}-t^{q+1}_j]-t_j(t^q_{j+1}-t^q_j)\right].$$

For the numerical solution of the Caputo-Fabrizio fractal-fractional model, we used the scheme above and present the graphical results in Figures 14.3–14.6. Next, we present the numerical procedure for the fractal-fractional in the sense of Atangana-Baleanu derivative in the following subsection:

14.5.2 NUMERICAL SCHEME FOR GENERALISED MITTAG-LEFFLER KERNEL USING LINEAR INTERPOLATIONS

We consider in this section a fractal-fractional Cauchy problem with the generalised Mittag-Leffler Kernel:

$$^{FFM}_0 D^{p,q}_t \mathbf{g}(t) = f(t,\mathbf{g}(t)), \text{ if } t > 0,$$
$$\mathbf{g}(0) = \mathbf{g}_0, \text{ if } t = 0.$$

We have for AB-RL case

$$^{ABR}_0 D^p_t \mathbf{g}(t) = qt^{q-1}f(t,\mathbf{g}(t)), \text{ if } t > 0,$$
$$\mathbf{g}(0) = \mathbf{g}_0, \text{ if } t = 0.$$

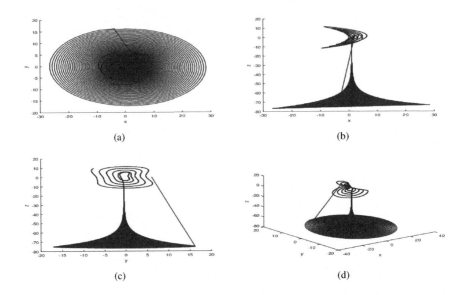

Figure 14.3 Model simulation of fractal-fractional Caputo-Fabrizio model when $p = 1, q = 1$.

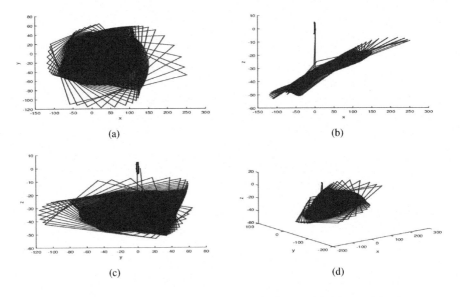

Figure 14.4 Model simulation of fractal-fractional Caputo-Fabrizio model when $p = 0.98, q = 1$.

$$\mathbf{g}(t) = (1-p)qt^{q-1}f(t,\mathbf{g}(t))$$

Application of Fractal-Fractional Operators to a Novel Chaotic Model

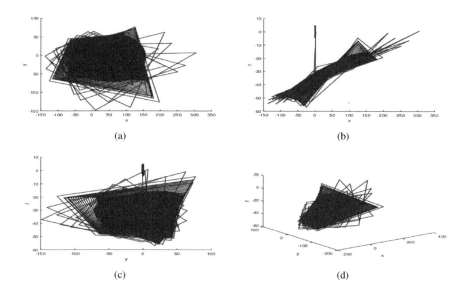

Figure 14.5 Model simulation of fractal-fractional Caputo-Fabrizio model when $p = q = 0.98$.

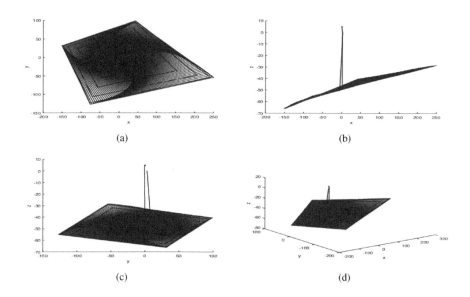

Figure 14.6 Model simulation of fractal-fractional Caputo-Fabrizio model when $p = 0.99, q = 0.75$.

$$+ \frac{pq}{\Gamma(p)} \int_0^t \tau^{q-1}(t-\tau)^{p-1} f(\tau, \mathbf{g}(\tau)) d\tau, \ \mathbf{g}(0) = \mathbf{g}_0.$$

At $t = t_{n+1}$, we have

$$\mathbf{g}(t_{n+1}) = (1-p)qt_{n+1}^{q-1}f(t_{n+1},\mathbf{g}(t_{n+1}))$$
$$+ \frac{pq}{\Gamma(p)} \int_0^{t_{n+1}} \tau^{q-1}(t_{n+1}-\tau)^{p-1} f(\tau,\mathbf{g}(\tau))d\tau.$$

$$\mathbf{g}(t_{n+1}) = (1-p)qt_{n+1}^{q-1}f(t_{n+1},\mathbf{g}(t_{n+1}))$$
$$+ \frac{pq}{\Gamma(p)} \sum_{j=0}^{n} \int_{t_j}^{t_{j+1}} (t_{n+1}-\tau)^{p-1} \tau^{q-1} f(\tau,\mathbf{g}(\tau))d\tau.$$

Approximating the function $f(\tau,\mathbf{g}(\tau))$ by polynomial approximations, we have

$$\mathbf{g}_{n+1} = qt_{n+1}^q (1-p)f(t_{n+1},\mathbf{g}_{n+1}^u)$$
$$+ \frac{pq}{\Gamma(p)} \sum_{j=0}^{n} f(t_j,\mathbf{g}_j) \left\{ t_{n+1}^{p+q-1} \left[B\left(\frac{t_{j+1}}{t_{n+1}}, q, p\right) \right. \right.$$
$$\left. \left. - B\left(\frac{t_j}{t_{n+1}}, q, p\right) \right] \right\}$$
$$+ \frac{pq}{\Gamma(p)} \sum_{j=0}^{n-1} \left(\frac{f(t_{j+1},\mathbf{g}_{j+1}) - f(t_j,\mathbf{g}_j)}{\Delta t} \right)$$
$$\left\{ t_{n+1}^{p+q} \left[B\left(\frac{t_{j+1}}{t_{n+1}}, q+1, p\right) - B\left(\frac{t_j}{t_{n+1}}, q+1, p\right) \right] \right.$$
$$\left. - t_j t_{n+1}^{p+q-1} \left[B\left(\frac{t_{j+1}}{t_{n+1}}, q, p\right) - B\left(\frac{t_j}{t_{n+1}}, q, p\right) \right] \right\}$$
$$+ \frac{pq}{\Gamma(p)} \left(\frac{f(t_n+h, \mathbf{g}_{n+1}^u) - f(t_n,\mathbf{g}_n)}{\Delta t} \right)$$
$$\times (\Delta t)^{p+q} \left\{ (n+1)^{p+q} \{ B(1,q+1,p) - B(\frac{t_n}{t_{n+1}}, q+1, p) \} \right.$$
$$\left. - n(n+1)^{p+q-1} \{ B(1,q,p) - B(\frac{t_n}{t_{n+1}}, q, p) \} \right\} \quad (14.17)$$

where

$$\mathbf{g}_{n+1}^u = (1-p)qf(t_n,\mathbf{g}_n)$$
$$+ \frac{pq}{\Gamma(p)} \sum_{j=0}^{n} f(t_j,\mathbf{g}_j)(n+1)^{p+q-1}$$
$$\times \left[B\left(\frac{t_{j+1}}{t_{n+1}}, q, p\right) - B\left(\frac{t_j}{t_{n+1}}, q, p\right) \right]. \quad (14.18)$$

The scheme for the case of Mittag-Leffler kernel can be applied to our system of equations as follows:

$$x_{n+1} = qt_{n+1}^q(1-p)f_1(t_{n+1}, x_{n+1}^u, y_{n+1}^u, z_{n+1}^u)$$

$$+\frac{pq}{\Gamma(p)}\sum_{j=0}^{n}f_1(t_j,x_j,y_j,z_j)\left\{t_{n+1}^{p+q-1}\left[B\left(\frac{t_{j+1}}{t_{n+1}},q,p\right)\right.\right.$$

$$\left.\left.-B\left(\frac{t_j}{t_{n+1}},q,p\right)\right\}\right]$$

$$+\frac{pq}{\Gamma(p)}\sum_{j=0}^{n-1}\left(\frac{f_1(t_{j+1},x_{j+1},y_{j+1},z_{j+1})-f_1(t_j,x_j,y_j,z_j)}{\Delta t}\right)$$

$$\left\{t_{n+1}^{p+q}\left[B\left(\frac{t_{j+1}}{t_{n+1}},q+1,p\right)-B\left(\frac{t_j}{t_{n+1}},q+1,p\right)\right]\right.$$

$$\left.-t_j t_{n+1}^{p+q-1}\left[B\left(\frac{t_{j+1}}{t_{n+1}},q,p\right)-B\left(\frac{t_j}{t_{n+1}},q,p\right)\right]\right\}$$

$$+\frac{pq}{\Gamma(p)}\left(\frac{f_1(t_n+h,x_{n+1}^u,y_{n+1}^u,z_{n+1}^u)-f_1(t_n,x_n,y_n,z_n)}{\Delta t}\right)$$

$$\times(\Delta t)^{p+q}\left\{(n+1)^{p+q}\{B(1,q+1,p)-B(\frac{t_n}{t_{n+1}},q+1,p)\}\right.$$

$$\left.-n(n+1)^{p+q-1}\{B(1,q,p)-B(\frac{t_n}{t_{n+1}},q,p)\}\right\},$$

$$y_{n+1} = qt_{n+1}^q(1-p)f_2(t_{n+1},x_{n+1}^u,y_{n+1}^u,z_{n+1}^u)$$

$$+\frac{pq}{\Gamma(p)}\sum_{j=0}^{n}f_2(t_j,x_j,y_j,z_j)\left\{t_{n+1}^{p+q-1}\left[B\left(\frac{t_{j+1}}{t_{n+1}},q,p\right)\right.\right.$$

$$\left.\left.-B\left(\frac{t_j}{t_{n+1}},q,p\right)\right\}\right]$$

$$+\frac{pq}{\Gamma(p)}\sum_{j=0}^{n-1}\left(\frac{f_2(t_{j+1},x_{j+1},y_{j+1},z_{j+1})-f_2(t_j,x_j,y_J,z_j)}{\Delta t}\right)$$

$$\left\{t_{n+1}^{p+q}\left[B\left(\frac{t_{j+1}}{t_{n+1}},q+1,p\right)-B\left(\frac{t_j}{t_{n+1}},q+1,p\right)\right]\right.$$

$$\left.-t_j t_{n+1}^{p+q-1}\left[B\left(\frac{t_{j+1}}{t_{n+1}},q,p\right)-B\left(\frac{t_j}{t_{n+1}},q,p\right)\right]\right\}$$

$$+\frac{pq}{\Gamma(p)}\left(\frac{f_2(t_n+h,x_{n+1}^u,y_{n+1}^u,z_{n+1}^u)-f_2(t_n,x_n,y_n,z_n)}{\Delta t}\right)$$

$$\times(\Delta t)^{p+q}\left\{(n+1)^{p+q}\{B(1,q+1,p)-B(\frac{t_n}{t_{n+1}},q+1,p)\}\right.$$

$$\left.-n(n+1)^{p+q-1}\{B(1,q,p)-B(\frac{t_n}{t_{n+1}},q,p)\}\right\},$$

$$z_{n+1} = qt_{n+1}^q(1-p)f_3(t_{n+1},x_{n+1}^u,y_{n+1}^u,z_{n+1}^u)$$

$$+\frac{pq}{\Gamma(p)}\sum_{j=0}^{n}f_3(t_j,x_j,y_j,z_j)\left\{t_{n+1}^{p+q-1}\left[B\left(\frac{t_{j+1}}{t_{n+1}},q,p\right)\right.\right.$$

$$-B\left(\frac{t_j}{t_{n+1}}, q, p\right)\bigg\}\bigg]$$
$$+\frac{pq}{\Gamma(p)}\sum_{j=0}^{n-1}\left(\frac{f_3(t_{j+1}, x_{j+1}, y_{j+1}, z_{j+1}) - f_2(t_j, x_j, y_j, z_j)}{\Delta t}\right)$$
$$\left\{t_{n+1}^{p+q}\left[B\left(\frac{t_{j+1}}{t_{n+1}}, q+1, p\right) - B\left(\frac{t_j}{t_{n+1}}, q+1, p\right)\right]\right.$$
$$\left.-t_j t_{n+1}^{p+q-1}\left[B\left(\frac{t_{j+1}}{t_{n+1}}, q, p\right) - B\left(\frac{t_j}{t_{n+1}}, q, p\right)\right]\right\}$$
$$+\frac{pq}{\Gamma(p)}\left(\frac{f_3(t_n+h, x_{n+1}^u, y_{n+1}^u, z_{n+1}^u) - f_3(t_n, x_n, y_n, z_n)}{\Delta t}\right)$$
$$\times (\Delta t)^{p+q}\left\{(n+1)^{p+q}\{B(1, q+1, p) - B(\frac{t_n}{t_{n+1}}, q+1, p)\}\right.$$
$$\left.-n(n+1)^{p+q-1}\{B(1, q, p) - B(\frac{t_n}{t_{n+1}}, q, p)\}\right\},$$

where

$$\begin{aligned}
x_{n+1}^u &= (1-p)qf_1(t_n, x_n, y_n, z_n) \\
&\quad +\frac{pq}{\Gamma(p)}\sum_{j=0}^{n} f_1(t_j, x_j, y_j, z_j)(n+1)^{p+q-1} \\
&\quad \times\left[B\left(\frac{t_{j+1}}{t_{n+1}}, q, p\right) - B\left(\frac{t_j}{t_{n+1}}, q, p\right)\right], \\
y_{n+1}^u &= (1-p)qf_2(t_n, x_n, y_n, z_n) \\
&\quad +\frac{pq}{\Gamma(p)}\sum_{j=0}^{n} f_2(t_j, x_j, y_j, z_j)(n+1)^{p+q-1} \\
&\quad \times\left[B\left(\frac{t_{j+1}}{t_{n+1}}, q, p\right) - B\left(\frac{t_j}{t_{n+1}}, q, p\right)\right], \\
z_{n+1}^u &= (1-p)qf_3(t_n, x_n, y_n, z_n) \\
&\quad +\frac{pq}{\Gamma(p)}\sum_{j=0}^{n} f_3(t_j, x_j, y_j, z_j)(n+1)^{p+q-1} \\
&\quad \times\left[B\left(\frac{t_{j+1}}{t_{n+1}}, q, p\right) - B\left(\frac{t_j}{t_{n+1}}, q, p\right)\right].
\end{aligned} \qquad (14.19)$$

With the help of the scheme shown above, we obtain the graphical results for the fractal-fractional model in the sense of Atangana-Baleanu, as shown in Figures 14.7–14.9.

Application of Fractal-Fractional Operators to a Novel Chaotic Model

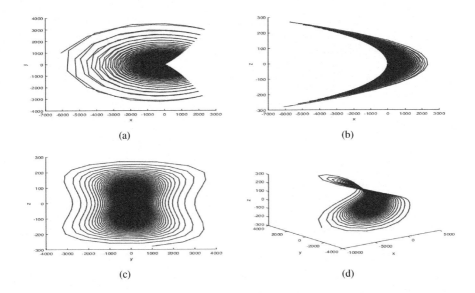

Figure 14.7 Model simulation of fractal-fractional Atangana-Baleanu model when $p = 1, q = 1$.

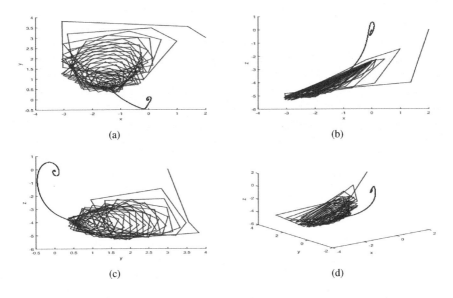

Figure 14.8 Model simulation of fractal-fractional Atangana-Baleanu model when $p = 0.9, q = 1$.

14.6 CONCLUSION

We constructed a new chaotic model in fractal-fractional operators which consists of three differential equations. We obtained reasonable initial conditions and

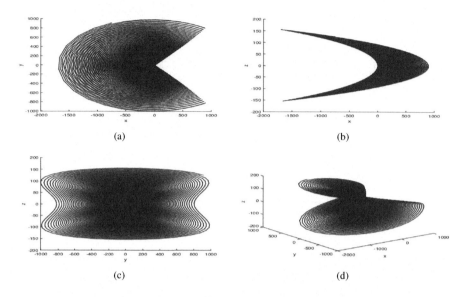

Figure 14.9 Model simulation of fractal-fractional Atangana-Baleanu model when $p = 1, q = 0.99$.

parameters values for that the model becomes chaotic. The model properties have been explored in detailed. Then, we applied the fractal-fractional operators and obtained the suitable and efficient numerical scheme for each operator based on the linear interpolations approach. The numerical results for each operator with different orders of fractal and fractional were shown graphically. We observed that the fractal and fractional order explicitly and jointly provide good results.

15 A 4D Chaotic System under Fractal-Fractional Operators

15.1 INTRODUCTION

Over the past few decades, the complexity science community has been deeply interested in studying chaos. Numerous significant advances in the study of chaos theory and its applications have been documented. Because of its distinctive random-like quality, the chaotic application has advanced significantly and permeated a variety of fields, from computer technology to life science. Since the discovery of the well-known Lorenz attractor, significant efforts have been made to reveal chaos in natural systems and develop chaotic systems for use in practical applications. Many nonlinear systems exhibit coexisting attractors also known as multistability, which is the coexistence of numerous attractors (or equilibrium states) for a constant set of parameters. It implies that the system's final state, when subjected to set parameters, is not unique and depends on the original circumstances. It largely makes it possible for the system to be more robust and flexible. The system may achieve the transition between several states to accommodate various work conditions when properly controlled. In order to further highlight the complexity of the system, there has recently been interesting in studying the coexisting attractors in chaotic systems. It was discovered that the butterfly attractor in Lorenz and Lorenz-type systems can be readily split into two separate symmetrical unusual attractors. Chaos in simple nonlinear dynamical systems is thought to be the complex, seemingly random, and frequently unexpected behaviour [73]. Widely used in electronic engineering, chaos in nonlinear dynamical systems has a distinctive kind of motion [80], information engineering [178], and some other areas [66, 141, 161]. These are all caused by the original conditions' sensitivity, their boundedness, and their inherent randomness [44]. Following the 1963 discovery of the first chaotic system [150], it attracted a lot of interest from the academic community and the creation of a new chaotic system began. A novel chaotic system with a different topology from the Lorenz system was created by Rossler in 1976 [51]. This is thought to be the most straightforward chaotic oscillation circuits. After that, Sprott proposed a number of chaotic systems in 1994 [221]. For the purpose of researching the anti-control of chaos, the two researchers Chen and Ueta created the Chen system in 1999 [45, 154]. The finding of Liu and Chen in 2003 led to the creation of the first chaotic four-wing butterfly attractor [147]. For more information on the use of chaotic systems in secure communication, see [102, 264].

Researcher interest in the coexistence of chaotic attractors has increased recently. [176, 225]. The dynamics of coexisting attractors are thought to be more complex

DOI: 10.1201/9781003359258-15

than those of generic chaotic attractors. See the outcomes in for more information about the coexisting attractors in detail [120, 130, 131, 187, 256, 284]. The present chapter studies the 4D chaotic system under different fractal-fractional operators. We give the existence and uniqueness of the system and show that the solution to the nonlinear system exists and has a unique system of solutions. We then study the model under the power law kernel, exponential kernel, and the Mittag-Leffler kernel by providing the numerical schemes based on linear interpolations. We derive the scheme for each case and present the numerical results.

15.2 MODEL DETAILS

We have given the following chaotic model that consists of three equations, modelled by the system of nonlinear ordinary differential equations:

$$\begin{cases} \frac{dx(t)}{dt} = y - x, \\ \frac{dy(t)}{dt} = -xz + w, \\ \frac{dz(t)}{dt} = xy - A, \\ \frac{dw(t)}{dt} = -By, \end{cases} \tag{15.1}$$

where $x(t)$, $y(t)$, $z(t)$, and $w(t)$ are the state variables while A and B are the parameters. This system is chaotic for the given values of the parameters $A = 5$, $B = 0.5$ with the initial conditions on state variables are $x(0) = 2$, $y(0) = 1$, and $z(0) = 1$ and $w(0) = 1$. With the fractal-fractional representation, the system given by (15.1) takes the form below:

$$\begin{cases} {}^{FF}D_{0,t}^{p,q}\left(x(t)\right) = y - x, \\ {}^{FF}D_{0,t}^{p,q}\left(y(t)\right) = -xz + w, \\ {}^{FF}D_{0,t}^{p,q}\left(z(t)\right) = xy - A, \\ {}^{FF}D_{0,t}^{p,q}\left(w(t)\right) = -By, \end{cases} \tag{15.2}$$

where p represents the fractional order while q is fractal order. On solving the model at the steady state, there is no possible equilibrium point of the model (15.2).

15.3 EXISTENCE AND UNIQUENESS

This section discusses the existence and uniqueness of the system (15.2). We assumed that for all $t \in [0,T]$, the functions $x(t)$, $y(t)$, $z(t)$, and $w(t)$ are bounded, i.e., $\|x\|_\infty \le M_x$, $\|y\|_\infty \le M_y$, $\|z\|_\infty \le M_z$, and $\|w\|_\infty \le M_w$. Further, the system (15.2) can be written in the form given by:

$$f_x(t,x,y,z,w) = y - x,$$

$$f_y(t,x,y,z,w) = -xz + w,$$

$$f_z(t,x,y,z,w) = xy - A,$$

$$f_w(t,x,y,z,w) = -By,$$

We can show in the following that these functions, f_x, f_y, f_z, and f_w hod the linear growth and Lipschitz conditions. We first prove that the functions satisfy the linear growth property.

$$\begin{aligned} \|f_x(t,x,y,z,w)\| &= \|y-x\| \leq \|y\| + \|x\|, \\ &\leq \|y\|_\infty + \|x\|_\infty, \\ &\leq M_y + M_x, \\ &\leq \infty. \end{aligned}$$

$$\begin{aligned} \|f_y(t,x,y,z,w)\| &= \|-xz+w\|, \\ &\leq \|x\|\|z\| + \|w\|, \\ &\leq \|x\|_\infty \|z\|_\infty + \|w\|_\infty, \\ &\leq M_x M_z + M_w, \\ &< \infty \end{aligned}$$

$$\begin{aligned} \|f_z(t,x,y,z,w)\| &= \|xy-A\|, \\ &\leq \|x\|\|y\| + A, \\ &\leq \|x\|_\infty \|y\|_\infty + A, \\ &\leq M_x M_y + A, \\ &\leq \infty. \end{aligned}$$

$$\begin{aligned} \|f_z(t,x,y,z,w)\| &= \|-By\|, \\ &\leq B\|y\|, \\ &\leq B\|y\|_\infty, \end{aligned}$$

$$\leq BM_y,$$

$$\leq \infty.$$

Now, we show the Lipschitz conditions:

$$|f_x(t,x_1,y,z,w) - f_x(t,x_2,y,z,w)| \leq |x_1 - x_2|,$$

$$|f_y(t,x,y_1,z,w) - f_y(t,x,y_2,z,w)| = 0 \leq |y_1 - y_2|,$$

$$|f_z(t,x,y,z_1,w) - f_z(t,x,y,z_2,w)| = 0 \leq |z_1 - z_2|,$$

$$|f_z(t,x,y,z,w_1) - f_z(t,x,y,z,w_2)| = 0 \leq |w_1 - w_2|.$$

It can be seen that the system satisfys the linear growth property and the Lipschitz conditions. Hence, the system admits a unique system of solution.

Besides the above proof, we can also satisfy the linear growth and Lipschitz conditions as,

$$|f_x(t,x,y,z,w)|^2 = |y-x|^2 = 2|y|^2 + 2|x|^2,$$

$$\leq 2M_y^2 + 2|x|^2,$$

$$\leq 2M_y^2(1 + \frac{1}{M_y^2}|x|^2),$$

if $\frac{1}{M_y^2} < 1$, then

$$|f_x(t,x,y,z,w)|^2 \leq K_x(1+|x|^2),$$

where $K_x = 2M_y^2$.

$$|f_y(t,x,y,z,w)|^2 = |-xz+w|^2 \leq 3(M_x^2 M_z^2 + M_w^2)(1+|y|^2),$$

$$\leq K_y(1+|y|^2),$$

where $K_y = 3(M_x^2 M_z^2 + M_w^2)$.

$$|f_z(t,x,y,z,w)|^2 \leq 2(M_x^2 M_y^2 + A^2)(1+|z|^2),$$

$$\leq K_z(1+|z|^2),$$

where $K_z = 2(M_x^2 M_y^2 + A^2)$.

$$|f_w(t,x,y,z,w)|^2 = B^2|y|^2 \leq B^2 M_y^2(1+|w|^2),$$

$$\leq K_z(1+|w|^2).$$

Snice indeed the system satisfies the Lipschitz condition, then if the inequality $1/M_x^2 < 1$, is obtained, then the system admits a unique system of solutions.

A 4D Chaotic System under Fractal-Fractional Operators

15.4 SCHEMES BASED ON LINEAR INTERPOLATIONS

The present section will explore the new schemes based on linear interpolations for the fractal-fractional ordinary differential equations. We present in the following schemes for each case of power, exponential, and MIttag-Leffler kernel.

15.4.1 NUMERICAL SCHEME FOR POWER LAW KERNEL USING LINEAR INTERPOLATIONS

In this case, we consider the following fractal-fractional Cauchy problem,

$$\begin{aligned} {}_{0}^{FFP}D_{t}^{p,q}\mathbf{g}(t) &= f(t,\mathbf{g}(t)), \text{ if } t>0, \\ \mathbf{g}(0) &= \mathbf{g}_0, \text{ if } t=0. \end{aligned} \quad (15.3)$$

It is assumed that $f(t,\mathbf{g}(t))$ is twice differentiable and that $\mathbf{g}(t)$ is at least continuous

$$\begin{aligned} {}_{0}^{RL}D_{t}^{p}\mathbf{g}(t) &= qt^{q-1}f(t,\mathbf{g}(t)), \text{ if } t>0, \\ \mathbf{g}(0) &= \mathbf{g}_0, \text{ if } t=0. \end{aligned} \quad (15.4)$$

We write further

$$\begin{aligned} \mathbf{g}(t) &= \frac{q}{\Gamma(p)} \int_0^t \tau^{q-1} f(\tau,\mathbf{g}(\tau))(t-\tau)^{p-1} d\tau, \text{ if } t>0, \\ \mathbf{g}(0) &= \mathbf{g}_0, \text{ if } t=0, \end{aligned} \quad (15.5)$$

at $t = t_{n+1}$, we have

$$\mathbf{g}(t_{n+1}) = \frac{q}{\Gamma(p)} \int_0^{t_{n+1}} \tau^{q-1} f(\tau,\mathbf{g}(\tau))(t_{n+1}-\tau)^{p-1} d\tau. \quad (15.6)$$

Also, we write

$$\mathbf{g}(t_{n+1}) = \frac{q}{\Gamma(p)} \sum_{j=0}^{n} \int_{t_j}^{t_{j+1}} \tau^{q-1} f(\tau,\mathbf{g}(\tau))(t_{n+1}-\tau)^{p-1} d\tau. \quad (15.7)$$

Thus, within $[t_j, t_{j+1}]$, we approximate the function

$$f(\tau,\mathbf{g}(\tau)) \approx P_j(\tau) = f(t_j,\mathbf{g}_j) + (t-t_j)\frac{f(t_{j+1},\mathbf{g}_{j+1}) - f(t_j,\mathbf{g}_j)}{\Delta t}, \quad (15.8)$$

replacing yields

$$\mathbf{g}(t_{n+1}) \approx \mathbf{g}_{n+1} = \frac{q}{\Gamma(p)} \sum_{j=0}^{n} \int_{t_j}^{t_{j+1}} \tau^{q-1} (t_{n+1}-\tau)^{p-1}$$

$$\times \left(f(t_j, \mathbf{g}_j) + (t - t_j) \frac{f(t_{j+1}, \mathbf{g}_{j+1}) - f(t_j, \mathbf{g}_j)}{\Delta t} \right) d\tau,$$

$$= \frac{q}{\Gamma(p)} \sum_{j=0}^{n} f(t_j, \mathbf{g}_j) \int_{t_j}^{t_{j+1}} \tau^{q-1}(t_{n+1} - \tau)^{p-1} d\tau$$

$$+ \frac{q}{\Gamma(p)} \sum_{j=0}^{n} \frac{f(t_{j+1}, \mathbf{g}_{j+1}) - f(t_j, \mathbf{g}_j)}{\Delta t}$$

$$\times \int_{t_j}^{t_{j+1}} (t_{n+1} - \tau)^{p-1}(t - t_j) d\tau. \tag{15.9}$$

Noting that

$$\int_{t_j}^{t_{j+1}} \tau^{q-1}(t_{n+1} - \tau)^{p-1} d\tau = \int_{0}^{t_{j+1}} \tau^{q-1}(t_{n+1} - \tau)^{p-1} d\tau$$
$$- \int_{0}^{t_j} \tau^{q-1}(t_{n+1} - \tau)^{p-1} d\tau, \tag{15.10}$$

where

$$\int_{0}^{t_{j+1}} \tau^{q-1}(t_{n+1} - \tau)^{p-1} d\tau = t_{n+1}^{p+q-1} \int_{0}^{\frac{t_{j+1}}{t_{n+1}}} \mathbf{g}^{q-1}(1 - \mathbf{g})^p \mathbf{g}^{-1} d\mathbf{g},$$
$$= t_{n+1}^{p+q-1} B\left(\frac{t_{j+1}}{t_{n+1}}, q, p\right). \tag{15.11}$$

$$\int_{0}^{t_j} \tau^{q-1}(t_{n+1} - \tau)^{p-1} d\tau = t_{n+1}^{p+q-1} B\left(\frac{t_j}{t_{n+1}}, q, p\right). \tag{15.12}$$

Therefore,

$$\int_{t_j}^{t_{j+1}} \tau^{q-1}(t_{n+1} - \tau)^{p-1}(\tau - t_j) d\tau = \int_{t_j}^{t_{j+1}} \tau^q (t_{n+1} - \tau)^{p-1} d\tau$$
$$- t_j \int_{t_j}^{t_{j+1}} \tau^{q-1}(t_{n+1} - \tau)^{p-1} d\tau,$$
$$= t_{n+1}^{p+q} \widehat{B}_{q+1} - t_j \widehat{B}_q t_{n+1}^{p+q-1},$$

where

$$\widehat{B}_{q+1} = \left\{ B\left(\frac{t_{j+1}}{t_{n+1}}, q+1, p\right) - B\left(\frac{t_j}{t_{n+1}}, q+1, p\right) \right\},$$
$$\widehat{B}_q = \left\{ B\left(\frac{t_{j+1}}{t_{n+1}}, q, p\right) - B\left(\frac{t_j}{t_{n+1}}, q, p\right) \right\}.$$

Replacing yields, and further simplifications give

$$\mathbf{g}_{n+1} = \frac{q}{\Gamma(p)} (\Delta t)^{p+q-1} \sum_{j=0}^{n} f(t_j, \mathbf{g}_j)(n+1)^{p+q-1}$$

A 4D Chaotic System under Fractal-Fractional Operators 217

$$\times \left\{ B\left(\frac{t_{j+1}}{t_{n+1}},q,p\right) - B\left(\frac{t_j}{t_{n+1}},q,p\right) \right\}$$

$$+ \frac{q}{\Gamma(p)}(\Delta t)^{p+q-1} \sum_{j=0}^{n-1} \left[f(t_{j+1},\mathbf{g}_{j+1}) - f(t_j,\mathbf{g}_j) \right]$$

$$\times \left\{ (n+1)^{p+q} B\left(\frac{t_{j+1}}{t_{n+1}},q+1,p\right) - n^{p+q} B\left(\frac{t_j}{t_{n+1}},q+1,p\right) \right.$$

$$\left. -j \left[B\left(\frac{t_{j+1}}{t_{n+1}},q,p\right) - B\left(\frac{t_j}{t_{n+1}},q,p\right) \right] (n+1)^{p+q-1} \right\}$$

$$+ \frac{q(\Delta t)^{p+q-1}}{\Gamma(p)} \left[f(t_n + h, \mathbf{g}_{n+1}^u) - f(t_n, \mathbf{g}_n) \right]$$

$$\times \left\{ (n+1)^{p+q} \left(B(1,q+1,p) - B\left(\frac{t_n}{t_{n+1}},q+1,p\right) \right) \right.$$

$$\left. - n \left(B(1,q,p) - B\left(\frac{t_n}{t_{n+1}},q,p\right) \right) (n+1)^{p+q-1} \right\} \qquad (15.13)$$

where

$$\mathbf{g}_{n+1}^u = \frac{q}{\Gamma(p+1)}(\Delta t)^{p+q-1} \sum_{j=0}^{n} f(t_j,\mathbf{g}_j)(n+1)^{p+q-1}$$

$$\times \left\{ B\left(\frac{t_{j+1}}{t_{n+1}},q,p\right) - B\left(\frac{t_j}{t_{n+1}},q,p\right) \right\}$$

For our proposed system, the scheme follows:

$$x_{n+1} = \frac{q}{\Gamma(p)}(\Delta t)^{p+q-1} \sum_{j=0}^{n} f_1(t_j,x_j,y_j,z_j,w_j)(n+1)^{p+q-1}$$

$$\times \left\{ B\left(\frac{t_{j+1}}{t_{n+1}},q,p\right) - B\left(\frac{t_j}{t_{n+1}},q,p\right) \right\} + \frac{q}{\Gamma(p)}(\Delta t)^{p+q-1}$$

$$\times \sum_{j=0}^{n-1} \left[f_1(t_{j+1},x_{j+1},y_{j+1},z_{j+1},w_{j+1}) \right.$$

$$\left. - f_1(t_j,x_j,y_j,z_j,w_j) \right]$$

$$\times \left\{ (n+1)^{p+q} B\left(\frac{t_{j+1}}{t_{n+1}},q+1,p\right) - n^{p+q} B\left(\frac{t_j}{t_{n+1}},q+1,p\right) \right.$$

$$\left. -j \left[B\left(\frac{t_{j+1}}{t_{n+1}},q,p\right) - B\left(\frac{t_j}{t_{n+1}},q,p\right) \right] (n+1)^{p+q-1} \right\}$$

$$+ \frac{q(\Delta t)^{p+q-1}}{\Gamma(p)} \left[f_1(t_n + h, x_{n+1}^u, y_{n+1}^u, z_{n+1}^u, w_{n+1}^u) \right.$$

$$\left. - f_1(t_n,x_n,y_n,z_n,w_n) \right]$$

$$\times \left\{ (n+1)^{p+q} \left(B(1,q+1,p) - B\left(\frac{t_n}{t_{n+1}},q+1,p\right) \right) \right.$$

$$-n\left(B(1,q,p)-B\left(\frac{t_n}{t_{n+1}},q,p\right)\right)(n+1)^{p+q-1}\bigg\},$$

$$\begin{aligned}
y_{n+1} =\ & \frac{q}{\Gamma(p)}(\Delta t)^{p+q-1}\sum_{j=0}^{n}f_2(t_j,x_j,y_j,z_j,w_j)(n+1)^{p+q-1} \\
& \times\left\{B\left(\frac{t_{j+1}}{t_{n+1}},q,p\right)-B\left(\frac{t_j}{t_{n+1}},q,p\right)\right\}+\frac{q}{\Gamma(p)}(\Delta t)^{p+q-1} \\
& \times\sum_{j=0}^{n-1}\Big[f_2(t_{j+1},x_{j+1},y_{j+1},z_{j+1},w_{j+1}) \\
& -f_2(t_j,x_j,y_j,z_j,w_j)\Big] \\
& \times\left\{(n+1)^{p+q}B\left(\frac{t_{j+1}}{t_{n+1}},q+1,p\right)-n^{p+q}B\left(\frac{t_j}{t_{n+1}},q+1,p\right)\right. \\
& \left.-j\left[B\left(\frac{t_{j+1}}{t_{n+1}},q,p\right)-B\left(\frac{t_j}{t_{n+1}},q,p\right)\right](n+1)^{p+q-1}\right\} \\
& +\frac{q(\Delta t)^{p+q-1}}{\Gamma(p)}\Big[f_2(t_n+h,x_{n+1}^u,y_{n+1}^u,z_{n+1}^u,w_{n+1}^u) \\
& -f_2(t_n,x_n,y_n,z_n,w_n)\Big] \\
& \times\left\{(n+1)^{p+q}\left(B(1,q+1,p)-B\left(\frac{t_n}{t_{n+1}},q+1,p\right)\right)\right. \\
& \left.-n\left(B(1,q,p)-B\left(\frac{t_n}{t_{n+1}},q,p\right)\right)(n+1)^{p+q-1}\right\}, \\
z_{n+1} =\ & \frac{q}{\Gamma(p)}(\Delta t)^{p+q-1}\sum_{j=0}^{n}f_3(t_j,x_j,y_j,z_j,w_j)(n+1)^{p+q-1} \\
& \times\left\{B\left(\frac{t_{j+1}}{t_{n+1}},q,p\right)-B\left(\frac{t_j}{t_{n+1}},q,p\right)\right\}+\frac{q}{\Gamma(p)}(\Delta t)^{p+q-1} \\
& \times\sum_{j=0}^{n-1}\Big[f_3(t_{j+1},x_{j+1},y_{j+1},z_{j+1},w_{j+1}) \\
& -f_3(t_j,x_j,y_j,z_j,w_j)\Big] \\
& \times\left\{(n+1)^{p+q}B\left(\frac{t_{j+1}}{t_{n+1}},q+1,p\right)-n^{p+q}B\left(\frac{t_j}{t_{n+1}},q+1,p\right)\right. \\
& \left.-j\left[B\left(\frac{t_{j+1}}{t_{n+1}},q,p\right)-B\left(\frac{t_j}{t_{n+1}},q,p\right)\right](n+1)^{p+q-1}\right\} \\
& +\frac{q(\Delta t)^{p+q-1}}{\Gamma(p)}\Big[f_3(t_n+h,x_{n+1}^u,y_{n+1}^u,z_{n+1}^u,w_{n+1}^u) \\
& -f_3(t_n,x_n,y_n,z_n,w_n)\Big] \\
& \times\left\{(n+1)^{p+q}\left(B(1,q+1,p)-B\left(\frac{t_n}{t_{n+1}},q+1,p\right)\right)\right.
\end{aligned}$$

$$-n\left(B(1,q,p) - B\left(\frac{t_n}{t_{n+1}},q,p\right)\right)(n+1)^{p+q-1}\bigg\},$$

$$\begin{aligned}w_{n+1} = & \frac{q}{\Gamma(p)}(\Delta t)^{p+q-1}\sum_{j=0}^{n} f_4(t_j,x_j,y_j,z_j,w_j)(n+1)^{p+q-1}\\ & \times\left\{B\left(\frac{t_{j+1}}{t_{n+1}},q,p\right) - B\left(\frac{t_j}{t_{n+1}},q,p\right)\right\} + \frac{q}{\Gamma(p)}(\Delta t)^{p+q-1}\\ & \times\sum_{j=0}^{n-1}\Big[f_4(t_{j+1},x_{j+1},y_{j+1},z_{j+1},w_{j+1})\\ & -f_4(t_j,x_j,y_j,z_j,w_j)\Big]\\ & \times\left\{(n+1)^{p+q}B\left(\frac{t_{j+1}}{t_{n+1}},q+1,p\right) - n^{p+q}B\left(\frac{t_j}{t_{n+1}},q+1,p\right)\right.\\ & \left. -j\left[B\left(\frac{t_{j+1}}{t_{n+1}},q,p\right) - B\left(\frac{t_j}{t_{n+1}},q,p\right)\right](n+1)^{p+q-1}\right\}\\ & +\frac{q(\Delta t)^{p+q-1}}{\Gamma(p)}\Big[f_4(t_n+h,x^u_{n+1},y^u_{n+1},z^u_{n+1},w^u_{n+1})\\ & -f_4(t_n,x_n,y_n,z_n,w_n)\Big]\\ & \times\left\{(n+1)^{p+q}\left(B(1,q+1,p) - B\left(\frac{t_n}{t_{n+1}},q+1,p\right)\right)\right.\\ & \left. -n\left(B(1,q,p) - B\left(\frac{t_n}{t_{n+1}},q,p\right)\right)(n+1)^{p+q-1}\right\},\end{aligned} \quad (15.14)$$

where

$$\begin{aligned}x^u_{n+1} = & \frac{q}{\Gamma(p+1)}(\Delta t)^{p+q-1}\sum_{j=0}^{n} f_1(t_j,x_j,y_j,z_j,w_j)\\ & \times(n+1)^{p+q-1}\left\{B\left(\frac{t_{j+1}}{t_{n+1}},q,p\right) - B\left(\frac{t_j}{t_{n+1}},q,p\right)\right\},\\ y^u_{n+1} = & \frac{q}{\Gamma(p+1)}(\Delta t)^{p+q-1}\sum_{j=0}^{n} f_2(t_j,x_j,y_j,z_j,w_j)\\ & \times(n+1)^{p+q-1}\left\{B\left(\frac{t_{j+1}}{t_{n+1}},q,p\right) - B\left(\frac{t_j}{t_{n+1}},q,p\right)\right\},\\ z^u_{n+1} = & \frac{q}{\Gamma(p+1)}(\Delta t)^{p+q-1}\sum_{j=0}^{n} f_3(t_j,x_j,y_j,z_j,w_j)\\ & \times(n+1)^{p+q-1}\left\{B\left(\frac{t_{j+1}}{t_{n+1}},q,p\right) - B\left(\frac{t_j}{t_{n+1}},q,p\right)\right\},\\ w^u_{n+1} = & \frac{q}{\Gamma(p+1)}(\Delta t)^{p+q-1}\sum_{j=0}^{n} f_4(t_j,x_j,y_j,z_j,w_j)\\ & \times(n+1)^{p+q-1}\left\{B\left(\frac{t_{j+1}}{t_{n+1}},q,p\right) - B\left(\frac{t_j}{t_{n+1}},q,p\right)\right\},\end{aligned}$$

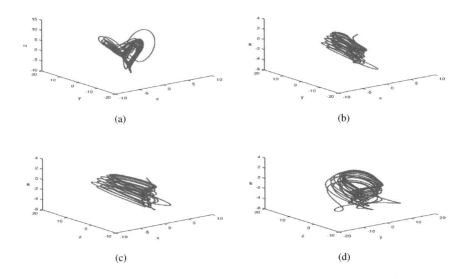

Figure 15.1 Model simulation of fractal-fractional Caputo model when $p = 1, q = 1$.

and

$$f_1(x,y,z,w,t) = y - x,$$
$$f_2(x,y,z,w,t) = -xz + w,$$
$$f_3(x,y,z,w,t) = xy - A,$$
$$f_4(x,y,z,w,t) = -By.$$

The numerical results shown in Figures 15.1–15.2 are presented for the fractal-fractional model in the sense of Caputo by using the approach presented above considering different orders of fractal and fractional operators.

15.4.2 NUMERICAL SCHEME FOR EXPONENTIAL DECAY KERNEL USING LINEAR INTERPOLATIONS

In this subsection, we consider the fractal-fractional Cauchy problem with exponential kernel, given by,

$$_0^{FFE}D_t^{p,q}\mathbf{g}(t) = f(t, \mathbf{g}(t)), \text{ if } t > 0,$$

A 4D Chaotic System under Fractal-Fractional Operators

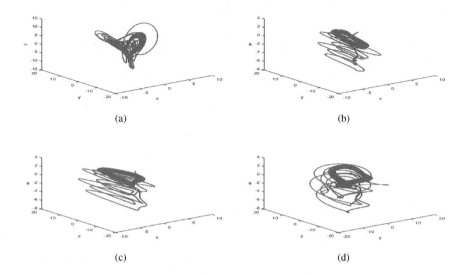

Figure 15.2 Model simulation of fractal-fractional Caputo model when $p = 0.98$, $q = 0.97$.

$$\mathbf{g}(0) = \mathbf{g}_0, \text{ if } t = 0.$$

where $\mathbf{g}(t) = (x(t), y(t), z(t), w(t))$ are the equations of the model proposed.

$$^{CF}_0D^p_t \mathbf{g}(t) = qt^{q-1} f(t, \mathbf{g}(t)), \text{ if } t > 0,$$

$$\mathbf{g}(0) = \mathbf{g}_0, \text{ if } t = 0.$$

$$\mathbf{g}(t) = (1-p)qt^{q-1} f(t, \mathbf{g}(t)) + pq \int_0^t \tau^{q-1} f(\tau, \mathbf{g}(\tau)) d\tau, \; y(0) = y_0,$$

at $t = t_{n+1}$, we have

$$\mathbf{g}(t_{n+1}) = (1-p)qt_{n+1}^{q-1} f(t_{n+1}, \mathbf{g}_{n+1}) + pq \int_0^{t_{n+1}} \tau^{q-1} f(\tau, \mathbf{g}(\tau)) d\tau,$$

at $t = t_n$, we have

$$\mathbf{g}(t_n) = (1-p)qt_n^{q-1} f(t_n, \mathbf{g}_n) + pq \int_0^{t_n} \tau^{q-1} f(\tau, \mathbf{g}(\tau)) d\tau.$$

Therefore,

$$\mathbf{g}(t_{n+1}) = \mathbf{g}(t_n) + (1-p)q \left[t_{n+1}^{q-1} f(t_{n+1}, \mathbf{g}_{n+1}) - t_n^{q-1} f(t_n, \mathbf{g}(t_n)) \right]$$

$$+pq \int_{t_n}^{t_{n+1}} \tau^{q-1} f(\tau, \mathbf{g}(\tau)) d\tau.$$

Using the polynomial approximations of $f(\tau, \mathbf{g}(\tau))$ within $[t_n, t_{n+1}]$ yields,

$$\begin{aligned} \mathbf{g}_{n+1} &= \mathbf{g}_n + (1-p)q \left[t_{n+1}^{q-1} f(t_{n+1}, \mathbf{g}_{n+1}^u) - t_n^{q-1} f(t_n, \mathbf{g}_n) \right] \\ &+ p \left[f(t_n, \mathbf{g}_n)(t_{n+1}^q - t_n^q) + \frac{f(t_{n+1}, \mathbf{g}_{n+1}^u) - f(t_n, \mathbf{g}_n)}{\Delta t} \right. \\ &\left. \left\{ \frac{q}{q+1}(t_{n+1}^{q+1} - t_n^{q+1}) - t_n(t_{n+1}^q - t_n^q) \right\} \right]. \end{aligned} \quad (15.15)$$

Further, we have

$$\begin{aligned} \mathbf{g}_{n+1} &= \mathbf{g}_n + (1-p)q(\Delta t)^{q-1} \\ &\quad \left\{ (n+1)^{q-1} f(t_n + h, \mathbf{g}_{n+1}^u) - n^{q-1} f(t_n, \mathbf{g}_n) \right\} \\ &\quad + p(\Delta t)^q \left\{ f(t_n, \mathbf{g}_n)((n+1)^q - n^q) \right. \\ &\quad + (f(t_n + h, \mathbf{g}_{n+1}^u) - f(t_n, \mathbf{g}_n)) \right\} \\ &\quad \times \left\{ \frac{q}{q+1}[(n+1)^{q+1} - n^{q+1}] - n\left((n+1)^q - n^q\right) \right\}, \end{aligned}$$

where

$$\begin{aligned} \mathbf{g}_{n+1}^u &= \mathbf{g}_0 + (1-p)q(\Delta t)^{q-1} n^{q-1} f(t_n, \mathbf{g}_n) \\ &\quad + p \sum_{j=0}^{n} f(t_j, \mathbf{g}_j) \left[\frac{q}{q+1}[t_{j+1}^{q+1} - t_j^{q+1}] - t_j(t_{j+1}^q - t_j^q) \right]. \end{aligned}$$

For our system, the scheme can be written as

$$\begin{aligned} x_{n+1} &= x_n + (1-p)q(\Delta t)^{q-1} \\ &\quad \left\{ (n+1)^{q-1} f_1(t_n + h, x_{n+1}^u, y_{n+1}^u, z_{n+1}^u, w_{n+1}^u) \right. \\ &\quad \left. - n^{q-1} f_1(t_n, x_n, y_n, z_n, w_n) \right\} \\ &\quad + p(\Delta t)^q \left\{ f_1(t_n, x_n, y_n, z_n, w_n)((n+1)^q - n^q) \right. \\ &\quad + (f_1(t_n + h, x_{n+1}^u, y_{n+1}^u, z_{n+1}^u, w_{n+1}^u) - f_1(t_n, x_n, y_n, z_n, w_n)) \right\} \\ &\quad \times \left\{ \frac{q}{q+1}[(n+1)^{q+1} - n^{q+1}] - n\left((n+1)^q - n^q\right) \right\}, \\ y_{n+1} &= y_n + (1-p)q(\Delta t)^{q-1} \end{aligned}$$

$$\begin{aligned}
&\left\{(n+1)^{q-1} f_2(t_n+h, x_{n+1}^u, y_{n+1}^u, z_{n+1}^u, w_{n+1}^u)\right.\\
&\left.-n^{q-1} f_2(t_n, x_n, y_n, z_n, w_n)\right\}\\
&+p(\Delta t)^q \left\{f_2(t_n, x_n, y_n, z_n, w_n)((n+1)^q - n^q)\right.\\
&\left.+(f_2(t_n+h, x_{n+1}^u, y_{n+1}^u, z_{n+1}^u, w_{n+1}^u) - f_2(t_n, x_n, y_n, z_n, w_n))\right\}\\
&\times \left\{\frac{q}{q+1}[(n+1)^{q+1} - n^{q+1}] - n\left((n+1)^q - n^q\right)\right\},
\end{aligned}$$

$$z_{n+1} = z_n + (1-p)q(\Delta t)^{q-1}$$
$$\left\{(n+1)^{q-1} f_3(t_n+h, x_{n+1}^u, y_{n+1}^u, z_{n+1}^u, w_{n+1}^u)\right.$$
$$\left.-n^{q-1} f_3(t_n, x_n, y_n, z_n, w_n)\right\}$$
$$+p(\Delta t)^q \left\{f_3(t_n, x_n, y_n, z_n, w_n)((n+1)^q - n^q)\right.$$
$$\left.+(f_3(t_n+h, x_{n+1}^u, y_{n+1}^u, z_{n+1}^u, w_{n+1}^u) - f_3(t_n, x_n, y_n, z_n, w_n))\right\}$$
$$\times \left\{\frac{q}{q+1}[(n+1)^{q+1} - n^{q+1}] - n\left((n+1)^q - n^q\right)\right\},$$

$$w_{n+1} = w_n + (1-p)q(\Delta t)^{q-1}$$
$$\left\{(n+1)^{q-1} f_4(t_n+h, x_{n+1}^u, y_{n+1}^u, z_{n+1}^u, w_{n+1}^u)\right.$$
$$\left.-n^{q-1} f_4(t_n, x_n, y_n, z_n, w_n)\right\}$$
$$+p(\Delta t)^q \left\{f_4(t_n, x_n, y_n, z_n, w_n)((n+1)^q - n^q)\right.$$
$$\left.+(f_4(t_n+h, x_{n+1}^u, y_{n+1}^u, z_{n+1}^u, w_{n+1}^u) - f_4(t_n, x_n, y_n, z_n, w_n))\right\}$$
$$\times \left\{\frac{q}{q+1}[(n+1)^{q+1} - n^{q+1}] - n\left((n+1)^q - n^q\right)\right\},$$

where

$$x_{n+1}^u = x_0 + (1-p)q(\Delta t)^{q-1} n^{q-1} f_1(t_n, x_n, y_n, z_n, w_n)$$
$$+ p \sum_{j=0}^{n} f_1(t_j, x_j, y_j, z_j, w_j) \left[\frac{q}{q+1}[t_{j+1}^{q+1} - t_j^{q+1}] - t_j(t_{j+1}^q - t_j^q)\right],$$

$$y_{n+1}^u = y_0 + (1-p)q(\Delta t)^{q-1} n^{q-1} f_2(t_n, x_n, y_n, z_n, w_n)$$
$$+ p \sum_{j=0}^{n} f_2(t_j, x_j, y_j, z_j, w_j) \left[\frac{q}{q+1}[t_{j+1}^{q+1} - t_j^{q+1}] - t_j(t_{j+1}^q - t_j^q)\right],$$

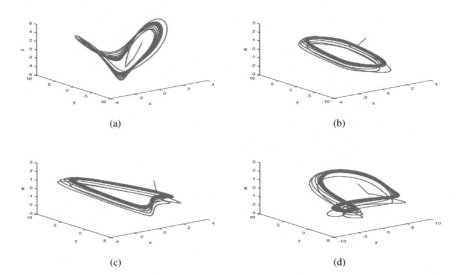

Figure 15.3 Model simulation of fractal-fractional Caputo-Fabrizio model when $p = 1$, $q = 1$.

$$z_{n+1}^u = z_0 + (1-p)q(\Delta t)^{q-1}n^{q-1}f_3(t_n, x_n, y_n, z_n, w_n)$$

$$+ p\sum_{j=0}^{n} f_3(t_j, x_j, y_j, z_j, w_j)\left[\frac{q}{q+1}[t_{j+1}^{q+1} - t_j^{q+1}] - t_j(t_{j+1}^q - t_j^q)\right],$$

$$w_{n+1}^u = w_0 + (1-p)q(\Delta t)^{q-1}n^{q-1}f_4(t_n, x_n, y_n, z_n, w_n)$$

$$+ p\sum_{j=0}^{n} f_4(t_j, x_j, y_j, z_j, w_j)\left[\frac{q}{q+1}[t_{j+1}^{q+1} - t_j^{q+1}] - t_j(t_{j+1}^q - t_j^q)\right].$$

The numerical results shown in Figures 15.3–15.11 have been obtained for the fractal-fractional model in the sense of Caputo-Fabrizio operator using the scheme above for different values of the fractal and fractional orders parameters.

15.4.3 NUMERICAL SCHEME FOR GENERALISED MITTAG-LEFFLER KERNEL USING LINEAR INTERPOLATIONS

We consider in this subsection, a fractal-fractional Cauchy problem with the generalised Mittag-Leffler Kernel:

$$_0^{FFM}D_t^{p,q}\mathbf{g}(t) = f(t, \mathbf{g}(t)), \text{ if } t > 0,$$

$$\mathbf{g}(0) = \mathbf{g}_0, \text{ if } t = 0.$$

A 4D Chaotic System under Fractal-Fractional Operators 225

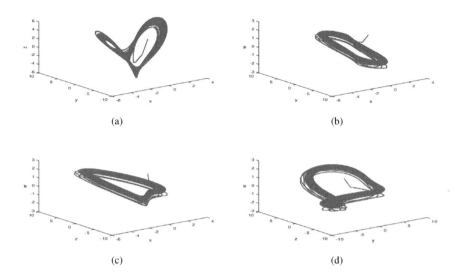

Figure 15.4 Model simulation of fractal-fractional Caputo-Fabrizio model when $p = 0.99, q = 1$.

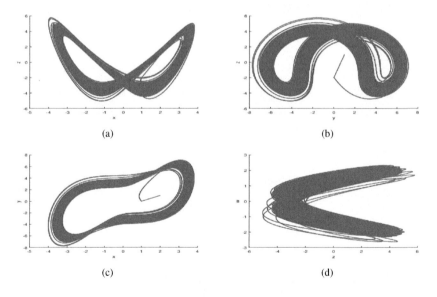

Figure 15.5 Model simulation of fractal-fractional Caputo-Fabrizio model when $p = 0.99, q = 1$.

We have for AB-RL case

$$_0^{ABR}D_t^p \mathbf{g}(t) = qt^{q-1}f(t,\mathbf{g}(t)), \text{ if } t > 0,$$

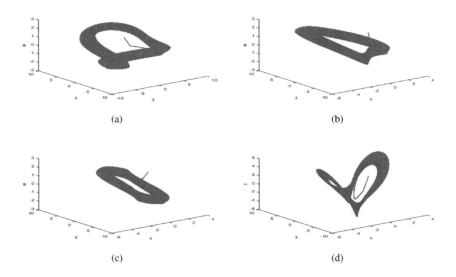

Figure 15.6 Model simulation of fractal-fractional Caputo-Fabrizio model when $p = 0.98, q = 1$.

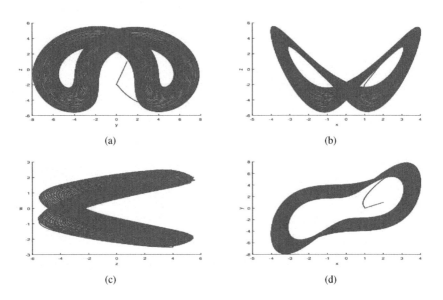

Figure 15.7 Model simulation of fractal-fractional Caputo-Fabrizio model when $p = 0.98, q = 1$.

$$\mathbf{g}(0) = \mathbf{g}_0, \text{ if } t = 0.$$

A 4D Chaotic System under Fractal-Fractional Operators 227

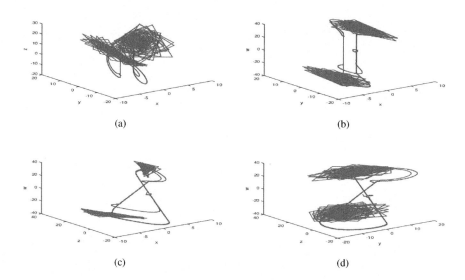

Figure 15.8 Model simulation of fractal-fractional Caputo-Fabrizio model when $p = 0.8, q = 1$.

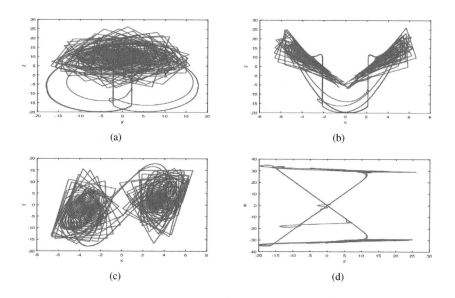

Figure 15.9 Model simulation of fractal-fractional Caputo-Fabrizio model when $p = 0.8, q = 1$.

$$\mathbf{g}(t) \;\;=\;\; (1-p)qt^{q-1}f(t,\mathbf{g}(t))$$

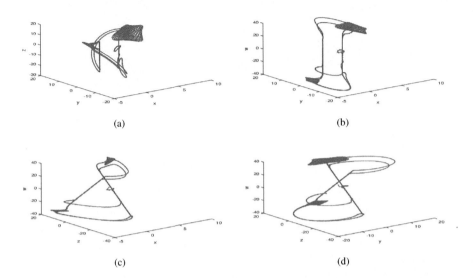

Figure 15.10 Model simulation of fractal-fractional Caputo-Fabrizio model when $p = 0.8, q = 0.99$.

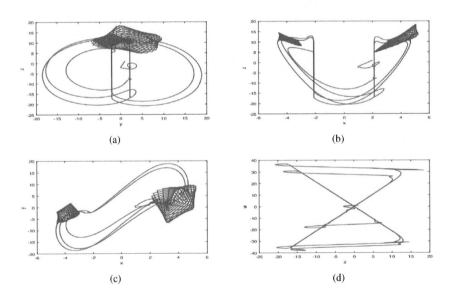

Figure 15.11 Model simulation of fractal-fractional Caputo-Fabrizio model when $p = 0.8, q = 0.99$.

$$+\frac{pq}{\Gamma(p)}\int_0^t \tau^{q-1}(t-\tau)^{p-1} f(\tau,\mathbf{g}(\tau))d\tau, \; \mathbf{g}(0) = \mathbf{g}_0.$$

A 4D Chaotic System under Fractal-Fractional Operators

At $t = t_{n+1}$, we have

$$\mathbf{g}(t_{n+1}) = (1-p)qt_{n+1}^{q-1}f(t_{n+1},\mathbf{g}(t_{n+1}))$$
$$+ \frac{pq}{\Gamma(p)}\int_0^{t_{n+1}} \tau^{q-1}(t_{n+1}-\tau)^{p-1}f(\tau,\mathbf{g}(\tau))d\tau.$$

$$\mathbf{g}(t_{n+1}) = (1-p)qt_{n+1}^{q-1}f(t_{n+1},\mathbf{g}(t_{n+1}))$$
$$+ \frac{pq}{\Gamma(p)}\sum_{j=0}^{n}\int_{t_j}^{t_{j+1}}(t_{n+1}-\tau)^{p-1}\tau^{q-1}f(\tau,\mathbf{g}(\tau))d\tau.$$

Approximating the function $f(\tau,\mathbf{g}(\tau))$ by polynomial approximations, we have

$$\mathbf{g}_{n+1} = qt_{n+1}^q(1-p)f(t_{n+1},\mathbf{g}_{n+1}^u)$$
$$+ \frac{pq}{\Gamma(p)}\sum_{j=0}^{n}f(t_j,\mathbf{g}_j)\left\{t_{n+1}^{p+q-1}\left[B\left(\frac{t_{j+1}}{t_{n+1}},q,p\right)\right.\right.$$
$$\left.\left.-B\left(\frac{t_j}{t_{n+1}},q,p\right)\right]\right\}$$
$$+ \frac{pq}{\Gamma(p)}\sum_{j=0}^{n-1}\left(\frac{f(t_{j+1},\mathbf{g}_{j+1}) - f(t_j,\mathbf{g}_j)}{\Delta t}\right)$$
$$\left\{t_{n+1}^{p+q}\left[B\left(\frac{t_{j+1}}{t_{n+1}},q+1,p\right) - B\left(\frac{t_j}{t_{n+1}},q+1,p\right)\right]\right.$$
$$\left. -t_j t_{n+1}^{p+q-1}\left[B\left(\frac{t_{j+1}}{t_{n+1}},q,p\right) - B\left(\frac{t_j}{t_{n+1}},q,p\right)\right]\right\}$$
$$+ \frac{pq}{\Gamma(p)}\left(\frac{f(t_n+h,\mathbf{g}_{n+1}^u) - f(t_n,\mathbf{g}_n)}{\Delta t}\right)$$
$$\times (\Delta t)^{p+q}\left\{(n+1)^{p+q}\{B(1,q+1,p) - B(\frac{t_n}{t_{n+1}},q+1,p)\}\right.$$
$$\left. -n(n+1)^{p+q-1}\{B(1,q,p) - B(\frac{t_n}{t_{n+1}},q,p)\}\right\} \quad (15.16)$$

where

$$\mathbf{g}_{n+1}^u = (1-p)qf(t_n,\mathbf{g}_n)$$
$$+ \frac{pq}{\Gamma(p)}\sum_{j=0}^{n}f(t_j,\mathbf{g}_j)(n+1)^{p+q-1}$$
$$\times \left[B\left(\frac{t_{j+1}}{t_{n+1}},q,p\right) - B\left(\frac{t_j}{t_{n+1}},q,p\right)\right]. \quad (15.17)$$

The scheme for the case of Mittag-Leffler kernel can be applied to our system of equations as follows:

$$x_{n+1} = qt_{n+1}^q(1-p)f_1(t_{n+1},x_{n+1}^u,y_{n+1}^u,z_{n+1}^u,w_{n+1}^u)$$

$$+\frac{pq}{\Gamma(p)}\sum_{j=0}^{n}f_1(t_j,x_j,y_j,z_j,w_j)\left\{t_{n+1}^{p+q-1}\left[B\left(\frac{t_{j+1}}{t_{n+1}},q,p\right)\right.\right.$$

$$\left.\left.-B\left(\frac{t_j}{t_{n+1}},q,p\right)\right\}\right]$$

$$+\frac{pq}{\Gamma(p)}\sum_{j=0}^{n-1}\left(\frac{f_1(t_{j+1},x_{j+1},y_{j+1},z_{j+1},w_{j+1})-f_1(t_j,x_j,y_j,z_j,w_j)}{\Delta t}\right)$$

$$\left\{t_{n+1}^{p+q}\left[B\left(\frac{t_{j+1}}{t_{n+1}},q+1,p\right)-B\left(\frac{t_j}{t_{n+1}},q+1,p\right)\right]\right.$$

$$\left.-t_j t_{n+1}^{p+q-1}\left[B\left(\frac{t_{j+1}}{t_{n+1}},q,p\right)-B\left(\frac{t_j}{t_{n+1}},q,p\right)\right]\right\}$$

$$+\frac{pq}{\Gamma(p)}\left(\frac{f_1(t_n+h,x_{n+1}^u,y_{n+1}^u,z_{n+1}^u,w_{n+1}^u)-f_1(t_n,x_n,y_n,z_n,w_n)}{\Delta t}\right)$$

$$\times(\Delta t)^{p+q}\left\{(n+1)^{p+q}\{B(1,q+1,p)-B(\frac{t_n}{t_{n+1}},q+1,p)\}\right.$$

$$\left.-n(n+1)^{p+q-1}\{B(1,q,p)-B(\frac{t_n}{t_{n+1}},q,p)\}\right\},$$

$$y_{n+1} = qt_{n+1}^q(1-p)f_2(t_{n+1},x_{n+1}^u,y_{n+1}^u,z_{n+1}^u,w_{n+1}^u)$$

$$+\frac{pq}{\Gamma(p)}\sum_{j=0}^{n}f_2(t_j,x_j,y_j,z_j,w_j)\left\{t_{n+1}^{p+q-1}\left[B\left(\frac{t_{j+1}}{t_{n+1}},q,p\right)\right.\right.$$

$$\left.\left.-B\left(\frac{t_j}{t_{n+1}},q,p\right)\right\}\right]$$

$$+\frac{pq}{\Gamma(p)}\sum_{j=0}^{n-1}\left(\frac{f_2(t_{j+1},x_{j+1},y_{j+1},z_{j+1},w_{j+1})-f_2(t_j,x_j,y_j,z_j,w_j)}{\Delta t}\right)$$

$$\left\{t_{n+1}^{p+q}\left[B\left(\frac{t_{j+1}}{t_{n+1}},q+1,p\right)-B\left(\frac{t_j}{t_{n+1}},q+1,p\right)\right]\right.$$

$$\left.-t_j t_{n+1}^{p+q-1}\left[B\left(\frac{t_{j+1}}{t_{n+1}},q,p\right)-B\left(\frac{t_j}{t_{n+1}},q,p\right)\right]\right\}$$

$$+\frac{pq}{\Gamma(p)}\left(\frac{f_2(t_n+h,x_{n+1}^u,y_{n+1}^u,z_{n+1}^u,w_{n+1}^u)-f_2(t_n,x_n,y_n,z_n,w_n)}{\Delta t}\right)$$

$$\times(\Delta t)^{p+q}\left\{(n+1)^{p+q}\{B(1,q+1,p)-B(\frac{t_n}{t_{n+1}},q+1,p)\}\right.$$

$$\left.-n(n+1)^{p+q-1}\{B(1,q,p)-B(\frac{t_n}{t_{n+1}},q,p)\}\right\},$$

$$z_{n+1} = qt_{n+1}^q(1-p)f_3(t_{n+1},x_{n+1}^u,y_{n+1}^u,z_{n+1}^u,w_{n+1}^u)$$

$$+\frac{pq}{\Gamma(p)}\sum_{j=0}^{n}f_3(t_j,x_j,y_j,z_j,w_j)\left\{t_{n+1}^{p+q-1}\left[B\left(\frac{t_{j+1}}{t_{n+1}},q,p\right)\right.\right.$$

A 4D Chaotic System under Fractal-Fractional Operators

$$-B\left(\frac{t_j}{t_{n+1}},q,p\right)\Big\}\Big]$$

$$+\frac{pq}{\Gamma(p)}\sum_{j=0}^{n-1}\left(\frac{f_3(t_{j+1},x_{j+1},y_{j+1},z_{j+1},w_{j+1})-f_3(t_j,x_j,y_j,z_j,w_j)}{\Delta t}\right)$$

$$\left\{t_{n+1}^{p+q}\left[B\left(\frac{t_{j+1}}{t_{n+1}},q+1,p\right)-B\left(\frac{t_j}{t_{n+1}},q+1,p\right)\right]\right.$$

$$\left.-t_j t_{n+1}^{p+q-1}\left[B\left(\frac{t_{j+1}}{t_{n+1}},q,p\right)-B\left(\frac{t_j}{t_{n+1}},q,p\right)\right]\right\}$$

$$+\frac{pq}{\Gamma(p)}\left(\frac{f_3(t_n+h,x_{n+1}^u,y_{n+1}^u,z_{n+1}^u,w_{n+1}^u)-f_3(t_n,x_n,y_n,z_n,w_n)}{\Delta t}\right)$$

$$\times(\Delta t)^{p+q}\left\{(n+1)^{p+q}\{B(1,q+1,p)-B(\frac{t_n}{t_{n+1}},q+1,p)\}\right.$$

$$\left.-n(n+1)^{p+q-1}\{B(1,q,p)-B(\frac{t_n}{t_{n+1}},q,p)\}\right\},$$

$$w_{n+1} = qt_{n+1}^q(1-p)f_4(t_{n+1},x_{n+1}^u,y_{n+1}^u,z_{n+1}^u,w_{n+1}^u)$$

$$+\frac{pq}{\Gamma(p)}\sum_{j=0}^{n}f_4(t_j,x_j,y_j,z_j,w_j)\left\{t_{n+1}^{p+q-1}\left[B\left(\frac{t_{j+1}}{t_{n+1}},q,p\right)\right.\right.$$

$$\left.\left.-B\left(\frac{t_j}{t_{n+1}},q,p\right)\right]\right\}$$

$$+\frac{pq}{\Gamma(p)}\sum_{j=0}^{n-1}\left(\frac{f_4(t_{j+1},x_{j+1},y_{j+1},z_{j+1},w_{j+1})-f_4(t_j,x_j,y_j,z_j,w_j)}{\Delta t}\right)$$

$$\left\{t_{n+1}^{p+q}\left[B\left(\frac{t_{j+1}}{t_{n+1}},q+1,p\right)-B\left(\frac{t_j}{t_{n+1}},q+1,p\right)\right]\right.$$

$$\left.-t_j t_{n+1}^{p+q-1}\left[B\left(\frac{t_{j+1}}{t_{n+1}},q,p\right)-B\left(\frac{t_j}{t_{n+1}},q,p\right)\right]\right\}$$

$$+\frac{pq}{\Gamma(p)}\left(\frac{f_4(t_n+h,x_{n+1}^u,y_{n+1}^u,z_{n+1}^u,w_{n+1}^u)-f_4(t_n,x_n,y_n,z_n,w_n)}{\Delta t}\right)$$

$$\times(\Delta t)^{p+q}\left\{(n+1)^{p+q}\{B(1,q+1,p)-B(\frac{t_n}{t_{n+1}},q+1,p)\}\right.$$

$$\left.-n(n+1)^{p+q-1}\{B(1,q,p)-B(\frac{t_n}{t_{n+1}},q,p)\}\right\},$$

where

$$x_{n+1}^u = (1-p)qf_1(t_n,x_n,y_n,z_n,w_n)$$

$$+\frac{pq}{\Gamma(p)}\sum_{j=0}^{n}f_1(t_j,x_j,y_j,z_j,w_j)(n+1)^{p+q-1}$$

$$\times\left[B\left(\frac{t_{j+1}}{t_{n+1}},q,p\right)-B\left(\frac{t_j}{t_{n+1}},q,p\right)\right],$$

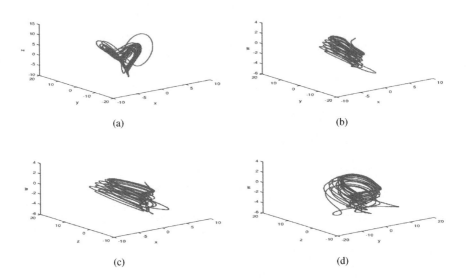

Figure 15.12 Model simulation of fractal-fractional Atangana-Baleanu model when $p = 1, q = 1$.

$$\begin{aligned}
y_{n+1}^u &= (1-p)qf_2(t_n, x_n, y_n, z_n, w_n) \\
&\quad + \frac{pq}{\Gamma(p)} \sum_{j=0}^{n} f_2(t_j, x_j, y_j, z_j, w_j)(n+1)^{p+q-1} \\
&\quad \times \left[B\left(\frac{t_{j+1}}{t_{n+1}}, q, p\right) - B\left(\frac{t_j}{t_{n+1}}, q, p\right) \right], \\
z_{n+1}^u &= (1-p)qf_3(t_n, x_n, y_n, z_n, w_n) \\
&\quad + \frac{pq}{\Gamma(p)} \sum_{j=0}^{n} f_3(t_j, x_j, y_j, z_j, w_j)(n+1)^{p+q-1} \\
&\quad \times \left[B\left(\frac{t_{j+1}}{t_{n+1}}, q, p\right) - B\left(\frac{t_j}{t_{n+1}}, q, p\right) \right], \\
w_{n+1}^u &= (1-p)qf_4(t_n, x_n, y_n, z_n, w_n) \\
&\quad + \frac{pq}{\Gamma(p)} \sum_{j=0}^{n} f_4(t_j, x_j, y_j, z_j, w_j)(n+1)^{p+q-1} \\
&\quad \times \left[B\left(\frac{t_{j+1}}{t_{n+1}}, q, p\right) - B\left(\frac{t_j}{t_{n+1}}, q, p\right) \right]. \quad (15.18)
\end{aligned}$$

The graphs shown in Figures 15.12–15.17 are obtained using the scheme based on linear interpolations for the case of Mittag-Leffler kernel.

A 4D Chaotic System under Fractal-Fractional Operators

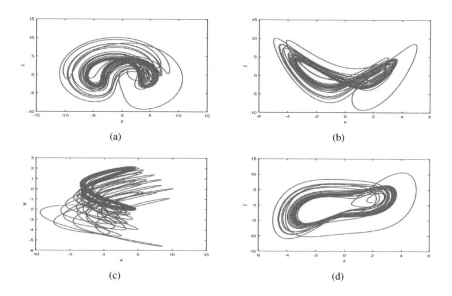

Figure 15.13 Model simulation of fractal-fractional Atangana-Baleanu model when $p = 1, q = 1$.

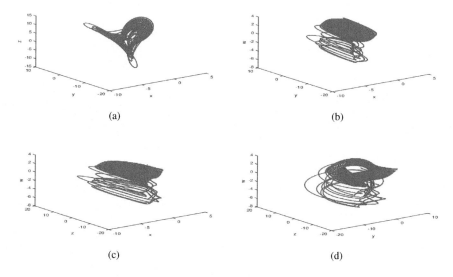

Figure 15.14 Model simulation of fractal-fractional Atangana-Baleanu model when $p = 0.99, q = 0.99$.

15.5 CONCLUSION

We studied a chaotic 4D model under different fractal-fractional operators in the sense of power, exponential, and Mittag-Leffler kernel. We found that system of

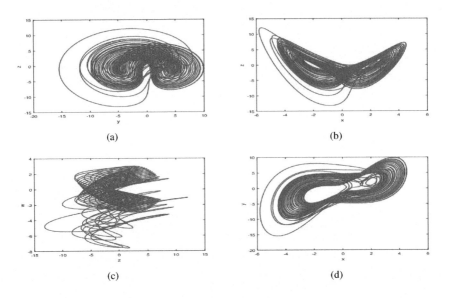

Figure 15.15 Model simulation of fractal-fractional Atangana-Baleanu model when $p = 0.99, q = 0.99$.

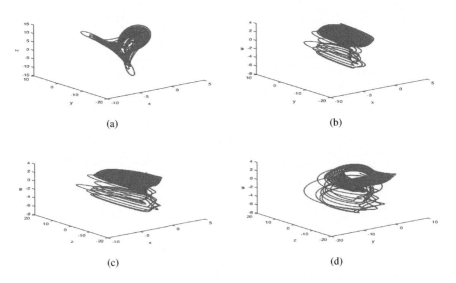

Figure 15.16 Model simulation of fractal-fractional Atangana-Baleanu model when $p = 0.98, q = 0.99$.

A 4D Chaotic System under Fractal-Fractional Operators

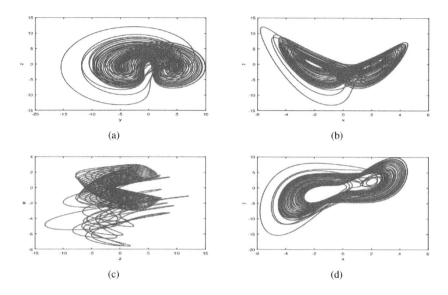

Figure 15.17 Model simulation of fractal-fractional Atangana-Baleanu model when $p = 0.98, q = 0.99$.

solution for the fractal-fractional model exists, and it is unique, through the theory of fixed point. We then presented some new numerical schemes that are based on linear interpolations for each case of fractal-fractional system and obtained the graphical results.

16 Self-Excited and Hidden Attractors through Fractal-Fractional Operators

16.1 INTRODUCTION

A concealed oscillation is a bounded oscillation that arises without impairing the stability of the stationary set. According to non-linear control theory, the emergence of a hidden oscillation in a bounded state time-invariant control system entails crossing a boundary in the parameter space where the local stability of the stationary states implies global stability. A hidden oscillation is referred to as a hidden attractor if it draws in all neighbouring oscillations. The emergence of a hidden attractor relates to a qualitative shift in behaviour from monostability to bi-stability for a dynamical system with a singular equilibrium point that is globally attractive. A hidden attractor is one that exists in the phase space of a dynamical system that, in the general case, turns out to be multistable and attracts all adjacent oscillations. The emergence of a hidden attractor relates to a qualitative shift in behaviour from monostability to bi-stability for a dynamical system with a singular equilibrium point that is globally attractive [102, 132, 225, 257]. A dynamical system could, in the general situation, end up being multistable and have coexisting local attractors in the phase space. Although trivial attractors, or stable equilibrium points, are simple to locate analytically or numerically, finding periodic and chaotic attractors can be difficult. In a physical or mathematical experiment, one must select an initial system state in the attractor's basin of attraction and observe how the system's state, starting from this initial state, following a transitory process, manifests the attractor. The distinction between hidden and self-excited attractors reflects the challenges in locating local attractors in phase space and disclosing basins of attraction. Because an unstable equilibrium is connected to the basin of attraction of a self-excited attractor, the self-excited attractors can be found numerically using a standard computational method. In this method, a trajectory that begins in the vicinity of unstable equilibrium is initially drawn to the state of oscillation before being drawn to it and then traced. Consequently, self-excited attractors can be easily discovered and represented mathematically. The attractor in the Lorenz system is self-excited with respect to all current equilibria and can be seen by any trajectory from their vicinity for classical parameter values; however, for some other parameter values, there are two trivial attractors coexisting with a chaotic attractor, which is a self-excited one with respect to the zero equilibrium only. In some locations in the phase space, hidden attractors have basins of attraction that are not connected to equilibria. Some important references on chaotic attractors can be found in [7,25,102,128,132,191,201,206,225,235,257]. We consider a self-cited hidden attractors model under the framework of different fractal-fractional operators. The model and their related results are given. We consider the approach based on linear interpolations, to provide schemes for the kernel of fractal-fractional chaotic system. The details of the graphical results for each case are shown.

16.2 CHAOTIC MODEL AND ITS DYNAMICAL BEHAVIOUR

In the present section, we consider a chaotic mathematical model presented in [172] to study its dynamics through fractal-fractional operators. The chaotic model we consider here is given by the following equations [172]:

$$\begin{cases} \frac{dx(t)}{dt} = -y, \\ \frac{dy(t)}{dt} = x + \delta yz, \\ \frac{dz(t)}{dt} = \sigma|(\sin(x))| - \beta y^2 - \varpi z - \mu. \end{cases} \quad (16.1)$$

In above model, $x(t)$, $y(t)$, and $z(t)$ are the state variables, whereas the parameters are δ, σ, β, ϖ, and μ. This system is chaotic for the given values of the parameters $\varpi = 0.2, \beta = 0.6, \sigma = 17, \mu = 0$, and $\delta = 0.63$ with the initial conditions on state variables $x(0) = 0.5$, $y(0) = -2$, and $z(0) = 0$. The fractal-fractional representation of the model (16.1) is given by

$$\begin{cases} {}^{FF}D_{0,t}^{p,q}\left(x(t)\right) = -y, \\ {}^{FF}D_{0,t}^{p,q}\left(y(t)\right) = x + \delta yz, \\ {}^{FF}D_{0,t}^{p,q}\left(z(t)\right) = -\sigma|(\sin(x))| - \beta y^2 - \varpi z - \mu, \end{cases} \quad (16.2)$$

where p represents the fractional order while q is the fractal order.

16.3 EXISTENCE AND UNIQUENESS

In the present section, we aim to study the existence and uniqueness of the model (16.2). It is assumed that for all $t \in [0,T]$, the functions $x(t)$, $y(t)$, and $z(t)$ are bounded, $\|x\|_\infty \leq M_x$, $\|y\|_\infty \leq M_y$ and $\|z\|_\infty \leq M_z$. Further, we write the model (16.2) in the form:

$$f_x(t,x,y,z) = -y,$$

$$f_y(t,x,y,z) = x + \delta yz,$$

$$f_z(t,x,y,z) = -\sigma|(\sin(x))| - \beta y^2 - \varpi z - \mu.$$

We show that the functions, f_x, f_y, and f_z satisfy the linear growth and Lipschitz conditions. We first show that the functions satisfy the linear growth property.

$$\|f_x(t,x,y,z)\| = \|-y\| \leq \|y\|,$$

$$\leq \|y\|_\infty,$$
$$\leq M_y,$$
$$\leq \infty.$$

$$\|f_y(t,x,y,z)\| = \|x+\delta yz\| \leq \|x\| + \delta\|y\|\|z\|,$$
$$\leq \|x\|_\infty + \delta\|y\|_\infty\|z\|_\infty,$$
$$\leq M_x + \delta M_y M_z,$$
$$< \infty$$

$$\|f_z(t,x,y,z)\| = \|-\sigma|(sin(x))| - \beta y^2 - \varpi z - \mu\|,$$
$$\leq \sigma\|sin(x)\| + \beta\|y\|^2 + \varpi\|z\| + \mu,$$
$$\leq \sigma\|sin(x)\| + \beta\|y\|^2 + \varpi\|z\| + \mu,$$
$$\leq \sigma\|sin(x)\|_\infty + \beta\|y\|_\infty^2 + \varpi\|z\|_\infty + \mu,$$
$$\leq \sigma M_x^1 + \beta M_y + \varpi M_z + \mu,$$
$$\leq \infty.$$

Now, we show the Lipschitz conditions:
$$|f_x(t,x_1,y,z) - f_x(t,x_2,y,z)| \leq |x_1 - x_2|,$$
$$|f_y(t,x,y_1,z) - f_y(t,x,y_2,z)| \leq \delta M_z|y_1 - y_2|,$$
$$|f_z(t,x,y,z_1) - f_z(t,x,y,z_2)| \leq D|z_1 - z_2|.$$

The system is a contraction if $\{M_x\delta, D\} < 1$; under the above conditions, then the system considered admits a system of unique solution.

16.4 EQUILIBRIUM POINTS ANALYSIS

The equilibrium points associated with the above model can be obtained by setting

$$\begin{cases} 0 = -y, \\ 0 = x + \delta yz, \\ 0 = -\sigma|(sin(x))| - \beta y^2 - \varpi z - \mu; \end{cases}$$

we have $E_1 = (0, 0, -\frac{\sigma}{\varpi})$. In order to study the stability of the given system at E_1, we have to evaluate the Jacobian matrix of the model at E_1, given by

$$J(E_1) = \begin{pmatrix} 0 & -1 & 0 \\ 1 & \frac{\delta\sigma}{\varpi} & 0 \\ 0 & 0 & -\varpi \end{pmatrix}$$

which gives the eigenvalues $\lambda_1 = -\varpi$ and $\lambda_{2,3} = \frac{\delta\sigma}{2\varpi} \pm \frac{\sqrt{\delta^2\sigma^2 - 4\varpi^2}}{2\varpi}$. Obviously, $\lambda_1 < 0$ while $\lambda_{2,3}$ will lead about the stability of the model for $\delta^2\sigma^2 < 4\varpi^2$ will gives unstable equilibrium and if $\delta^2\sigma^2 > 4\varpi^2$ will give a saddle point.

16.5 NUMERICAL PROCEDURE FOR THE CHAOTIC MODEL

We present the numerical schemes for the solution of fractal-fractional model considered above, using the concept of linear interpolations. For each kernel, we have the following scheme.

16.6 NUMERICAL SCHEME FOR POWER LAW KERNEL

In this case, we consider the following fractal-fractional Cauchy problem:

$$_0^{FFP}D_t^{p,q}\mathbf{g}(t) = f(t, \mathbf{g}(t)), \text{ if } t > 0,$$

$$\mathbf{g}(0) = \mathbf{g}_0, \text{ if } t = 0. \tag{16.3}$$

It is assumed that $f(t, \mathbf{g}(t))$ is twice differentiable and that $\mathbf{g}(t)$ is at least continuous

$$_0^{RL}D_t^p\mathbf{g}(t) = qt^{q-1}f(t, \mathbf{g}(t)), \text{ if } t > 0,$$

$$\mathbf{g}(0) = \mathbf{g}_0, \text{ if } t = 0. \tag{16.4}$$

We write further,

$$\mathbf{g}(t) = \frac{q}{\Gamma(p)} \int_0^t \tau^{q-1} f(\tau, \mathbf{g}(\tau))(t-\tau)^{p-1} d\tau, \text{ if } t > 0;$$

$$\mathbf{g}(0) = \mathbf{g}_0, \text{ if } t = 0, \tag{16.5}$$

at $t = t_{n+1}$, we have

$$\mathbf{g}(t_{n+1}) = \frac{q}{\Gamma(p)} \int_0^{t_{n+1}} \tau^{q-1} f(\tau, \mathbf{g}(\tau))(t_{n+1}-\tau)^{p-1} d\tau. \tag{16.6}$$

Also, we write

$$\mathbf{g}(t_{n+1}) = \frac{q}{\Gamma(p)} \sum_{j=0}^n \int_{t_j}^{t_{j+1}} \tau^{q-1} f(\tau, \mathbf{g}(\tau))(t_{n+1}-\tau)^{p-1} d\tau. \tag{16.7}$$

Thus, within $[t_j, t_{j+1}]$, we approximate the function

$$f(\tau, \mathbf{g}(\tau)) \approx P_j(\tau) = f(t_j, \mathbf{g}_j) + (t - t_j)\frac{f(t_{j+1}, \mathbf{g}_{j+1}) - f(t_j, \mathbf{g}_j)}{\Delta t}, \quad (16.8)$$

replacing yields

$$\begin{aligned}
\mathbf{g}(t_{n+1}) \approx \mathbf{g}_{n+1} &= \frac{q}{\Gamma(p)} \sum_{j=0}^{n} \int_{t_j}^{t_{j+1}} \tau^{q-1}(t_{n+1} - \tau)^{p-1} \\
&\quad \times \left(f(t_j, \mathbf{g}_j) + (t - t_j)\frac{f(t_{j+1}, \mathbf{g}_{j+1}) - f(t_j, \mathbf{g}_j)}{\Delta t} \right) d\tau, \\
&= \frac{q}{\Gamma(p)} \sum_{j=0}^{n} f(t_j, \mathbf{g}_j) \int_{t_j}^{t_{j+1}} \tau^{q-1}(t_{n+1} - \tau)^{p-1} d\tau \\
&\quad + \frac{q}{\Gamma(p)} \sum_{j=0}^{n} \frac{f(t_{j+1}, \mathbf{g}_{j+1}) - f(t_j, \mathbf{g}_j)}{\Delta t} \\
&\quad \times \int_{t_j}^{t_{j+1}} (t_{n+1} - \tau)^{p-1}(t - t_j) d\tau, \quad (16.9)
\end{aligned}$$

noting that

$$\int_{t_j}^{t_{j+1}} \tau^{q-1}(t_{n+1} - \tau)^{p-1} d\tau = \int_{0}^{t_{j+1}} \tau^{q-1}(t_{n+1} - \tau)^{p-1} d\tau$$
$$- \int_{0}^{t_j} \tau^{q-1}(t_{n+1} - \tau)^{p-1} d\tau, \quad (16.10)$$

where

$$\begin{aligned}
\int_{0}^{t_{j+1}} \tau^{q-1}(t_{n+1} - \tau)^{p-1} d\tau &= t_{n+1}^{p+q-1} \int_{0}^{\frac{t_{j+1}}{t_{n+1}}} \mathbf{g}^{q-1}(1 - \mathbf{g})^p \mathbf{g}^{-1} d\mathbf{g}, \\
&= t_{n+1}^{p+q-1} B\left(\frac{t_{j+1}}{t_{n+1}}, q, p\right), \quad (16.11)
\end{aligned}$$

$$\int_{0}^{t_j} \tau^{q-1}(t_{n+1} - \tau)^{p-1} d\tau = t_{n+1}^{p+q-1} B\left(\frac{t_j}{t_{n+1}}, q, p\right). \quad (16.12)$$

Therefore,

$$\begin{aligned}
\int_{t_j}^{t_{j+1}} \tau^{q-1}(t_{n+1} - \tau)^{p-1}(\tau - t_j) d\tau &= \int_{t_j}^{t_{j+1}} \tau^q (t_{n+1} - \tau)^{p-1} d\tau \\
&\quad - t_j \int_{t_j}^{t_{j+1}} \tau^{q-1}(t_{n+1} - \tau)^{p-1} d\tau, \\
&= t_{n+1}^{p+q} \overbrace{B}^{q+1} - t_j \overbrace{B}^{q} t_{n+1}^{p+q-1},
\end{aligned}$$

where

$$\widehat{B}_{q+1} = \left\{ B\left(\frac{t_{j+1}}{t_{n+1}}, q+1, p\right) - B\left(\frac{t_j}{t_{n+1}}, q+1, p\right) \right\},$$

$$\widehat{B}_{q} = \left\{ B\left(\frac{t_{j+1}}{t_{n+1}}, q, p\right) - B\left(\frac{t_j}{t_{n+1}}, q, p\right) \right\}.$$

Replacing yields and further simplifications give

$$\begin{aligned}
\mathbf{g}_{n+1} = {} & \frac{q}{\Gamma(p)} (\Delta t)^{p+q-1} \sum_{j=0}^{n} f(t_j, \mathbf{g}_j)(n+1)^{p+q-1} \\
& \times \left\{ B\left(\frac{t_{j+1}}{t_{n+1}}, q, p\right) - B\left(\frac{t_j}{t_{n+1}}, q, p\right) \right\} \\
& + \frac{q}{\Gamma(p)} (\Delta t)^{p+q-1} \sum_{j=0}^{n-1} \left[f(t_{j+1}, \mathbf{g}_{j+1}) - f(t_j, \mathbf{g}_j) \right] \\
& \times \left\{ (n+1)^{p+q} B\left(\frac{t_{j+1}}{t_{n+1}}, q+1, p\right) - n^{p+q} B\left(\frac{t_j}{t_{n+1}}, q+1, p\right) \right. \\
& \left. - j \left[B\left(\frac{t_{j+1}}{t_{n+1}}, q, p\right) - B\left(\frac{t_j}{t_{n+1}}, q, p\right) \right] (n+1)^{p+q-1} \right\} \\
& + \frac{q(\Delta t)^{p+q-1}}{\Gamma(p)} \left[f(t_n + h, \mathbf{g}^u_{n+1}) - f(t_n, \mathbf{g}_n) \right] \\
& \times \left\{ (n+1)^{p+q} \left(B(1, q+1, p) - B\left(\frac{t_n}{t_{n+1}}, q+1, p\right) \right) \right. \\
& \left. - n \left(B(1, q, p) - B\left(\frac{t_n}{t_{n+1}}, q, p\right) \right) (n+1)^{p+q-1} \right\}
\end{aligned} \tag{16.13}$$

where

$$\mathbf{g}^u_{n+1} = \frac{q}{\Gamma(p+1)} (\Delta t)^{p+q-1} \sum_{j=0}^{n} f(t_j, \mathbf{g}_j)(n+1)^{p+q-1} \\
\times \left\{ B\left(\frac{t_{j+1}}{t_{n+1}}, q, p\right) - B\left(\frac{t_j}{t_{n+1}}, q, p\right) \right\}.$$

For our proposed system, the scheme follows

$$\begin{aligned}
x_{n+1} = {} & \frac{q}{\Gamma(p)} (\Delta t)^{p+q-1} \sum_{j=0}^{n} f_1(t_j, x_j, y_j, z_j)(n+1)^{p+q-1} \\
& \times \left\{ B\left(\frac{t_{j+1}}{t_{n+1}}, q, p\right) - B\left(\frac{t_j}{t_{n+1}}, q, p\right) \right\} + \frac{q}{\Gamma(p)} (\Delta t)^{p+q-1} \\
& \times \sum_{j=0}^{n-1} \left[f_1(t_{j+1}, x_{j+1}, y_{j+1}, z_{j+1}) \right. \\
& \left. - f_1(t_j, x_j, y_j, z_j) \right]
\end{aligned}$$

$$\times \left\{ (n+1)^{p+q} B\left(\frac{t_{j+1}}{t_{n+1}}, q+1, p\right) - n^{p+q} B\left(\frac{t_j}{t_{n+1}}, q+1, p\right) \right.$$
$$\left. -j\left[B\left(\frac{t_{j+1}}{t_{n+1}}, q, p\right) - B\left(\frac{t_j}{t_{n+1}}, q, p\right) \right](n+1)^{p+q-1} \right\}$$
$$+ \frac{q(\Delta t)^{p+q-1}}{\Gamma(p)} \left[f_1(t_n + h, x^u_{n+1}, y^u_{n+1}, z^u_{n+1}) \right.$$
$$\left. - f_1(t_n, x_n, y_n, z_n) \right]$$
$$\times \left\{ (n+1)^{p+q} \left(B(1, q+1, p) - B\left(\frac{t_n}{t_{n+1}}, q+1, p\right) \right) \right.$$
$$\left. -n \left(B(1, q, p) - B\left(\frac{t_n}{t_{n+1}}, q, p\right) \right)(n+1)^{p+q-1} \right\},$$

$$y_{n+1} = \frac{q}{\Gamma(p)} (\Delta t)^{p+q-1} \sum_{j=0}^{n} f_2(t_j, x_j, y_j, z_j)(n+1)^{p+q-1}$$
$$\times \left\{ B\left(\frac{t_{j+1}}{t_{n+1}}, q, p\right) - B\left(\frac{t_j}{t_{n+1}}, q, p\right) \right\} + \frac{q}{\Gamma(p)} (\Delta t)^{p+q-1}$$
$$\times \sum_{j=0}^{n-1} \left[f_2(t_{j+1}, x_{j+1}, y_{j+1}, z_{j+1}) \right.$$
$$\left. - f_2(t_j, x_j, y_j, z_j) \right]$$
$$\times \left\{ (n+1)^{p+q} B\left(\frac{t_{j+1}}{t_{n+1}}, q+1, p\right) - n^{p+q} B\left(\frac{t_j}{t_{n+1}}, q+1, p\right) \right.$$
$$\left. -j\left[B\left(\frac{t_{j+1}}{t_{n+1}}, q, p\right) - B\left(\frac{t_j}{t_{n+1}}, q, p\right) \right](n+1)^{p+q-1} \right\}$$
$$+ \frac{q(\Delta t)^{p+q-1}}{\Gamma(p)} \left[f_2(t_n + h, x^u_{n+1}, y^u_{n+1}, z^u_{n+1}) \right.$$
$$\left. - f_2(t_n, x_n, y_n, z_n) \right]$$
$$\times \left\{ (n+1)^{p+q} \left(B(1, q+1, p) - B\left(\frac{t_n}{t_{n+1}}, q+1, p\right) \right) \right.$$
$$\left. -n \left(B(1, q, p) - B\left(\frac{t_n}{t_{n+1}}, q, p\right) \right)(n+1)^{p+q-1} \right\},$$

$$z_{n+1} = \frac{q}{\Gamma(p)} (\Delta t)^{p+q-1} \sum_{j=0}^{n} f_3(t_j, x_j, y_j, z_j)(n+1)^{p+q-1}$$
$$\times \left\{ B\left(\frac{t_{j+1}}{t_{n+1}}, q, p\right) - B\left(\frac{t_j}{t_{n+1}}, q, p\right) \right\} + \frac{q}{\Gamma(p)} (\Delta t)^{p+q-1}$$
$$\times \sum_{j=0}^{n-1} \left[f_3(t_{j+1}, x_{j+1}, y_{j+1}, z_{j+1}) \right.$$
$$\left. - f_3(t_j, x_j, y_j, z_j) \right]$$

$$\times \left\{ (n+1)^{p+q} B\left(\frac{t_{j+1}}{t_{n+1}}, q+1, p\right) - n^{p+q} B\left(\frac{t_j}{t_{n+1}}, q+1, p\right) \right.$$
$$\left. - j\left[B\left(\frac{t_{j+1}}{t_{n+1}}, q, p\right) - B\left(\frac{t_j}{t_{n+1}}, q, p\right) \right](n+1)^{p+q-1} \right\}$$
$$+ \frac{q(\Delta t)^{p+q-1}}{\Gamma(p)} \left[f_3(t_n + h, x^u_{n+1}, y^u_{n+1}, z^u_{n+1}) \right.$$
$$\left. - f_3(t_n, x_n, y_n, z_n) \right]$$
$$\times \left\{ (n+1)^{p+q} \left(B(1, q+1, p) - B\left(\frac{t_n}{t_{n+1}}, q+1, p\right) \right) \right.$$
$$\left. - n \left(B(1, q, p) - B\left(\frac{t_n}{t_{n+1}}, q, p\right) \right)(n+1)^{p+q-1} \right\}, \tag{16.14}$$

where

$$x^u_{n+1} = \frac{q}{\Gamma(p+1)} (\Delta t)^{p+q-1} \sum_{j=0}^{n} f_1(t_j, x_j, y_j, z_j)$$
$$\times (n+1)^{p+q-1} \left\{ B\left(\frac{t_{j+1}}{t_{n+1}}, q, p\right) - B\left(\frac{t_j}{t_{n+1}}, q, p\right) \right\},$$
$$y^u_{n+1} = \frac{q}{\Gamma(p+1)} (\Delta t)^{p+q-1} \sum_{j=0}^{n} f_2(t_j, x_j, y_j, z_j)$$
$$\times (n+1)^{p+q-1} \left\{ B\left(\frac{t_{j+1}}{t_{n+1}}, q, p\right) - B\left(\frac{t_j}{t_{n+1}}, q, p\right) \right\},$$
$$z^u_{n+1} = \frac{q}{\Gamma(p+1)} (\Delta t)^{p+q-1} \sum_{j=0}^{n} f_3(t_j, x_j, y_j, z_j)$$
$$\times (n+1)^{p+q-1} \left\{ B\left(\frac{t_{j+1}}{t_{n+1}}, q, p\right) - B\left(\frac{t_j}{t_{n+1}}, q, p\right) \right\},$$

where

$$f_1(x, y, z, t) = -y,$$
$$f_2(x, y, z, t) = x + \delta y z,$$
$$f_3(x, y, z, t) = -\sigma|(\sin(x))| - \beta y^2 - \varpi z - \mu.$$

The scheme presented above has been applied successfully to the proposed fractal-fractional model in the sense of Caputo operator by considering the different values of the fractal-fractional order parameters. Some suitable results are presented in Figures 16.1–16.2.

Self-Excited and Hidden Attractors through Fractal-Fractional Operators

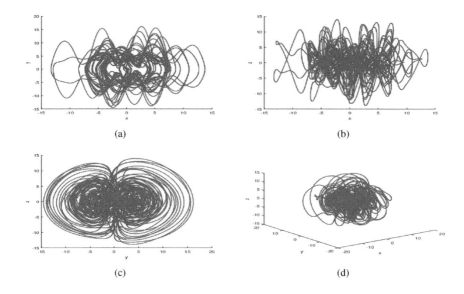

Figure 16.1 Model simulation of fractal-fractional Caputo model when $p = 1, q = 1$.

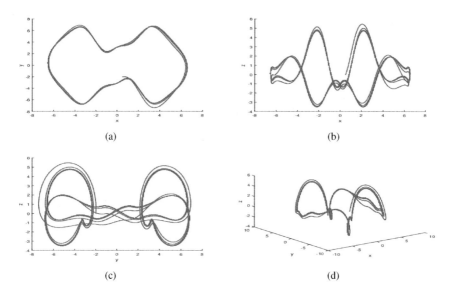

Figure 16.2 Model simulation of fractal-fractional Caputo model when $p = 0.96$ and $q = 0.99$.

16.6.1 NUMERICAL SCHEME FOR EXPONENTIAL DECAY KERNEL USING LINEAR INTERPOLATIONS

In this subsection, we consider the fractal-fractional Cauchy problem with exponential kernel, given by

$$_0^{FFE}D_t^{p,q}\mathbf{g}(t) = f(t,\mathbf{g}(t)), \text{ if } t > 0,$$

$$\mathbf{g}(0) = \mathbf{g}_0, \text{ if } t = 0,$$

where $\mathbf{g}(t) = (x(t), y(t), z(t))$ are the equations of the model proposed.

$$_0^{CF}D_t^p \mathbf{g}(t) = qt^{q-1}f(t, \mathbf{g}(t)), \text{ if } t > 0,$$

$$\mathbf{g}(0) = \mathbf{g}_0, \text{ if } t = 0.$$

$$\mathbf{g}(t) = (1-p)qt^{q-1}f(t,\mathbf{g}(t)) + pq\int_0^t \tau^{q-1}f(\tau,\mathbf{g}(\tau))d\tau, \quad y(0) = y_0;$$

at $t = t_{n+1}$, we have

$$\mathbf{g}(t_{n+1}) = (1-p)qt_{n+1}^{q-1}f(t_{n+1},\mathbf{g}_{n+1}) + pq\int_0^{t_{n+1}} \tau^{q-1}f(\tau,\mathbf{g}(\tau))d\tau;$$

at $t = t_n$, we have

$$\mathbf{g}(t_n) = (1-p)qt_n^{q-1}f(t_n,\mathbf{g}_n) + pq\int_0^{t_n} \tau^{q-1}f(\tau,\mathbf{g}(\tau))d\tau.$$

Therefore,

$$\mathbf{g}(t_{n+1}) = \mathbf{g}(t_n) + (1-p)q\left[t_{n+1}^{q-1}f(t_{n+1},\mathbf{g}_{n+1}) - t_n^{q-1}f(t_n,\mathbf{g}(t_n))\right]$$
$$+ pq\int_{t_n}^{t_{n+1}} \tau^{q-1}f(\tau,\mathbf{g}(\tau))d\tau.$$

Using the polynomial approximations of $f(\tau, \mathbf{g}(\tau))$ within $[t_n, t_{n+1}]$ yields,

$$\mathbf{g}_{n+1} = \mathbf{g}_n + (1-p)q\left[t_{n+1}^{q-1}f(t_{n+1},\mathbf{g}_{n+1}^u) - t_n^{q-1}f(t_n,\mathbf{g}_n)\right]$$
$$+ p\left[f(t_n,\mathbf{g}_n)(t_{n+1}^q - t_n^q) + \frac{f(t_{n+1},\mathbf{g}_{n+1}^u) - f(t_n,\mathbf{g}_n)}{\Delta t}\right.$$
$$\left.\left\{\frac{q}{q+1}(t_{n+1}^{q+1} - t_n^{q+1}) - t_n(t_{n+1}^q - t_n^q)\right\}\right]. \tag{16.15}$$

Further, we have

$$\mathbf{g}_{n+1} = \mathbf{g}_n + (1-p)q(\Delta t)^{q-1}$$
$$\left\{(n+1)^{q-1}f(t_n+h,\mathbf{g}_{n+1}^u) - n^{q-1}f(t_n,\mathbf{g}_n)\right\}$$
$$+ p(\Delta t)^q \left\{f(t_n,\mathbf{g}_n)((n+1)^q - n^q)\right.$$
$$\left. + (f(t_n+h,\mathbf{g}_{n+1}^u) - f(t_n,\mathbf{g}_n))\right\}$$

$$\times \left\{ \frac{q}{q+1}[(n+1)^{q+1} - n^{q+1}] - n\left((n+1)^q - n^q\right) \right\},$$

where

$$\mathbf{g}_{n+1}^u = \mathbf{g}_0 + (1-p)q(\Delta t)^{q-1}n^{q-1}f(t_n, \mathbf{g}_n)$$

$$+ p\sum_{j=0}^{n} f(t_j, \mathbf{g}_j)\left[\frac{q}{q+1}[t_{j+1}^{q+1} - t_j^{q+1}] - t_j(t_{j+1}^q - t_j^q)\right].$$

For our system, the scheme can be written as

$$x_{n+1} = x_n + (1-p)q(\Delta t)^{q-1}$$
$$\left\{(n+1)^{q-1}f_1(t_n + h, x_{n+1}^u, y_{n+1}^u, z_{n+1}^u)\right.$$
$$\left. - n^{q-1}f_1(t_n, x_n, y_n, z_n)\right\}$$
$$+ p(\Delta t)^q \left\{ f_1(t_n, x_n, y_n, z_n)((n+1)^q - n^q) \right.$$
$$\left. + (f_1(t_n + h, x_{n+1}^u, y_{n+1}^u, z_{n+1}^u) - f_1(t_n, x_n, y_n, z_n)) \right\}$$
$$\times \left\{ \frac{q}{q+1}[(n+1)^{q+1} - n^{q+1}] - n\left((n+1)^q - n^q\right) \right\},$$

$$y_{n+1} = y_n + (1-p)q(\Delta t)^{q-1}$$
$$\left\{(n+1)^{q-1}f_2(t_n + h, x_{n+1}^u, y_{n+1}^u, z_{n+1}^u)\right.$$
$$\left. - n^{q-1}f_2(t_n, x_n, y_n, z_n)\right\}$$
$$+ p(\Delta t)^q \left\{ f_2(t_n, x_n, y_n, z_n)((n+1)^q - n^q) \right.$$
$$\left. + (f_2(t_n + h, x_{n+1}^u, y_{n+1}^u, z_{n+1}^u) - f_2(t_n, x_n, y_n, z_n)) \right\}$$
$$\times \left\{ \frac{q}{q+1}[(n+1)^{q+1} - n^{q+1}] - n\left((n+1)^q - n^q\right) \right\},$$

$$z_{n+1} = z_n + (1-p)q(\Delta t)^{q-1}$$
$$\left\{(n+1)^{q-1}f_3(t_n + h, x_{n+1}^u, y_{n+1}^u, z_{n+1}^u)\right.$$
$$\left. - n^{q-1}f_3(t_n, x_n, y_n, z_n)\right\}$$
$$+ p(\Delta t)^q \left\{ f_3(t_n, x_n, y_n, z_n)((n+1)^q - n^q) \right.$$
$$\left. + (f_3(t_n + h, x_{n+1}^u, y_{n+1}^u, z_{n+1}^u) - f_3(t_n, x_n, y_n, z_n)) \right\}$$

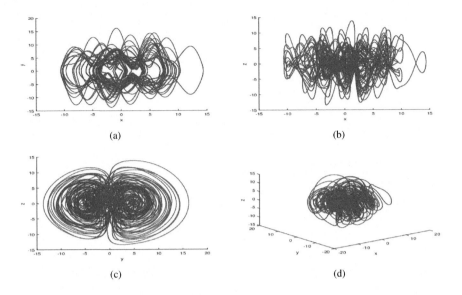

Figure 16.3 Model simulation of fractal-fractional Caputo-Fabrizio model when $p = 1$ and $q = 1$.

$$\times \left\{ \frac{q}{q+1}[(n+1)^{q+1} - n^{q+1}] - n\left((n+1)^q - n^q\right) \right\},$$

where

$$x_{n+1}^u = x_0 + (1-p)q(\Delta t)^{q-1}n^{q-1}f_1(t_n, x_n, y_n, z_n)$$
$$+ p\sum_{j=0}^{n} f_1(t_j, x_j, y_j, z_j)\left[\frac{q}{q+1}[t_{j+1}^{q+1} - t_j^{q+1}] - t_j(t_{j+1}^q - t_j^q)\right],$$

$$y_{n+1}^u = y_0 + (1-p)q(\Delta t)^{q-1}n^{q-1}f_2(t_n, x_n, y_n, z_n)$$
$$+ p\sum_{j=0}^{n} f_2(t_j, x_j, y_j, z_j)\left[\frac{q}{q+1}[t_{j+1}^{q+1} - t_j^{q+1}] - t_j(t_{j+1}^q - t_j^q)\right],$$

$$z_{n+1}^u = z_0 + (1-p)q(\Delta t)^{q-1}n^{q-1}f_3(t_n, x_n, y_n, z_n)$$
$$+ p\sum_{j=0}^{n} f_3(t_j, x_j, y_j, z_j)\left[\frac{q}{q+1}[t_{j+1}^{q+1} - t_j^{q+1}] - t_j(t_{j+1}^q - t_j^q)\right].$$

The graphical results shown in Figures 16.3 and 16.4 are obtained for the fractal-fractional model in the sense of Caputo-Fabrizio operator considering many values of the fractal and fractional order parameters. These graphs are some selected graphs although there exist many graphical results for this model with different order of fractal and fractional.

Self-Excited and Hidden Attractors through Fractal-Fractional Operators

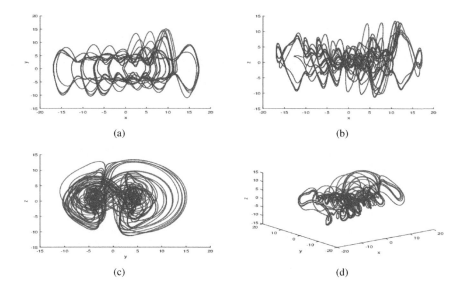

Figure 16.4 Model simulation of fractal-fractional Caputo-Fabrizio model when $p = 0.98$ and $q = 0.96$.

16.6.2 NUMERICAL SCHEME FOR GENERALISED MITTAG-LEFFLER KERNEL USING LINEAR INTERPOLATIONS

We consider in this section a fractal-fractional Cauchy problem with the generalised Mittag-Leffler kernel:

$$_{0}^{FFM}D_{t}^{p,q}\mathbf{g}(t) = f(t,\mathbf{g}(t)), \text{ if } t > 0,$$

$$\mathbf{g}(0) = \mathbf{g}_0, \text{ if } t = 0.$$

We have for AB-RL case

$$_{0}^{ABR}D_{t}^{p}\mathbf{g}(t) = qt^{q-1}f(t,\mathbf{g}(t)), \text{ if } t > 0,$$

$$\mathbf{g}(0) = \mathbf{g}_0, \text{ if } t = 0.$$

$$\mathbf{g}(t) = (1-p)qt^{q-1}f(t,\mathbf{g}(t))$$
$$+ \frac{pq}{\Gamma(p)}\int_{0}^{t}\tau^{q-1}(t-\tau)^{p-1}f(\tau,\mathbf{g}(\tau))d\tau, \ \mathbf{g}(0) = \mathbf{g}_0.$$

At $t = t_{n+1}$, we have

$$\mathbf{g}(t_{n+1}) = (1-p)qt_{n+1}^{q-1}f(t_{n+1},\mathbf{g}(t_{n+1}))$$

$$+\frac{pq}{\Gamma(p)}\int_0^{t_{n+1}} \tau^{q-1}(t_{n+1}-\tau)^{p-1}f(\tau,\mathbf{g}(\tau))d\tau.$$

$$\mathbf{g}(t_{n+1}) = (1-p)qt_{n+1}^{q-1}f(t_{n+1},\mathbf{g}(t_{n+1}))$$
$$+\frac{pq}{\Gamma(p)}\sum_{j=0}^{n}\int_{t_j}^{t_{j+1}}(t_{n+1}-\tau)^{p-1}\tau^{q-1}f(\tau,\mathbf{g}(\tau))d\tau.$$

Approximating the function $f(\tau,\mathbf{g}(\tau))$ by polynomial approximations, we have

$$\mathbf{g}_{n+1} = qt_{n+1}^q(1-p)f(t_{n+1},\mathbf{g}_{n+1}^u)$$
$$+\frac{pq}{\Gamma(p)}\sum_{j=0}^{n}f(t_j,\mathbf{g}_j)\left\{t_{n+1}^{p+q-1}\left[B\left(\frac{t_{j+1}}{t_{n+1}},q,p\right)\right.\right.$$
$$\left.\left.-B\left(\frac{t_j}{t_{n+1}},q,p\right)\right]\right\}$$
$$+\frac{pq}{\Gamma(p)}\sum_{j=0}^{n-1}\left(\frac{f(t_{j+1},\mathbf{g}_{j+1})-f(t_j,\mathbf{g}_j)}{\Delta t}\right)$$
$$\left\{t_{n+1}^{p+q}\left[B\left(\frac{t_{j+1}}{t_{n+1}},q+1,p\right)-B\left(\frac{t_j}{t_{n+1}},q+1,p\right)\right]\right.$$
$$\left.-t_j t_{n+1}^{p+q-1}\left[B\left(\frac{t_{j+1}}{t_{n+1}},q,p\right)-B\left(\frac{t_j}{t_{n+1}},q,p\right)\right]\right\}$$
$$+\frac{pq}{\Gamma(p)}\left(\frac{f(t_n+h,\mathbf{g}_{n+1}^u)-f(t_n,\mathbf{g}_n)}{\Delta t}\right)$$
$$\times (\Delta t)^{p+q}\left\{(n+1)^{p+q}\{B(1,q+1,p)-B(\frac{t_n}{t_{n+1}},q+1,p)\}\right.$$
$$\left.-n(n+1)^{p+q-1}\{B(1,q,p)-B(\frac{t_n}{t_{n+1}},q,p)\}\right\} \quad (16.16)$$

where

$$\mathbf{g}_{n+1}^u = (1-p)qf(t_n,\mathbf{g}_n)$$
$$+\frac{pq}{\Gamma(p)}\sum_{j=0}^{n}f(t_j,\mathbf{g}_j)(n+1)^{p+q-1}$$
$$\times\left[B\left(\frac{t_{j+1}}{t_{n+1}},q,p\right)-B\left(\frac{t_j}{t_{n+1}},q,p\right)\right]. \quad (16.17)$$

The scheme for the case of Mittag-Leffler kernel can be applied to our system of equations as follows:

$$x_{n+1} = qt_{n+1}^q(1-p)f_1(t_{n+1},x_{n+1}^u,y_{n+1}^u,z_{n+1}^u)$$

Self-Excited and Hidden Attractors through Fractal-Fractional Operators

$$+ \frac{pq}{\Gamma(p)} \sum_{j=0}^{n} f_1(t_j, x_j, y_j, z_j) \left\{ t_{n+1}^{p+q-1} \left[B\left(\frac{t_{j+1}}{t_{n+1}}, q, p\right) \right. \right.$$

$$\left. \left. - B\left(\frac{t_j}{t_{n+1}}, q, p\right) \right] \right\}$$

$$+ \frac{pq}{\Gamma(p)} \sum_{j=0}^{n-1} \left(\frac{f_1(t_{j+1}, x_{j+1}, y_{j+1}, z_{j+1}) - f_1(t_j, x_j, y_j, z_j)}{\Delta t} \right)$$

$$\left\{ t_{n+1}^{p+q} \left[B\left(\frac{t_{j+1}}{t_{n+1}}, q+1, p\right) - B\left(\frac{t_j}{t_{n+1}}, q+1, p\right) \right] \right.$$

$$\left. - t_j t_{n+1}^{p+q-1} \left[B\left(\frac{t_{j+1}}{t_{n+1}}, q, p\right) - B\left(\frac{t_j}{t_{n+1}}, q, p\right) \right] \right\}$$

$$+ \frac{pq}{\Gamma(p)} \left(\frac{f_1(t_n + h, x_{n+1}^u, y_{n+1}^u, z_{n+1}^u) - f_1(t_n, x_n, y_n, z_n)}{\Delta t} \right)$$

$$\times (\Delta t)^{p+q} \left\{ (n+1)^{p+q} \{ B(1, q+1, p) - B(\frac{t_n}{t_{n+1}}, q+1, p) \} \right.$$

$$\left. - n(n+1)^{p+q-1} \{ B(1, q, p) - B(\frac{t_n}{t_{n+1}}, q, p) \} \right\},$$

$$y_{n+1} = q t_{n+1}^q (1-p) f_2(t_{n+1}, x_{n+1}^u, y_{n+1}^u, z_{n+1}^u)$$

$$+ \frac{pq}{\Gamma(p)} \sum_{j=0}^{n} f_2(t_j, x_j, y_j, z_j) \left\{ t_{n+1}^{p+q-1} \left[B\left(\frac{t_{j+1}}{t_{n+1}}, q, p\right) \right. \right.$$

$$\left. \left. - B\left(\frac{t_j}{t_{n+1}}, q, p\right) \right] \right\}$$

$$+ \frac{pq}{\Gamma(p)} \sum_{j=0}^{n-1} \left(\frac{f_2(t_{j+1}, x_{j+1}, y_{j+1}, z_{j+1}) - f_2(t_j, x_j, y_j, z_j)}{\Delta t} \right)$$

$$\left\{ t_{n+1}^{p+q} \left[B\left(\frac{t_{j+1}}{t_{n+1}}, q+1, p\right) - B\left(\frac{t_j}{t_{n+1}}, q+1, p\right) \right] \right.$$

$$\left. - t_j t_{n+1}^{p+q-1} \left[B\left(\frac{t_{j+1}}{t_{n+1}}, q, p\right) - B\left(\frac{t_j}{t_{n+1}}, q, p\right) \right] \right\}$$

$$+ \frac{pq}{\Gamma(p)} \left(\frac{f_2(t_n + h, x_{n+1}^u, y_{n+1}^u, z_{n+1}^u) - f_2(t_n, x_n, y_n, z_n)}{\Delta t} \right)$$

$$\times (\Delta t)^{p+q} \left\{ (n+1)^{p+q} \{ B(1, q+1, p) - B(\frac{t_n}{t_{n+1}}, q+1, p) \} \right.$$

$$\left. - n(n+1)^{p+q-1} \{ B(1, q, p) - B(\frac{t_n}{t_{n+1}}, q, p) \} \right\},$$

$$z_{n+1} = q t_{n+1}^q (1-p) f_3(t_{n+1}, x_{n+1}^u, y_{n+1}^u, z_{n+1}^u)$$

$$+ \frac{pq}{\Gamma(p)} \sum_{j=0}^{n} f_3(t_j, x_j, y_j, z_j) \left\{ t_{n+1}^{p+q-1} \left[B\left(\frac{t_{j+1}}{t_{n+1}}, q, p\right) \right. \right.$$

$$\left. \left. - B\left(\frac{t_j}{t_{n+1}}, q, p\right) \right] \right\}$$

$$+\frac{pq}{\Gamma(p)}\sum_{j=0}^{n-1}\left(\frac{f_3(t_{j+1},x_{j+1},y_{j+1},z_{j+1})-f_2(t_j,x_j,y_j,z_j)}{\Delta t}\right)$$
$$\left\{t_{n+1}^{p+q}\left[B\left(\frac{t_{j+1}}{t_{n+1}},q+1,p\right)-B\left(\frac{t_j}{t_{n+1}},q+1,p\right)\right]\right.$$
$$\left.-t_j t_{n+1}^{p+q-1}\left[B\left(\frac{t_{j+1}}{t_{n+1}},q,p\right)-B\left(\frac{t_j}{t_{n+1}},q,p\right)\right]\right\}$$
$$+\frac{pq}{\Gamma(p)}\left(\frac{f_3(t_n+h,x_{n+1}^u,y_{n+1}^u,z_{n+1}^u)-f_3(t_n,x_n,y_n,z_n)}{\Delta t}\right)$$
$$\times(\Delta t)^{p+q}\left\{(n+1)^{p+q}\{B(1,q+1,p)-B(\frac{t_n}{t_{n+1}},q+1,p)\}\right.$$
$$\left.-n(n+1)^{p+q-1}\{B(1,q,p)-B(\frac{t_n}{t_{n+1}},q,p)\}\right\},$$

where

$$x_{n+1}^u = (1-p)qf_1(t_n,x_n,y_n,z_n)$$
$$+\frac{pq}{\Gamma(p)}\sum_{j=0}^{n}f_1(t_j,x_j,y_j,z_j)(n+1)^{p+q-1}$$
$$\times\left[B\left(\frac{t_{j+1}}{t_{n+1}},q,p\right)-B\left(\frac{t_j}{t_{n+1}},q,p\right)\right],$$
$$y_{n+1}^u = (1-p)qf_2(t_n,x_n,y_n,z_n)$$
$$+\frac{pq}{\Gamma(p)}\sum_{j=0}^{n}f_2(t_j,x_j,y_j,z_j)(n+1)^{p+q-1}$$
$$\times\left[B\left(\frac{t_{j+1}}{t_{n+1}},q,p\right)-B\left(\frac{t_j}{t_{n+1}},q,p\right)\right],$$
$$z_{n+1}^u = (1-p)qf_3(t_n,x_n,y_n,z_n)$$
$$+\frac{pq}{\Gamma(p)}\sum_{j=0}^{n}f_3(t_j,x_j,y_j,z_j)(n+1)^{p+q-1}$$
$$\times\left[B\left(\frac{t_{j+1}}{t_{n+1}},q,p\right)-B\left(\frac{t_j}{t_{n+1}},q,p\right)\right]. \tag{16.18}$$

The scheme described above is used to obtain the graphical results for the model in the sense of AB derivative. We use various values of the fractal and fractional order parameters and present graphical results for the fractal-fractional model in the sense of Atangana-Baleanu derivative; see (16.5–16.7).

16.7 CONCLUSION

We discussed a self-cited chaotic attractors model based on fractal-fractional differential equations. We presented the mathematical results for the model. We presented

Self-Excited and Hidden Attractors through Fractal-Fractional Operators 253

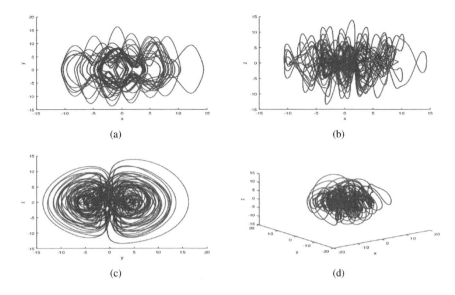

Figure 16.5 Model simulation of fractal-fractional Atangana-Baleanu model when $p = 1$ and $q = 1$.

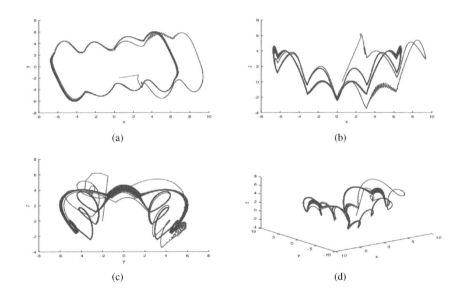

Figure 16.6 Model simulation of fractal-fractional Atangana-Baleanu model when $p = 0.8$ and $q = 0.99$.

the fractal-fractional model using three different concepts of operators, and for each operator, a numerical scheme was provided. The numerical schemes based on linear

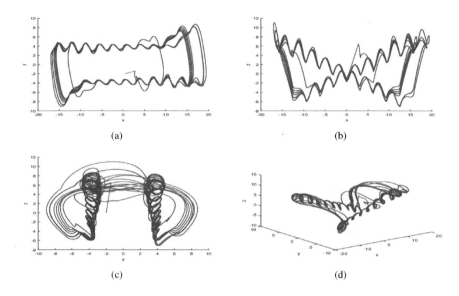

Figure 16.7 Model simulation of fractal-fractional Atangana-Baleanu model when $p = 0.8$ and $q = 0.9$.

interpolations for the power law, exponential law, and Mittag-Leffler have been used, and we obtained the numerical results graphically.

17 Dynamical Analysis of a Chaotic Model in Fractal-Fractional Operators

17.1 INTRODUCTION

Non-linear chaotic systems have complex dynamic behaviour and are sensitive to the beginning conditions. A chaotic attractor is further described as a chaotic set that a dynamic system likes to grow towards. Numerous engineering and non-engineering domains, including intelligent controls, power systems, secure communication, biology, and mathematics, use chaotic systems in a variety of ways [91, 191, 266]. In addition, Lorenz found chaotic attractors while researching atmospheric convection, and in 1963, he presented the first three-dimensional chaotic system. Rossler continued his study on the dissipative dynamical system after that, and in 1976, he suggested a new chaotic system. Since then, more research has been done; for instance, in 1999 Chen developed a new three-dimensional attractor that is not topologically identical to the Lorenz system [43, 61, 62, 90, 123–128, 173, 174, 188, 189, 192, 213]. Furthermore, there has recently been a lot of interest in researching hidden attractors in areas where the basin of attraction does not meet with the tiny neighbourhood of equilibria [52, 98, 99, 219]. The field of chaos theory and its applications would benefit greatly by the suggestion of new chaotic attractors with novel shapes and dynamics. This study introduces a novel four-dimensional chaotic system that consists primarily of four simple terms and four multiplier terms. For instance, one proposal suggests a fresh four-dimensional system with a different structure and topology from those already in use. Equilibrium points, their stabilities, bifurcation diagrams, Poincare maps, and power spectra are used to analyse the fundamental properties of the new system, demonstrating how the four-dimensional chaotic system opened up new avenues for study within the framework of chaotic theories [129, 171, 223, 224, 237, 259, 270], with [101, 121, 214] and [26–29]. In this chapter, we take a look at a set of four non-linear differential equations that can simulate chaotic behaviour. The present chapter investigates the dynamical analysis of a chaotic model under fractal-fractional operators. Appropriate initial and parameter values are suggested for which the model becomes chaotic, and then we have shown some related results of the model. We give the existence and uniqueness results for the model. Further, we give new numerical schemes based on middle-point interpolations for each kernel, such as power, exponential and Mittag-Leffler law. Some

DOI: 10.1201/9781003359258-17

simulation results for different fractal-fractional orders are given for each numerical scheme.

17.2 MODEL DESCRIPTIONS

The present section describes the formulation and the results involved in the chaotic model. The chaotic model we consider here is given by the following equations [36]:

$$\begin{cases} \frac{dx(t)}{dt} = y, \\ \frac{dy(t)}{dt} = z, \\ \frac{dz(t)}{dt} = z + ayw - zw, \\ \frac{dw(t)}{dt} = xy + byz, \end{cases} \quad (17.1)$$

where $x(t)$, $y(t)$, $z(t)$, and $w(t)$ are the state variables, while a and b are the two parameters. This system is chaotic for the given values of the parameters $a = -3$ and $b = 1$ with the initial conditions on state variables $x(0) = 2.5$, $y(0) = 0.1$, $z(0) = -1$, and $w(0) = -0.4$. The fractal-fractional representation of the model (17.1) is given by

$$\begin{cases} {}^{FF}D_{0,t}^{p,\,q}\bigl(x(t)\bigr) = y, \\ {}^{FF}D_{0,t}^{p,\,q}\bigl(y(t)\bigr) = z, \\ {}^{FF}D_{0,t}^{p,\,q}\bigl(z(t)\bigr) = z + ayw - zw, \\ {}^{FF}D_{0,t}^{p,\,q}\bigl(w(t)\bigr) = xy + byz, \end{cases} \quad (17.2)$$

where p represents the fractional order, while q is the fractal order.

17.3 EXISTENCE AND UNIQUENESS

In the present section, we aim to study the existence and uniqueness of the model (17.2). It is assumed that for all $t \in [0,T]$, the functions $x(t)$, $y(t)$, $z(t)$, and $w(t)$ are bounded, $\|x\|_\infty \le M_x$, $\|y\|_\infty \le M_y$, $\|z\|_\infty \le M_z$, and $\|w\|_\infty \le M_w$. For simplicity, we let

$$\begin{aligned} f_x(t,x,y,z,w) &= y, \\ f_y(t,x,y,z,w) &= z, \\ f_z(t,x,y,z,w) &= z + ayw - zw, \\ f_w(t,x,y,z,w) &= xy + byz. \end{aligned}$$

We show that the functions, f_x, f_y, f_z, and f_w satisfy the linear growth and Lipschitz conditions. We first show that the functions satisfy the linear growth property.

$$\begin{aligned} |f_x(t,x,y,z,w)| &= |y| \le \sup_{t \in D_y} |y(t)| = M_y < \infty, \\ |f_y(t,x,y,z,w)| &= |z| \le \sup_{t \in D_z} |z(t)| = M_z < \infty, \end{aligned}$$

Dynamical Analysis of a Chaotic Model in Fractal-Fractional Operators

$$|f_z(t,x,y,z,w)| = |z+ayw-zw|,$$
$$\leq |z|+a|y|+|z||w|,$$

$M_z + aM_yM_w + M_w|z|,$

$$\leq (M_z + aM_yM_w)\left(1 + \frac{M_w}{M_z + aM_yM_w}|Z|\right),$$
$$\leq K_z^1(1+|z|),$$

if $\frac{M_w}{M_z+aM_yM_w} < 1$. But

$$|f_z(t,x,y,z,w)| = |z+ayw-zw| \leq M_z + aM_yM_w + M_wM_z < \infty.$$

$$|f_w(t,x,y,z,w)| = |xy+byz|,$$
$$\leq 2M_xM_y + bM_yM_z < \infty.$$

But

$$|f_x(t,x,y,z,w)|^2 = |y|^2 \leq M_y^2(1+|x|^2),$$
$$|f_y(t,x,y,z,w)|^2 = |z|^2 \leq M_z^2(1+|y|^2),$$
$$|f_z(t,x,y,z,w)|^2 = |z+ayw-zw|^2,$$
$$\leq 3|z|^2 + 3a^2|y|^2|w|^2 + 3|z|^2|w|^2,$$
$$\leq 3|z|^2 + 3a^2M_y^2M_w^2 + 3|z|^2M_w^2,$$
$$\leq 3a^2M_y^2M_w^2\left(1 + \frac{1+M_w^2}{a^2M_y^2M_w^2}|z|^2\right),$$
$$\leq K_z(1+|z|^2),$$

if $K_z = 3a^2M_y^2M_w^2$ and $frac1+M_w^2 a^2M_y^2M_w^2 < 1$.

$$|f_w(t,x,y,z,w)|^2 = |xy+byz|^2,$$
$$\leq 2|x|^2|y|^2 + 2b^2|y|^2||z|^2,$$
$$\leq 2M_x^2M_y^2 + 2b^2M_y^2M_z^2(1+|w|^2),$$
$$\leq K_w(1+|w|^2),$$

where $K_w = 2M_x^2 M_y^2 + 2b^2 M_y^2 M_z^2$. Indeed the system satisfies the Lipschitz conditions; therefore, under the conditions that $\frac{1+M_w^2}{a^2 M_y^2 M_w^2} < 1$, then the system admits a unique system of solutions.

17.3.1 MODEL ANALYSIS

The above system has the following equilibrium point, $E = (0,0,0,w^*)$. The Jacobian matrix associated with the model (17.2) is given by

$$J = \begin{pmatrix} 0 & 1 & 0 & 0 \\ 0 & 0 & 1 & 0 \\ 0 & aw & 1-w & ay-z \\ y & x+bz & by & 0 \end{pmatrix};$$

when $a = -1$ and $b = 1$, the Jacobian matrix becomes

$$J = \begin{pmatrix} 0 & 1 & 0 & 0 \\ 0 & 0 & 1 & 0 \\ 0 & -w & 1-w & 0 \\ 0 & x & 0 & 0 \end{pmatrix}.$$

We obtain a set of eigenvalues based on w, which can be seen in detail in [36]; there exist three stable saddle focuses for $w = 1.1, 2.25,$ and 5.23 while for $w = 0.85$ and 0.28 there exist two unstable saddle focuses. To study the numerical solution and identify more interesting graphical behaviour of the proposed system (17.2), we provide the effective numerical algorithms and its solution in the coming sections.

17.4 NUMERICAL SCHEMES BASED ON MIDDLE-POINT INTERPOLATIONS

The aim of this section is to present the new numerical schemes based on middle-point interpolation method for the fractal-fractional differential equations using power, exponential, and Mittag-Leffler kernels. We explain the procedure for each kernel and present some graphical results for illustrations. We begin with the power law case.

17.4.1 NUMERICAL SCHEME FOR POWER LAW CASE

In this subsection, we consider the fractal-fractional Cauchy problem for the power law case,

$$\begin{aligned} {}_0^{FFP}D_t^{p,q}\mathbf{g}(t) &= f(t,\mathbf{g}(t)), \text{ if } t > 0, \\ \mathbf{g}(0) &= \mathbf{g}_0, \text{ if } t = 0, \end{aligned}$$

where $\mathbf{g}(t) = (x(t), y(t), z(t), w(t))$ defines the equations of the model. We can write further

$$_0^{RL}D_t^p \mathbf{g}(t) = qt^{q-1} f(t, \mathbf{g}(t)), \quad \text{if } t > 0,$$

$$\mathbf{g}(0) = \mathbf{g}_0, \quad \text{if } t = 0.$$

Further, we write

$$\mathbf{g}(t) = \frac{q}{\Gamma(p)} \int_0^t \tau^{q-1} f(\tau, \mathbf{g}(\tau))(t - \tau)^{p-1} d\tau,$$

$$\mathbf{g}(0) = \mathbf{g}_0.$$

At $t = t_{n+1}$, we have

$$\mathbf{g}(t_{n+1}) = \frac{q}{\Gamma(p)} \int_0^{t_{n+1}} \tau^{q-1} f(\tau, \mathbf{g}(\tau))(t_{n+1} - \tau)^{p-1} d\tau,$$

$$\mathbf{g}(t_{n+1}) = \frac{q}{\Gamma(p)} \sum_{j=0}^{n} \int_{t_j}^{t_{j+1}} \tau^{q-1} f(\tau, \mathbf{g}(\tau))(t_{n+1} - \tau)^{p-1} d\tau.$$

Within $[t_j, t_{j+1}]$, we approximate the function $f(\tau, \mathbf{g}(\tau))$ using the middle-point method. Thus,

$$\mathbf{g}_{n+1} = \frac{q}{\Gamma(p)} \sum_{j=0}^{n} \int_{t_j}^{t_{j+1}} (t_{n+1} - \tau)^{p-1} \tau^{q-1} f\left(t_j + \frac{\Delta t}{2}, \frac{\mathbf{g}_j + \mathbf{g}_{j+1}}{2}\right) d\tau.$$

Further, we have

$$\mathbf{g}_{n+1} = \frac{q}{\Gamma(p)} \sum_{j=0}^{n} f\left(t_j + \frac{\Delta t}{2}, \frac{\mathbf{g}_j + \mathbf{g}_{j+1}}{2}\right) \int_{t_j}^{t_{j+1}} \tau^{q-1} (t_{n+1} - \tau)^{p-1} d\tau,$$

noting that

$$\int_{t_j}^{t_{j+1}} \tau^{q-1} (t_{n+1} - \tau)^{p-1} d\tau = \int_0^{t_{j+1}} \tau^{q-1} (t_{n+1} - \tau)^{p-1} d\tau$$

$$- \int_0^{t_j} \tau^{q-1} (t_{n+1} - \tau)^{p-1} d\tau,$$

where

$$\int_0^{t_{j+1}} \tau^{q-1} (t_{n+1} - \tau)^{p-1} d\tau = t_{n+1}^{p+q-1} B\left(\frac{t_{j+1}}{t_{n+1}}, q, p\right),$$

$$\int_0^{t_j} \tau^{q-1} (t_{n+1} - \tau)^{p-1} d\tau = t_{n+1}^{p+q-1} B\left(\frac{t_j}{t_{n+1}}, q, p\right).$$

Therefore,

$$\begin{aligned}\mathbf{g}_{n+1} &= \frac{q}{\Gamma(p)} t_{n+1}^{p+q-1} \sum_{j=0}^{n-1} f\left(t_j + \frac{\Delta t}{2}, \frac{\mathbf{g}_j + \mathbf{g}_{j+1}}{2}\right) \left\{ B\left(\frac{t_{j+1}}{t_{n+1}}, q, p\right) - B\left(\frac{t_j}{t_{n+1}}, q, p\right) \right\} \\ &+ \frac{q}{\Gamma(p)} t_{n+1}^{p+q-1} f\left(t_n + \frac{\Delta t}{2}, \frac{\mathbf{g}_n + \mathbf{g}_{n+1}^p}{2}\right) \left\{ B(1, q, p) - B\left(\frac{t_n}{t_{n+1}}, q, p\right) \right\}.\end{aligned}$$

Now, for our system (17.2), we can write the scheme finally for the case of power law kernel, given by

$$\begin{aligned}x_{n+1} &= \frac{q}{\Gamma(p)} t_{n+1}^{p+q-1} \sum_{j=0}^{n-1} f_1\left(t_j + \frac{\Delta t}{2}, \frac{x_j + x_{j+1}}{2}, \frac{y_j + y_{j+1}}{2}, \frac{z_j + z_{j+1}}{2}, \frac{w_j + w_{j+1}}{2}\right) \\ &\times \left\{ B\left(\frac{t_{j+1}}{t_{n+1}}, q, p\right) - B\left(\frac{t_j}{t_{n+1}}, q, p\right) \right\} \\ &+ \frac{q}{\Gamma(p)} t_{n+1}^{p+q-1} f_1\left(t_n + \frac{\Delta t}{2}, \frac{x_n + x_{n+1}^u}{2}, \frac{y_n + y_{n+1}^u}{2}, \frac{z_n + z_{n+1}^u}{2}, \frac{w_n + w_{n+1}^u}{2}\right) \\ &\times \left\{ B(1, q, p) - B\left(\frac{t_n}{t_{n+1}}, q, p\right) \right\},\end{aligned}$$

$$\begin{aligned}y_{n+1} &= \frac{q}{\Gamma(p)} t_{n+1}^{p+q-1} \sum_{j=0}^{n-1} f_2\left(t_j + \frac{\Delta t}{2}, \frac{x_j + x_{j+1}}{2}, \frac{y_j + y_{j+1}}{2}, \frac{z_j + z_{j+1}}{2}, \frac{w_j + w_{j+1}}{2}\right) \\ &\times \left\{ B\left(\frac{t_{j+1}}{t_{n+1}}, q, p\right) - B\left(\frac{t_j}{t_{n+1}}, q, p\right) \right\} \\ &+ \frac{q}{\Gamma(p)} t_{n+1}^{p+q-1} f_2\left(t_n + \frac{\Delta t}{2}, \frac{x_n + x_{n+1}^u}{2}, \frac{y_n + y_{n+1}^u}{2}, \frac{z_n + z_{n+1}^u}{2}, \frac{w_n + w_{n+1}^u}{2}\right) \\ &\times \left\{ B(1, q, p) - B\left(\frac{t_n}{t_{n+1}}, q, p\right) \right\},\end{aligned}$$

$$\begin{aligned}z_{n+1} &= \frac{q}{\Gamma(p)} t_{n+1}^{p+q-1} \sum_{j=0}^{n-1} f_3\left(t_j + \frac{\Delta t}{2}, \frac{x_j + x_{j+1}}{2}, \frac{y_j + y_{j+1}}{2}, \frac{z_j + z_{j+1}}{2}, \frac{w_j + w_{j+1}}{2}\right) \\ &\times \left\{ B\left(\frac{t_{j+1}}{t_{n+1}}, q, p\right) - B\left(\frac{t_j}{t_{n+1}}, q, p\right) \right\} \\ &+ \frac{q}{\Gamma(p)} t_{n+1}^{p+q-1} f_3\left(t_n + \frac{\Delta t}{2}, \frac{x_n + x_{n+1}^u}{2}, \frac{y_n + y_{n+1}^u}{2}, \frac{z_n + z_{n+1}^u}{2}, \frac{w_n + w_{n+1}^u}{2}\right) \\ &\times \left\{ B(1, q, p) - B\left(\frac{t_n}{t_{n+1}}, q, p\right) \right\},\end{aligned}$$

$$\begin{aligned}w_{n+1} &= \frac{q}{\Gamma(p)} t_{n+1}^{p+q-1} \sum_{j=0}^{n-1} f_4\left(t_j + \frac{\Delta t}{2}, \frac{x_j + x_{j+1}}{2}, \frac{y_j + y_{j+1}}{2}, \frac{z_j + z_{j+1}}{2}, \frac{w_j + w_{j+1}}{2}\right) \\ &\times \left\{ B\left(\frac{t_{j+1}}{t_{n+1}}, q, p\right) - B\left(\frac{t_j}{t_{n+1}}, q, p\right) \right\} \\ &+ \frac{q}{\Gamma(p)} t_{n+1}^{p+q-1} f_4\left(t_n + \frac{\Delta t}{2}, \frac{x_n + x_{n+1}^u}{2}, \frac{y_n + y_{n+1}^u}{2}, \frac{z_n + z_{n+1}^u}{2}, \frac{w_n + w_{n+1}^u}{2}\right)\end{aligned}$$

Dynamical Analysis of a Chaotic Model in Fractal-Fractional Operators

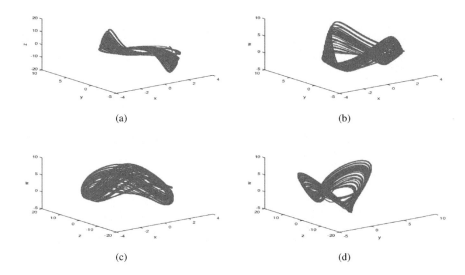

Figure 17.1 Model simulation of fractal-fractional Caputo model when $p = 1$ and $q = 1$.

$$\times \left\{ B(1,q,p) - B\left(\frac{t_n}{t_{n+1}}, q, p\right) \right\},$$

where

$$f_1(t,x,y,z,w) = y,$$

$$f_2(t,x,y,z,w) = -z,$$

$$f_3(t,x,y,z,w) = z + ayw - zw,$$

$$f_4(t,x,y,z,w) = xy + byz.$$

We have the graphical results for the fractal-fractional model using the numerical scheme based on middle-point interpolation for power law kernel case; we obtained the results graphically in Figures 17.1 and 17.2.

17.4.2 NUMERICAL SCHEME BASED ON MIDDLE-POINT INTERPOLATION FOR EXPONENTIAL CASE

We consider in this section the following fractal-fractional Cauchy problem,

$${}_0^{FFE}D_t^{p,q}\mathbf{g}(t) = f(t,\mathbf{g}(t)), \text{ if } t > 0,$$

$$\mathbf{g}(0) = \mathbf{g}_0, \text{ if } t = 0.$$

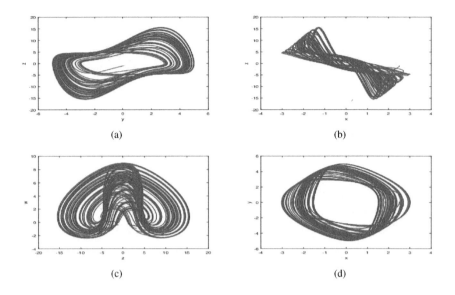

Figure 17.2 Model simulation of fractal-fractional Caputo model when $p = 1$ and $q = 1$.

We can write further

$$^{CFR}_{0}D^p_t\mathbf{g}(t) = qt^{q-1}f(t,\mathbf{g}(t)), \text{ if } t > 0,$$

$$\mathbf{g}(0) = \mathbf{g}_0, \text{ if } t = 0.$$

We have further the following

$$\mathbf{g}(t) = qt^{q-1}(1-p)f(t,\mathbf{g}(t))t^{q-1} + pq\int_0^t \tau^{q-1}f(\tau,\mathbf{g}(\tau))d\tau, \text{ if } t > 0,$$

$$\mathbf{g}(0) = \mathbf{g}_0, \text{ if } t = 0.$$

At $t = t_{n+1}$ and $t = t_n$, we have

$$\mathbf{g}(t_{n+1}) = (1-p)qt_{n+1}^{q-1}f(t_{n+1},\mathbf{g}^p_{n+1}) + pq\int_0^{t_{n+1}} \tau^{q-1}f(\tau,\mathbf{g}(\tau))d\tau$$

and

$$\mathbf{g}(t_n) = (1-p)qt_n^{q-1}f(t_n,\mathbf{g}_n) + pq\int_0^{t_n} \tau^{q-1}f(\tau,\mathbf{g}(\tau))d\tau.$$

Therefore,

$$\mathbf{g}(t_{n+1}) = \mathbf{g}_n + (1-p)q\left(t_{n+1}^{q-1}f(t_{n+1},\mathbf{g}^u_{n+1}) - t_n^{q-1}f(t_n,\mathbf{g}_n)\right)$$

Dynamical Analysis of a Chaotic Model in Fractal-Fractional Operators

$$+ pq \int_{t_n}^{t_{n+1}} \tau^{q-1} f(\tau, \mathbf{g}(\tau)) d\tau,$$

$$\mathbf{g}_{n+1} = \mathbf{g}_n + (1-p)q \left[t_{n+1}^{q-1} f(t_{n+1}, \mathbf{g}_{n+1}^u) - t_n^{q-1} f(t_n, \mathbf{g}_n) \right]$$

$$+ p(\Delta t)^q f\left(t_n + \frac{\Delta t}{2}, \mathbf{g}_n + (\Delta t)^q f(t_n, \mathbf{g}_n) \{(n+1)^q - n^q\} \right),$$

where

$$\mathbf{g}_{n+1}^u = \mathbf{g}_n + p(\Delta t)^q f\left(t_n + \frac{\Delta t}{2}, \mathbf{g}_n + (\Delta t)^q f(t_n, \mathbf{g}_n) \{(n+1)^q - n^q\} \right).$$

Finally, the scheme for our proposed model can be shown by the following form:

$$x_{n+1} = x_n + (1-p)q \left[t_{n+1}^{q-1} f_1(t_{n+1}, x_{n+1}^u, y_{n+1}^u, z_{n+1}^u, w_{n+1}^u) \right.$$

$$\left. - t_n^{q-1} f_1(t_n, x_n, y_n, z_n, w_n) \right]$$

$$+ p(\Delta t)^q f_1\left(t_n + \frac{\Delta t}{2}, x_n + (\Delta t)^q f_1(t_n, x_n, y_n, z_n, w_n) \right.$$

$$\times \{(n+1)^q - n^q\},$$

$$y_n + (\Delta t)^q f_1(t_n, x_n, y_n, z_n, w_n)\{(n+1)^q - n^q\}, z_n$$

$$\left. + (\Delta t)^q f_1(t_n, x_n, y_n, z_n, w_n)\{(n+1)^q - n^q\} \right),$$

$$y_{n+1} = x_n + (1-p)q \left[t_{n+1}^{q-1} f_2(t_{n+1}, x_{n+1}^u, y_{n+1}^u, z_{n+1}^u, w_{n+1}^u) \right.$$

$$\left. - t_n^{q-1} f_2(t_n, x_n, y_n, z_n, w_n) \right]$$

$$+ p(\Delta t)^q f_2\left(t_n + \frac{\Delta t}{2}, x_n + (\Delta t)^q f_2(t_n, x_n, y_n, z_n, w_n) \right.$$

$$\times \{(n+1)^q - n^q\},$$

$$y_n + (\Delta t)^q f_2(t_n, x_n, y_n, z_n, w_n)\{(n+1)^q - n^q\}, z_n$$

$$\left. + (\Delta t)^q f_2(t_n, x_n, y_n, z_n, w_n)\{(n+1)^q - n^q\} \right),$$

$$z_{n+1} = z_n + (1-p)q \left[t_{n+1}^{q-1} f_3(t_{n+1}, x_{n+1}^u, y_{n+1}^u, z_{n+1}^u, w_{n+1}^u) \right.$$

$$\left. - t_n^{q-1} f_3(t_n, x_n, y_n, z_n, w_n) \right]$$

$$+ p(\Delta t)^q f_3\left(t_n + \frac{\Delta t}{2}, x_n + (\Delta t)^q f_3(t_n, x_n, y_n, z_n, w_n) \right.$$

$$\times \{(n+1)^q - n^q\},$$

$$y_n + (\Delta t)^q f_3(t_n,x_n,y_n,z_n,w_n)\{(n+1)^q - n^q\}, z_n$$
$$+ (\Delta t)^q f_3(t_n,x_n,y_n,z_n,w_n)\{(n+1)^q - n^q\}\Big),$$

$$w_{n+1} = w_n + (1-p)q\left[t_{n+1}^{q-1} f_4(t_{n+1},x_{n+1}^u,y_{n+1}^u,z_{n+1}^u,w_{n+1}^u)\right.$$
$$\left. - t_n^{q-1} f_4(t_n,x_n,y_n,z_n,w_n)\right]$$
$$+ p(\Delta t)^q f_3\left(t_n + \frac{\Delta t}{2}, x_n + (\Delta t)^q f_4(t_n,x_n,y_n,z_n,w_n)\right)$$
$$\times \{(n+1)^q - n^q\},$$
$$y_n + (\Delta t)^q f_4(t_n,x_n,y_n,z_n,w_n)\{(n+1)^q - n^q\}, z_n$$
$$+ (\Delta t)^q f_4(t_n,x_n,y_n,z_n,w_n)\{(n+1)^q - n^q\}\Big),$$

where

$$x_{n+1}^u = x_n + p(\Delta t)^q f_1\left(t_n + \frac{\Delta t}{2}, x_n + (\Delta t)^q f_1(t_n,x_n,y_n,z_n,w_n)\{(n+1)^q - n^q\},\right.$$
$$y_n + (\Delta t)^q f_1(t_n,x_n,y_n,z_n,w_n)\{(n+1)^q - n^q\}, z_n + (\Delta t)^q f_1(t_n,x_n,y_n,z_n,w_n)$$
$$\left. \times \{(n+1)^q - n^q\}\right),$$

$$y_{n+1}^u = y_n + p(\Delta t)^q f_2\left(t_n + \frac{\Delta t}{2}, x_n + (\Delta t)^q f_2(t_n,x_n,y_n,z_n,w_n)\{(n+1)^q - n^q\},\right.$$
$$y_n + (\Delta t)^q f_2(t_n,x_n,y_n,z_n,w_n)\{(n+1)^q - n^q\}, z_n + (\Delta t)^q f_2(t_n,x_n,y_n,z_n,w_n)$$
$$\left. \times \{(n+1)^q - n^q\}\right),$$

$$z_{n+1}^u = z_n + p(\Delta t)^q f_3\left(t_n + \frac{\Delta t}{2}, x_n + (\Delta t)^q f_3(t_n,x_n,y_n,z_n,w_n)\{(n+1)^q - n^q\},\right.$$
$$y_n + (\Delta t)^q f_3(t_n,x_n,y_n,z_n,w_n)$$
$$\left. \times \{(n+1)^q - n^q\}, z_n + (\Delta t)^q f_3(t_n,x_n,y_n,z_n,w_n)\{(n+1)^q - n^q\}\right),$$

$$w_{n+1}^u = w_n + p(\Delta t)^q f_3\left(t_n + \frac{\Delta t}{2}, x_n + (\Delta t)^q f_4(t_n,x_n,y_n,z_n,w_n)\{(n+1)^q - n^q\},\right.$$
$$y_n + (\Delta t)^q f_4(t_n,x_n,y_n,z_n,w_n)\{(n+1)^q - n^q\}, z_n + (\Delta t)^q f_4(t_n,x_n,y_n,z_n,w_n)$$
$$\left. \times \{(n+1)^q - n^q\}\right).$$

For the case of fractal-fractional model based on exponential kernel, we considered the scheme above based on middle-point interpolation and obtained the results shown in Figures 17.3–17.6 for different p and q values.

Dynamical Analysis of a Chaotic Model in Fractal-Fractional Operators

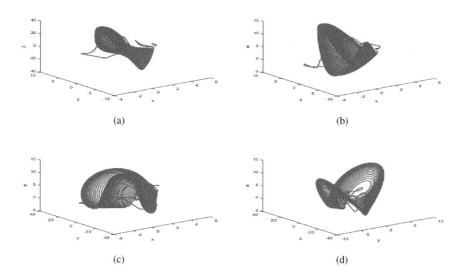

Figure 17.3 Model simulation of fractal-fractional Caputo-Fabrizio model when $p = 1$ and $q = 1$.

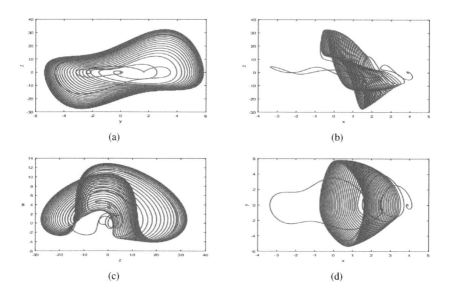

Figure 17.4 Model simulation of fractal-fractional Caputo-Fabrizio model when $p = 1$ and $q = 1$.

17.4.3 NUMERICAL SCHEME FOR THE MITTAG-LEFFLER CASE

This subsection presents the numerical scheme for Mittag-Leffler law using middle-point interpolation. We consider the following fractal-fractional Cauchy problem for

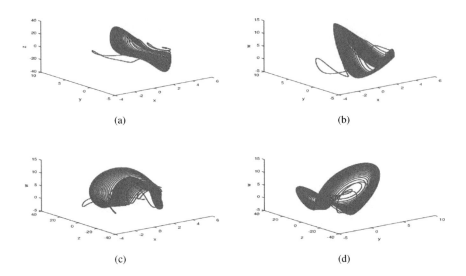

Figure 17.5 Model simulation of fractal-fractional Caputo-Fabrizio model when $p = 1$ and $q = 0.8$.

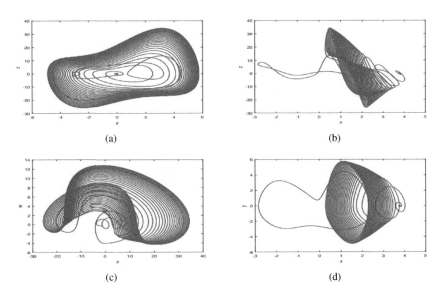

Figure 17.6 Model simulation of fractal-fractional Caputo-Fabrizio model when $p = 01$ and $q = 0.8$.

the Mittag-Leffler case,

$$_{0}^{FFM}D_{t}^{p,q}\mathbf{g}(t) = f(t, \mathbf{g}(t)), \text{ if } t > 0,$$

Dynamical Analysis of a Chaotic Model in Fractal-Fractional Operators

$$\mathbf{g}(0) = \mathbf{g}_0, \text{ if } t = 0.$$

We can write further

$${}_0^{ABR}D_t^p \mathbf{g}(t) = qt^{q-1}f(t,\mathbf{g}(t)), \text{ if } t > 0,$$

$$\mathbf{g}(0) = \mathbf{g}_0, \text{ if } t = 0.$$

Further, we write

$$\mathbf{g}(t) = (1-p)qt^{q-1}f(t,\mathbf{g}(t)) + \frac{pq}{\Gamma(p)}\int_0^t \tau^{q-1}f(\tau,\mathbf{g}(\tau))(t-\tau)^{p-1}d\tau,$$

$$\mathbf{g}(0) = \mathbf{g}_0.$$

For $t = t_{n+1}$, we have

$$\mathbf{g}(t_{n+1}) = (1-p)qt_{n+1}^{q-1}f(t_{n+1},\mathbf{g}^u(t_{n+1}))$$
$$+ \frac{pq}{\Gamma(p)}\sum_{j=0}^{n}\int_{t_j}^{t_{j+1}} \tau^{q-1}f(\tau,\mathbf{g}(\tau))(t_{n+1}-\tau)^{p-1}d\tau,$$

whereas

$$\mathbf{g}^u(t_{n+1}) = \frac{q}{\Gamma(p)}\int_0^{t_{n+1}} \tau^{q-1}(t_{n+1}-\tau)^{p-1}f(\tau,\mathbf{g}(\tau))d\tau,$$
$$= \frac{q}{\Gamma(p)}\sum_{j=0}^{n}\int_{t_j}^{t_{j+1}} \tau^{q-1}(t_{n+1}-\tau)^{p-1}f(\tau,\mathbf{g}(\tau))d\tau.$$

Hence, we use the Euler approximations,

$$\mathbf{g}_{n+1}^u = \frac{q}{\Gamma(p)}\sum_{j=0}^{n}\int_{t_j}^{t_{j+1}} f(t_j,\mathbf{g}_j)(t_{n+1}-\tau)^{p-1}\tau^{q-1}d\tau,$$
$$= \frac{q}{\Gamma(p)}\sum_{j=0}^{n} f(t_j,\mathbf{g}_j)t_{n+1}^{p+q-1}\left[B\left(\frac{t_{j+1}}{t_{n+1}},q,p\right) - B\left(\frac{t_j}{t_{n+1}},q,p\right)\right].$$

Therefore, the numerical scheme is given as

$$\mathbf{g}_{n+1} = \frac{q}{\Gamma(p)}t_{n+1}^{p+q-1}\sum_{j=0}^{n-1} f\left(t_j + \frac{\Delta t}{2}, \frac{\mathbf{g}_j+\mathbf{g}_{j+1}}{2}\right)\left[B\left(\frac{t_{j+1}}{t_{n+1}},q,p\right) - B\left(\frac{t_j}{t_{n+1}},q,p\right)\right]$$
$$+ \frac{q}{\Gamma(p)}t_{n+1}^{p+q-1} f\left(t_n + \frac{\Delta t}{2}, \frac{\mathbf{g}_n}{2} + \frac{\mathbf{g}_{n+1}^u}{2}\right)\left\{B(1,q,p) - B\left(\frac{t_n}{t_{n+1}},q,p\right)\right\},$$

and

$$\mathbf{g}_{(t_{n+1})} \approx \mathbf{g}_{n+1} = (1-p)qt_{n+1}^{q-1}f(t_{n+1},\mathbf{g}_{n+1}^{u})$$
$$+ \frac{pq}{\Gamma(p)}\sum_{j=0}^{n-1}f\left(t_j + \frac{h}{2}, \frac{\mathbf{g}_j + \mathbf{g}_{j+1}}{2}\right)t_{n+1}^{p+q-1}$$
$$\times \left[B\left(\frac{t_{j+1}}{t_{n+1}},q,p\right) - B\left(\frac{t_j}{t_{n+1}},q,p\right)\right]$$
$$+ \frac{pq}{\Gamma(p)}t_{n+1}^{p+q-1}\left\{B(1,q,p) - B\left(\frac{t_n}{t_{n+1}},q,p\right)\right\},$$

where,

$$\mathbf{g}_{n+1}^{u} = \frac{q}{\Gamma(p)}\sum_{j=0}^{n}f(t_j,\mathbf{g}_j)t_{n+1}^{p+q-1}\left[B\left(\frac{t_{j+1}}{t_{n+1}},q,p\right) - B\left(\frac{t_j}{t_{n+1}},q,p\right)\right].$$

We now write the scheme for our system given by

$$x_{n+1} = (1-p)qt_{n+1}^{q-1}f_1(t_{n+1},x_{n+1}^{u},y_{n+1}^{u},z_{n+1}^{u},w_{n+1}^{u})$$
$$+ \frac{pq}{\Gamma(p)}\sum_{j=0}^{n-1}f_1\left(t_j + \frac{h}{2}, \frac{x_j+x_{j+1}}{2}, \frac{y_j+y_{j+1}}{2}, \frac{z_j+z_{j+1}}{2}, \frac{w_j+w_{j+1}}{2}\right)$$
$$\times t_{n+1}^{p+q-1}\left[B\left(\frac{t_{j+1}}{t_{n+1}},q,p\right) - B\left(\frac{t_j}{t_{n+1}},q,p\right)\right]$$
$$+ \frac{pq}{\Gamma(p)}t_{n+1}^{p+q-1}\left\{B(1,q,p) - B\left(\frac{t_n}{t_{n+1}},q,p\right)\right\},$$

$$y_{n+1} = (1-p)qt_{n+1}^{q-1}f_2(t_{n+1},x_{n+1}^{u},y_{n+1}^{u},z_{n+1}^{u},w_{n+1}^{u})$$
$$+ \frac{pq}{\Gamma(p)}\sum_{j=0}^{n-1}f_2\left(t_j + \frac{h}{2}, \frac{x_j+x_{j+1}}{2}, \frac{y_j+y_{j+1}}{2}, \frac{z_j+z_{j+1}}{2}, \frac{w_j+w_{j+1}}{2}\right)$$
$$\times t_{n+1}^{p+q-1}\left[B\left(\frac{t_{j+1}}{t_{n+1}},q,p\right) - B\left(\frac{t_j}{t_{n+1}},q,p\right)\right]$$
$$+ \frac{pq}{\Gamma(p)}t_{n+1}^{p+q-1}\left\{B(1,q,p) - B\left(\frac{t_n}{t_{n+1}},q,p\right)\right\},$$

$$z_{n+1} = (1-p)qt_{n+1}^{q-1}f_3(t_{n+1},x_{n+1}^{u},y_{n+1}^{u},z_{n+1}^{u},w_{n+1}^{u})$$
$$+ \frac{pq}{\Gamma(p)}\sum_{j=0}^{n-1}f_3\left(t_j + \frac{h}{2}, \frac{x_j+x_{j+1}}{2}, \frac{y_j+y_{j+1}}{2}, \frac{z_j+z_{j+1}}{2}, \frac{w_j+w_{j+1}}{2}\right)$$
$$\times t_{n+1}^{p+q-1}\left[B\left(\frac{t_{j+1}}{t_{n+1}},q,p\right) - B\left(\frac{t_j}{t_{n+1}},q,p\right)\right]$$
$$+ \frac{pq}{\Gamma(p)}t_{n+1}^{p+q-1}\left\{B(1,q,p) - B\left(\frac{t_n}{t_{n+1}},q,p\right)\right\},$$

$$w_{n+1} = (1-p)qt_{n+1}^{q-1}f_4(t_{n+1},x_{n+1}^{u},y_{n+1}^{u},z_{n+1}^{u},w_{n+1}^{u})$$

Dynamical Analysis of a Chaotic Model in Fractal-Fractional Operators

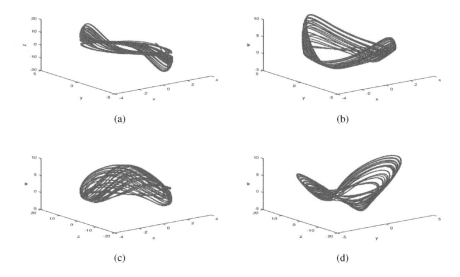

Figure 17.7 Model simulation of fractal-fractional Atangana-Baleanu model when $p = 1$ and $q = 1$.

$$+ \frac{pq}{\Gamma(p)} \sum_{j=0}^{n-1} f_4\left(t_j + \frac{h}{2}, \frac{x_j + x_{j+1}}{2}, \frac{y_j + y_{j+1}}{2}, \frac{z_j + z_{j+1}}{2}, \frac{w_j + w_{j+1}}{2}\right)$$
$$\times t_{n+1}^{p+q-1} \left[B\left(\frac{t_{j+1}}{t_{n+1}}, q, p\right) - B\left(\frac{t_j}{t_{n+1}}, q, p\right) \right]$$
$$+ \frac{pq}{\Gamma(p)} t_{n+1}^{p+q-1} \left\{ B(1, q, p) - B\left(\frac{t_n}{t_{n+1}}, q, p\right) \right\},$$

where

$$x_{n+1}^u = \frac{q}{\Gamma(p)} \sum_{j=0}^{n} f_1(t_j, x_j, y_j, z_j, w_j) t_{n+1}^{p+q-1} \left[B\left(\frac{t_{j+1}}{t_{n+1}}, q, p\right) - B\left(\frac{t_j}{t_{n+1}}, q, p\right) \right],$$

$$y_{n+1}^u = \frac{q}{\Gamma(p)} \sum_{j=0}^{n} f_2(t_j, x_j, y_j, z_j, w_j) t_{n+1}^{p+q-1} \left[B\left(\frac{t_{j+1}}{t_{n+1}}, q, p\right) - B\left(\frac{t_j}{t_{n+1}}, q, p\right) \right],$$

$$z_{n+1}^u = \frac{q}{\Gamma(p)} \sum_{j=0}^{n} f_3(t_j, x_j, y_j, z_j, w_j) t_{n+1}^{p+q-1} \left[B\left(\frac{t_{j+1}}{t_{n+1}}, q, p\right) - B\left(\frac{t_j}{t_{n+1}}, q, p\right) \right],$$

$$w_{n+1}^u = \frac{q}{\Gamma(p)} \sum_{j=0}^{n} f_4(t_j, x_j, y_j, z_j, w_j) t_{n+1}^{p+q-1} \left[B\left(\frac{t_{j+1}}{t_{n+1}}, q, p\right) - B\left(\frac{t_j}{t_{n+1}}, q, p\right) \right].$$

Figures 17.7–17.10 are obtained using the above schemes based on middle-point interpolations for Mittag-Leffler kernel.

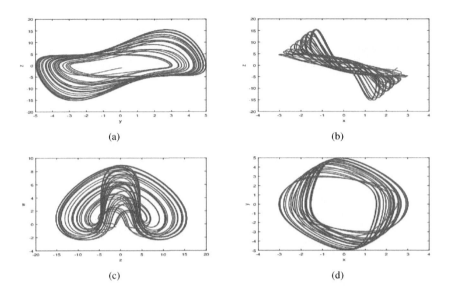

Figure 17.8 Model simulation of fractal-fractional Atangana-Baleanu model when $p = 01$ and $q = 1$.

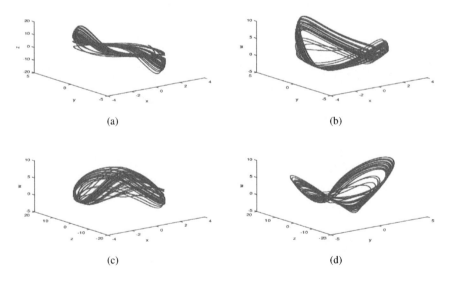

Figure 17.9 Model simulation of fractal-fractional Atangana-Baleanu model when $p = 1$ and $q = 0.99$.

17.5 CONCLUSION

We examined the numerical analysis of a chaotic model under various fractal-fractional operators. Fractal-fractional operators were considered in the sense of

Dynamical Analysis of a Chaotic Model in Fractal-Fractional Operators

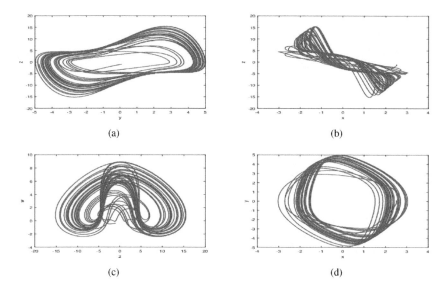

Figure 17.10 Model simulation of fractal-fractional Atangana-Baleanu model when $p = 1$ and $q = 0.99$.

power law, exponential law, and the Mittag-Leffler law. We studied the existence and uniqueness of the system and found that the model has a system of unique solutions and exists. Then, we presented their equilibrium points and some related discussion has been made. Further, we used the idea of middle-point interpolation and presented numerical schemes for each kernel. We have also given some illustrations in the form of figures for each fractal-fractional-order-based scheme.

18 A Chaotic Cancer Model in Fractal-Fractional Operators

18.1 INTRODUCTION

Any of the several illnesses characterised by the growth of aberrant cells that divide out of control and have the capacity to invade and destroy healthy bodily tissue are referred to as cancers. The propensity of cancer to spread throughout your body is common. The second greatest cause of death worldwide is cancer. On the other hand, because of advancements in cancer detection, treatment, and prevention, survival rates are rising for many cancer types. Changes or alterations to the DNA within cells are the primary cause of cancer. A cell's DNA is organised into numerous distinct genes, each of which carries a set of instructions directing the cell's performance of certain tasks as well as its growth and division. Incorrect instructions can make a cell cease functioning normally and even give it the chance to develop cancer. A healthy cell may be instructed by a gene mutation to promote rapid growth. A gene mutation may instruct a cell to divide and develop more quickly. This results in the creation of numerous additional cells with the same mutation. failure to inhibit unchecked cell growth Normal cells are aware of when to cease growing so that there proper proportions of each type of cell present. The mechanisms that instruct cancer cells when to stop growing are lost. A tumour suppressor gene mutation permits cancer cells to keep multiplying and accumulating and make blunders when fixing DNA mistakes. DNA repair genes scan a cell's DNA for faults and correct them. Gene mutations typically take place during healthy cell growth. However, cells have a system in place that may detect errors and correct them. On occasion, a mistake gets overlooked. This might result in a cell developing cancer. Every year, cancer kills people, making it a serious danger to human health as well as a difficult infectious disease. Some forms of treatment, such as surgery, are used to treat cancer and are seen to be somewhat effective. The latest medical innovations include immunotherapy and gene therapy. The given therapy is one of these healing methods that effectively treats the cancer in the patient and ensures their survival. In some situations, the treatment not only destroys the tumour cells but also kills some healthy tissues or severely damages them. To destroy the most tumour cells while causing the least damage to healthy tissue, some caution must be taken when adjusting the amount of the therapy [281]. But it will take some time for these strategies to mature. But the failure of these strategies is frequently noted. Additional details on this matter are available in [38, 118, 280]. In a theoretical perspective, tumour-immune dynamics have a long history.

The mathematical modelling of tumour growth dynamics is thought to be an active study subject among biologists, mathematicians, and engineers. Researchers have always employed a variety of methods to simulate cancer dynamics. Mathematical modeling is thought to be one of the potentially potent methods for better tumour treatment. The cellular automata are used to study tumour growth models, where efforts are made to accurately represent the patient, treatment, and tumour properties in the model [160, 207]. The work of Anderson and Chaplain [37] and Endearing et al. [46] also described the tumour in PDEs using the cellular automata. Some other related work, see [63], where generic tumour growth model Using the ordinary differential equation to be considered, where Through the number of tumours, one may see the dynamics of the tumour growth. immunological and healthy cells. Notably, some of these mathematical simulations of the actual mechanisms underlying cancer in the human body result in chaos. Thus, the idea of a fractal-fractional framework will be introduced to a mathematical model with a classical differential operator in this chapter. In this chapter, a chaotic cancer model under the framework of farctal-fractional operators is proposed. We give details of the existence and uniqueness for the proposed cancer model. Using the idea of middle-point interpolations, we present the numerical schemes for power law, exponential law, and the Mittag-Leffler law. For each case, we give some graphs for different fractal and fractional orders.

18.2 MODEL FRAMEWORK

The chaotic cancer model that described the dynamics of a tumour considered here is given by the following equations [156]:

$$\begin{cases} \frac{dx(t)}{dt} = ax(1-y)(1+z) - x^2 y, \\ \frac{dy(t)}{dt} = \beta y(1-z)(1+x) - y^2 z, \\ \frac{dz(t)}{dt} = \gamma z(1-x)(1+y) - z^2 x, \end{cases} \tag{18.1}$$

where $x(t)$, $y(t)$, and $z(t)$ are the state variables, whereas a, β, and γ denote the parameters of the system. This system is chaotic for the given values of the parameters $a = 0.3440$, $\beta = 0.5337$ and $\gamma = 0.6278$ with the initial conditions on state variables $x(0) = 01$, $y(0) = 01$, and $z(0) = 01$. The fractal-fractional representation of the model (18.1) is given by

$$\begin{cases} {}^{FF}D_{0,t}^{p,\,q}\bigl(x(t)\bigr) = ax(1-y)(1+z) - x^2 y, \\ {}^{FF}D_{0,t}^{p,\,q}\bigl(y(t)\bigr) = \beta y(1-z)(1+x) - y^2 z, \\ {}^{FF}D_{0,t}^{p,\,q}\bigl(z(t)\bigr) = \gamma z(1-x)(1+y) - z^2 x, \end{cases} \tag{18.2}$$

where p represents the fractional order, while q is the fractal order.

18.3 EXISTENCE AND UNIQUENESS

In the present section, we aim to study the existence and uniqueness of the model (18.2). It is assumed that for all $t \in [0,T]$, the functions $x(t)$, $y(t)$, and $z(t)$ are bounded, $\|x\|_\infty \leq M_x$, $\|y\|_\infty \leq M_y$, and $\|z\|_\infty \leq M_z$. For simplicity, we let

$$f_x(t,x,y,z) = ax(1-y)(1+z) - x^2 y,$$

$$f_y(t,x,y,z) = \beta y(1-z)(1+x) - y^2 z,$$

$$f_z(t,x,y,z) = \gamma z(1-x)(1+y) - z^2 x.$$

The system is highly non-linear and therefore cannot be solved analytically; we shall rely on the numerical scheme. However, we shall show that the system admits a unique system of solutions under the obtained conditions. If x, y, and z are bounded, then f_x, f_y, and f_z are also bounded.

Proof 2 *Assuming that x, y, and z are bounded, then*

$$|f_x(t,x,y,z)| = |ax(1-y)(1+z) - x^2 y|,$$

$$\leq a|x||1+z-y-yz| + |x^2||y|,$$

$$\leq aM_x(1 + M_x + M_y + M_y M_z) + M_x^2 M_y < \infty,$$

$$|f_y(t,x,y,z)| = |\beta y(1-z)(1+x) - y^2 z|,$$

$$\leq \beta |y||1-z||1+x| + |y^2||z|,$$

$$\leq \beta M_y(1 + M_z + M_x + M_x M_z) + M_y^2 M_z < \infty,$$

$$|f_z(t,x,y,z)| = |\gamma z(1-x)(1+y) - z^2 x|,$$

$$\leq \gamma |z||1-x||1+y| + |z^2||x|,$$

$$\leq \gamma M_z(1 + M_x + M_y + M_x M_y) + M_z^2 M_x < \infty.$$

On the other hand, we have that

$$|f_x(t,x,y,z)|^2 = |ax(1-y)(1+z) - x^2 y|^2,$$

$$\leq 2|ax|1-y||1+z||^2 + 2|x^4||y|^2,$$

$$\leq 2a^2|x|^2(1 + |y| + |z| + |y||z|) + 2|x|^2|x|^2|y|^2.$$

$$\leq 2a^2|x|^2(1+M_y+M_z+M_yM_z)+2|x|^2M_x^2M_y^2,$$
$$\leq (2a^2(1+M_y+M_z+M_yM_z+2M_x^2M_y^2))|x|^2,$$
$$\leq \overline{K}_x(1+|x|^2),$$

where $\overline{K}_x = (2a^2(1+M_y+M_z+M_yM_z+2M_x^2M_y^2))$. Following the same routine as presented earlier, we conclude that

$$|f_y(t,x,y,z)|^2 \leq \overline{K}_y(1+|y|^2),$$
$$\leq |f_z(t,x,y,z)|^2,$$
$$\leq \overline{K}_z(1+|z|^2),$$

where $\overline{K}_y = 2\beta^2(1+M_x+M_z+M_xM_z)+2M_y^2M_z^2$ and $\overline{K}_z = 2\gamma^2(1+M_x+M_y+M_xM_y)+2M_z^2M_x^2$. Similarly we have

$$|f_x(t,x_1,y,z)-f_x(t,x_2,y,z)| \leq |a|(|x_1-x_2|(|1-y||1+z|))+|x_1-x_2|^2|y|,$$
$$\leq \left[|a|(1+M_y+M_z+M_yM_z)+M_y(2M_{x_1}^2+2M_{x_2}^2)\right]|x_1-x_2|$$

$$|f_y(t,x,y_1,z)-f_y(t,x,y_2,z)| \leq |\beta||y_1-y_2||1-z+x-xz|+|z||y_1^2-y_2^2|,$$
$$\leq \left(\beta(1+M_z+M_x+M_xM_z)+M_z(M_{y_1}+M_{y_2})\right)|y_1-y_2|.$$

Using the similar process, f_z can be verified easily.

So, the system admits a unique solution.

18.3.1 EQUILIBRIUM POINTS

In the present section, we investigate the local stability of the cancer chaotic model, given that by (18.2) setting

$$\begin{cases} 0 = ax(1-y)(1+z)-x^2y, \\ 0 = \beta y(1-z)(1+x)-y^2z, \\ 0 = \gamma z(1-x)(1+y)-z^2x, \end{cases}$$

we get the set of equilibrium points:

$$\begin{cases} P_0 = (0,0,0), \\ P_1 = (0,-1,\frac{\beta}{\beta-1}), \text{ if } \beta \neq 1, \\ P_2 = (\frac{\gamma}{\gamma-1},0,-1), \text{ if } \gamma \neq 1, \\ P_3 = (-1,\frac{\alpha}{\alpha-1},0), \text{ if } \alpha \neq 1. \end{cases} \quad (18.3)$$

A Chaotic Cancer Model in Fractal-Fractional Operators

We now present the stability analysis of the above fixed points. The Jacobian matrix of the model (18.2) at P_0 is given by

$$J(P_0) = \begin{pmatrix} \alpha & 0 & 0 \\ 0 & \beta & 0 \\ 0 & 0 & \gamma \end{pmatrix}.$$

For the case when α, and β, and γ are negative, then the Jacobian matrix $J(P_0)$ becomes locally asymptotically stable.

At the equilibrium point P_1, the Jacobian matrix obtained is as follows:

$$J(P_1) = \begin{pmatrix} 2\alpha\left(\frac{\beta}{\beta-1}+1\right) & 0 & 0 \\ -\beta\left(1-\frac{\beta}{\beta-1}\right) & \left(1-\frac{\beta}{\beta-1}\right)\beta + \frac{2\beta}{\beta-1} & \beta-1 \\ -\frac{\beta^2}{(\beta-1)^2} & \frac{\beta\gamma}{\beta-1} & 0 \end{pmatrix}.$$

We have possibly three eigenvalues with negative real parts if

$$\alpha > 0.5, \ \beta \in \left(\frac{2\alpha-1}{4\alpha-1}, 0.5\right), \ \gamma \in \left(0, \frac{2\beta-1}{\beta(\beta-1)}\right),$$

for which the system becomes locally asymptotically stable.

$$J(P_2) = \begin{pmatrix} 0 & -\frac{\gamma^2}{(\gamma-1)^2} & \frac{\alpha\gamma}{\gamma-1} \\ 0 & 2\beta\left(\frac{\gamma}{\gamma-1}+1\right) & 0 \\ \gamma-1 & -\gamma\left(1-\frac{\gamma}{\gamma-1}\right) & \left(1-\frac{\gamma}{\gamma-1}\right)\gamma + \frac{2\gamma}{\gamma-1} \end{pmatrix}.$$

If the conidian $\beta = \frac{1}{2}$, $\gamma < \frac{\alpha+1}{2\alpha} - \frac{1}{2}\sqrt{\frac{\alpha^2+4}{\alpha^2}}$ fulfills, then the equilibrium point P_2 is locally asymptotically stable.

18.4 NUMERICAL PROCEDURE FOR THE CHAOTIC MODEL

We give the numerical schemes for the solution of the fractal-fractional model given in (18.2) by using the idea of middle-point interpolations. We consider in the following three different kernels, such as power law, exponential law and the Mittag-Leffler law. Also, the simulation results for various values of fractal-fractional order are presented.

18.4.1 NUMERICAL SCHEME FOR POWER LAW CASE

In this subsection, we consider the fractal-fractional Cauchy problem for the power law case,

$$_{0}^{FFP}D_{t}^{p,q}\mathbf{g}(t) = f(t, \mathbf{g}(t)), \text{ if } t > 0,$$

$$\mathbf{g}(0) = \mathbf{g}_0, \text{ if } t = 0,$$

where $\mathbf{g}(t) = (x(t), y(t), z(t))$ defines the equations of the model. We can write further

$$^{RL}_0 D^p_t \mathbf{g}(t) = qt^{q-1} f(t, \mathbf{g}(t)), \text{ if } t > 0,$$

$$\mathbf{g}(0) = \mathbf{g}_0, \text{ if } t = 0.$$

Further, we write,

$$\mathbf{g}(t) = \frac{q}{\Gamma(p)} \int_0^t \tau^{q-1} f(\tau, \mathbf{g}(\tau))(t-\tau)^{p-1} d\tau,$$

$$\mathbf{g}(0) = \mathbf{g}_0.$$

At $t = t_{n+1}$, we have

$$\mathbf{g}(t_{n+1}) = \frac{q}{\Gamma(p)} \int_0^{t_{n+1}} \tau^{q-1} f(\tau, \mathbf{g}(\tau))(t_{n+1}-\tau)^{p-1} d\tau,$$

$$\mathbf{g}(t_{n+1}) = \frac{q}{\Gamma(p)} \sum_{j=0}^{n} \int_{t_j}^{t_{j+1}} \tau^{q-1} f(\tau, \mathbf{g}(\tau))(t_{n+1}-\tau)^{p-1} d\tau.$$

Within $[t_j, t_{j+1}]$, we approximate the function $f(\tau, \mathbf{g}(\tau))$ using the middle-point method. Thus,

$$\mathbf{g}_{n+1} = \frac{q}{\Gamma(p)} \sum_{j=0}^{n} \int_{t_j}^{t_{j+1}} (t_{n+1}-\tau)^{p-1} \tau^{q-1} f\left(t_j + \frac{\Delta t}{2}, \frac{\mathbf{g}_j + \mathbf{g}_{j+1}}{2}\right) d\tau.$$

Further, we have

$$\mathbf{g}_{n+1} = \frac{q}{\Gamma(p)} \sum_{j=0}^{n} f\left(t_j + \frac{\Delta t}{2}, \frac{\mathbf{g}_j + \mathbf{g}_{j+1}}{2}\right) \int_{t_j}^{t_{j+1}} \tau^{q-1}(t_{n+1}-\tau)^{p-1} d\tau,$$

noting that

$$\int_{t_j}^{t_{j+1}} \tau^{q-1}(t_{n+1}-\tau)^{p-1} d\tau = \int_0^{t_{j+1}} \tau^{q-1}(t_{n+1}-\tau)^{p-1} d\tau$$
$$- \int_0^{t_j} \tau^{q-1}(t_{n+1}-\tau)^{p-1} d\tau,$$

where

$$\int_0^{t_{j+1}} \tau^{q-1}(t_{n+1}-\tau)^{p-1} d\tau = t_{n+1}^{p+q-1} B\left(\frac{t_{j+1}}{t_{n+1}}, q, p\right),$$

# A Chaotic Cancer Model in Fractal-Fractional Operators

$$\int_0^{t_j} \tau^{q-1}(t_{n+1}-\tau)^{p-1} d\tau = t_{n+1}^{p+q-1} B\left(\frac{t_j}{t_{n+1}}, q, p\right).$$

Therefore,

$$g_{n+1} = \frac{q}{\Gamma(p)} t_{n+1}^{p+q-1} \sum_{j=0}^{n-1} f\left(t_j + \frac{\Delta t}{2}, \frac{g_j + g_{j+1}}{2}\right) \left\{ B\left(\frac{t_{j+1}}{t_{n+1}}, q, p\right) - B\left(\frac{t_j}{t_{n+1}}, q, p\right) \right\}$$
$$+ \frac{q}{\Gamma(p)} t_{n+1}^{p+q-1} f\left(t_n + \frac{\Delta t}{2}, \frac{g_n + g_{n+1}^p}{2}\right) \left\{ B(1, q, p) - B\left(\frac{t_n}{t_{n+1}}, q, p\right) \right\}.$$

Now, for our system (18.2), we can write the scheme finally for the case of power law kernel, given by

$$x_{n+1} = \frac{q}{\Gamma(p)} t_{n+1}^{p+q-1} \sum_{j=0}^{n-1} f_1\left(t_j + \frac{\Delta t}{2}, \frac{x_j+x_{j+1}}{2}, \frac{y_j+y_{j+1}}{2}, \frac{z_j+z_{j+1}}{2}\right) \widehat{B}_1$$
$$+ \frac{q}{\Gamma(p)} t_{n+1}^{p+q-1} f_1\left(t_n + \frac{\Delta t}{2}, \frac{x_n+x_{n+1}^u}{2}, \frac{y_n+y_{n+1}^u}{2}, \frac{z_n+z_{n+1}^u}{2}\right) \widehat{B}_2,$$

$$y_{n+1} = \frac{q}{\Gamma(p)} t_{n+1}^{p+q-1} \sum_{j=0}^{n-1} f_2\left(t_j + \frac{\Delta t}{2}, \frac{x_j+x_{j+1}}{2}, \frac{y_j+y_{j+1}}{2}, \frac{z_j+z_{j+1}}{2}\right) \widehat{B}_1$$
$$+ \frac{q}{\Gamma(p)} t_{n+1}^{p+q-1} f_2\left(t_n + \frac{\Delta t}{2}, \frac{x_n+x_{n+1}^u}{2}, \frac{y_n+y_{n+1}^u}{2}, \frac{z_n+z_{n+1}^u}{2}\right) \widehat{B}_2,$$

$$z_{n+1} = \frac{q}{\Gamma(p)} t_{n+1}^{p+q-1} \sum_{j=0}^{n-1} f_3\left(t_j + \frac{\Delta t}{2}, \frac{x_j+x_{j+1}}{2}, \frac{y_j+y_{j+1}}{2}, \frac{z_j+z_{j+1}}{2}\right) \widehat{B}_1$$
$$+ \frac{q}{\Gamma(p)} t_{n+1}^{p+q-1} f_3\left(t_n + \frac{\Delta t}{2}, \frac{x_n+x_{n+1}^u}{2}, \frac{y_n+y_{n+1}^u}{2}, \frac{z_n+z_{n+1}^u}{2}\right) \widehat{B}_2,$$

where

$$f_1(t,x,y,z) = ax(1-y)(1+z) - x^2 y,$$
$$f_2(t,x,y,z) = \beta y(1-z)(1+x) - y^2 z,$$
$$f_3(t,x,y,z) = \gamma z(1-x)(1+y) - z^2 x,$$

and

$$\widehat{B}_1 = \left\{ B\left(\frac{t_{j+1}}{t_{n+1}}, q, p\right) - B\left(\frac{t_j}{t_{n+1}}, q, p\right) \right\},$$
$$\widehat{B}_2 = \left\{ B(1, q, p) - B\left(\frac{t_n}{t_{n+1}}, q, p\right) \right\}.$$

Using the middle-point interpolation formula for the case of fractal-fractional power law, we present the graphical results in Figures 18.1–18.2 using arbitrary values for fractal-fractional orders q and p.

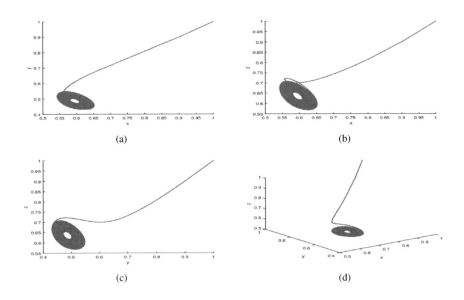

Figure 18.1 Model simulation of fractal-fractional Caputo model when $p = 1$ and $q = 1$.

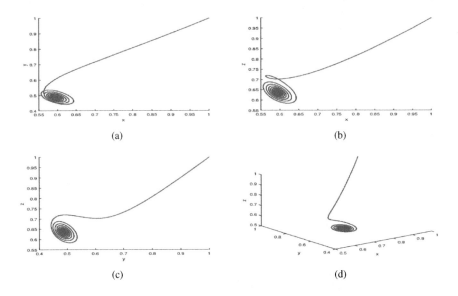

Figure 18.2 Model simulation of fractal-fractional Caputo model when $p = 0.98$ and $q = 0.99$.

18.4.2 NUMERICAL SCHEME FOR EXPONENTIAL CASE

We consider in this subsection the following fractal-fractional Cauchy problem:

$${}_{0}^{FFE}D_{t}^{p,q}\mathbf{g}(t) = f(t,\mathbf{g}(t)), \text{ if } t > 0,$$

A Chaotic Cancer Model in Fractal-Fractional Operators

$$\mathbf{g}(0) = \mathbf{g}_0, \text{ if } t = 0.$$

We can write further

$$_0^{CFR}D_t^p \mathbf{g}(t) = qt^{q-1}f(t,\mathbf{g}(t)), \text{ if } t > 0,$$

$$\mathbf{g}(0) = \mathbf{g}_0, \text{ if } t = 0.$$

We have further the following

$$\mathbf{g}(t) = qt^{q-1}(1-p)f(t,\mathbf{g}(t))t^{q-1} + pq\int_0^t \tau^{q-1}f(\tau,\mathbf{g}(\tau))d\tau, \text{ if } t > 0,$$

$$\mathbf{g}(0) = \mathbf{g}_0, \text{ if } t = 0.$$

At $t = t_{n+1}$, and $t = t_n$, we have

$$\mathbf{g}(t_{n+1}) = (1-p)qt_{n+1}^{q-1}f(t_{n+1},\mathbf{g}_{n+1}^p) + pq\int_0^{t_{n+1}} \tau^{q-1}f(\tau,\mathbf{g}(\tau))d\tau,$$

and

$$\mathbf{g}(t_n) = (1-p)qt_n^{q-1}f(t_n,\mathbf{g}_n) + pq\int_0^{t_n} \tau^{q-1}f(\tau,\mathbf{g}(\tau))d\tau.$$

Therefore,

$$\mathbf{g}(t_{n+1}) = \mathbf{g}_n + (1-p)q\left(t_{n+1}^{q-1}f(t_{n+1},\mathbf{g}_{n+1}^u) - t_n^{q-1}f(t_n,\mathbf{g}_n)\right)$$
$$+ pq\int_{t_n}^{t_{n+1}} \tau^{q-1}f(\tau,\mathbf{g}(\tau))d\tau,$$

$$\mathbf{g}_{n+1} = \mathbf{g}_n + (1-p)q\left[t_{n+1}^{q-1}f(t_{n+1},\mathbf{g}_{n+1}^u) - t_n^{q-1}f(t_n,\mathbf{g}_n)\right]$$
$$+ p(\Delta t)^q f\left(t_n + \frac{\Delta t}{2}, \mathbf{g}_n + (\Delta t)^q f(t_n,\mathbf{g}_n)\{(n+1)^q - n^q\}\right),$$

where

$$\mathbf{g}_{n+1}^u = \mathbf{g}_n + p(\Delta t)^q f\left(t_n + \frac{\Delta t}{2}, \mathbf{g}_n + (\Delta t)^q f(t_n,\mathbf{g}_n)\{(n+1)^q - n^q\}\right).$$

Finally, the scheme for our proposed model can be shown by the following form:

$$x_{n+1} = x_n + (1-p)q\left[t_{n+1}^{q-1}f_1(t_{n+1},x_{n+1}^u,y_{n+1}^u,z_{n+1}^u) - t_n^{q-1}f_1(t_n,x_n,y_n,z_n)\right]$$
$$+ p(\Delta t)^q f_1\left(t_n + \frac{\Delta t}{2}, x_n + (\Delta t)^q f_1(t_n,x_n,y_n,z_n)\{(n+1)^q - n^q\},$$

$$y_n + (\Delta t)^q f_1(t_n, x_n, y_n, z_n)$$
$$\times \{(n+1)^q - n^q\}, z_n + (\Delta t)^q f_1(t_n, x_n, y_n, z_n)\{(n+1)^q - n^q\}\Big),$$

$$y_{n+1} = x_n + (1-p)q\left[t_{n+1}^{q-1} f_2(t_{n+1}, x_{n+1}^u, y_{n+1}^u, z_{n+1}^u) - t_n^{q-1} f_2(t_n, x_n, y_n, z_n)\right]$$
$$+ p(\Delta t)^q f_2\left(t_n + \frac{\Delta t}{2}, x_n + (\Delta t)^q f_2(t_n, x_n, y_n, z_n)\{(n+1)^q - n^q\},\right.$$
$$y_n + (\Delta t)^q f_2(t_n, x_n, y_n, z_n)\{(n+1)^q - n^q\}, z_n + (\Delta t)^q f_2(t_n, x_n, y_n, z_n)$$
$$\times \{(n+1)^q - n^q\}\Big),$$

$$z_{n+1} = z_n + (1-p)q\left[t_{n+1}^{q-1} f_3(t_{n+1}, x_{n+1}^u, y_{n+1}^u, z_{n+1}^u) - t_n^{q-1} f_3(t_n, x_n, y_n, z_n)\right]$$
$$+ p(\Delta t)^q f_3\left(t_n + \frac{\Delta t}{2}, x_n + (\Delta t)^q f_3(t_n, x_n, y_n, z_n)\{(n+1)^q - n^q\},\right.$$
$$y_n + (\Delta t)^q f_3(t_n, x_n, y_n, z_n)\{(n+1)^q - n^q\}, z_n + (\Delta t)^q f_3(t_n, x_n, y_n, z_n)$$
$$\times \{(n+1)^q - n^q\}\Big),$$

where

$$x_{n+1}^u = x_n + p(\Delta t)^q f_1\left(t_n + \frac{\Delta t}{2}, x_n + (\Delta t)^q f_1(t_n, x_n, y_n, z_n)\{(n+1)^q - n^q\},\right.$$
$$y_n + (\Delta t)^q f_1(t_n, x_n, y_n, z_n)\{(n+1)^q - n^q\}, z_n + (\Delta t)^q f_1(t_n, x_n, y_n, z_n)$$
$$\times \{(n+1)^q - n^q\}\Big),$$

$$y_{n+1}^u = y_n + p(\Delta t)^q f_2\left(t_n + \frac{\Delta t}{2}, x_n + (\Delta t)^q f_2(t_n, x_n, y_n, z_n)\{(n+1)^q - n^q\},\right.$$
$$y_n + (\Delta t)^q f_2(t_n, x_n, y_n, z_n)\{(n+1)^q - n^q\}, z_n + (\Delta t)^q f_2(t_n, x_n, y_n, z_n)$$
$$\times \{(n+1)^q - n^q\}\Big),$$

$$z_{n+1}^u = z_n + p(\Delta t)^q f_3\left(t_n + \frac{\Delta t}{2}, x_n + (\Delta t)^q f_3(t_n, x_n, y_n, z_n)\{(n+1)^q - n^q\},\right.$$
$$y_n + (\Delta t)^q f_3(t_n, x_n, y_n, z_n)\{(n+1)^q - n^q\}, z_n + (\Delta t)^q f_3(t_n, x_n, y_n, z_n)$$
$$\times \{(n+1)^q - n^q\}\Big).$$

The numerical results obtained in Figures 18.3 and 18.4 are based on the above fractal-fractional-order middle-point interpolation method. We considered some arbitrary fractal and fractional-order to get the numerical results.

A Chaotic Cancer Model in Fractal-Fractional Operators

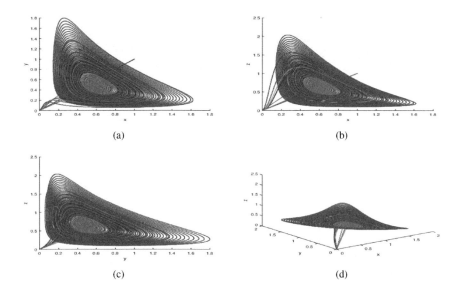

Figure 18.3 Model simulation of fractal-fractional Caputo-Fabrizio model when $q = 1$ and for different values of p.

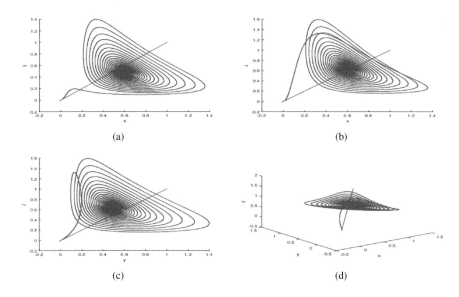

Figure 18.4 Model simulation of fractal-fractional Caputo-Fabrizio model when $p = 0.98$ and $q = 0.99$.

18.4.3 NUMERICAL SCHEME FOR THE MITTAG-LEFFLER CASE

This subsection presents the numerical scheme for Mittag-Leffler law using middle-point interpolation. We consider the following fractal-fractional Cauchy problem for the Mittag-Leffler case,

$$_0^{FFM}D_t^{p,q}\mathbf{g}(t) = f(t,\mathbf{g}(t)), \text{ if } t > 0,$$

$$\mathbf{g}(0) = \mathbf{g}_0, \text{ if } t = 0.$$

We can write further

$$_0^{ABR}D_t^p\mathbf{g}(t) = qt^{q-1}f(t,\mathbf{g}(t)), \text{ if } t > 0,$$

$$\mathbf{g}(0) = \mathbf{g}_0, \text{ if } t = 0.$$

Further, we write

$$\mathbf{g}(t) = (1-p)qt^{q-1}f(t,\mathbf{g}(t)) + \frac{pq}{\Gamma(p)}\int_0^t \tau^{q-1}f(\tau,\mathbf{g}(\tau))(t-\tau)^{p-1}d\tau,$$

$$\mathbf{g}(0) = \mathbf{g}_0.$$

For $t = t_{n+1}$, we have

$$\mathbf{g}(t_{n+1}) = (1-p)qt_{n+1}^{q-1}f(t_{n+1},\mathbf{g}^u(t_{n+1}))$$

$$+ \frac{pq}{\Gamma(p)}\sum_{j=0}^n \int_{t_j}^{t_{j+1}} \tau^{q-1}f(\tau,\mathbf{g}(\tau))(t_{n+1}-\tau)^{p-1}d\tau,$$

whereas

$$\mathbf{g}^u(t_{n+1}) = \frac{q}{\Gamma(p)}\int_0^{t_{n+1}} \tau^{q-1}(t_{n+1}-\tau)^{p-1}f(\tau,\mathbf{g}(\tau))d\tau,$$

$$= \frac{q}{\Gamma(p)}\sum_{j=0}^n \int_{t_j}^{t_{j+1}} \tau^{q-1}(t_{n+1}-\tau)^{p-1}f(\tau,\mathbf{g}(\tau))d\tau.$$

Hence, we use the Euler approximations,

$$\mathbf{g}_{n+1}^u = \frac{q}{\Gamma(p)}\sum_{j=0}^n \int_{t_j}^{t_{j+1}} f(t_j,\mathbf{g}_j)(t_{n+1}-\tau)^{p-1}\tau^{q-1}d\tau,$$

$$= \frac{q}{\Gamma(p)}\sum_{j=0}^n f(t_j,\mathbf{g}_j)t_{n+1}^{p+q-1}\left[B\left(\frac{t_{j+1}}{t_{n+1}},q,p\right) - B\left(\frac{t_j}{t_{n+1}},q,p\right)\right].$$

Therefore, the numerical scheme is given as

$$\mathbf{g}_{n+1} = \frac{q}{\Gamma(p)}t_{n+1}^{p+q-1}\sum_{j=0}^{n-1} f\left(t_j + \frac{\Delta t}{2}, \frac{\mathbf{g}_j + \mathbf{g}_{j+1}}{2}\right)\left[B\left(\frac{t_{j+1}}{t_{n+1}},q,p\right) - B\left(\frac{t_j}{t_{n+1}},q,p\right)\right]$$

$$+ \frac{q}{\Gamma(p)}t_{n+1}^{p+q-1}f\left(t_n + \frac{\Delta t}{2}, \frac{\mathbf{g}_n}{2} + \frac{\mathbf{g}_{n+1}^u}{2}\right)\left\{B(1,q,p) - B\left(\frac{t_n}{t_{n+1}},q,p\right)\right\},$$

and

$$g_{(t_{n+1})} \approx g_{n+1} = (1-p)qt_{n+1}^{q-1}f(t_{n+1}, g_{n+1}^u)$$
$$+ \frac{pq}{\Gamma(p)}\sum_{j=0}^{n-1} f\left(t_j + \frac{h}{2}, \frac{g_j + g_{j+1}}{2}\right) t_{n+1}^{p+q-1}$$
$$\times \left[B\left(\frac{t_{j+1}}{t_{n+1}}, q, p\right) - B\left(\frac{t_j}{t_{n+1}}, q, p\right)\right]$$
$$+ \frac{pq}{\Gamma(p)} t_{n+1}^{p+q-1}\left\{B(1,q,p) - B\left(\frac{t_n}{t_{n+1}}, q, p\right)\right\},$$

where,

$$g_{n+1}^u = \frac{q}{\Gamma(p)}\sum_{j=0}^{n} f(t_j, g_j) t_{n+1}^{p+q-1}\left[B\left(\frac{t_{j+1}}{t_{n+1}}, q, p\right) - B\left(\frac{t_j}{t_{n+1}}, q, p\right)\right].$$

We now write the scheme for our system given by

$$x_{n+1} = (1-p)qt_{n+1}^{q-1}f_1(t_{n+1}, x_{n+1}^u, y_{n+1}^u, z_{n+1}^u)$$
$$+ \frac{pq}{\Gamma(p)}\sum_{j=0}^{n-1} f_1\left(t_j + \frac{h}{2}, \frac{x_j + x_{j+1}}{2}, \frac{y_j + y_{j+1}}{2}, \frac{z_j + z_{j+1}}{2}\right)$$
$$\times t_{n+1}^{p+q-1}\left[B\left(\frac{t_{j+1}}{t_{n+1}}, q, p\right) - B\left(\frac{t_j}{t_{n+1}}, q, p\right)\right]$$
$$+ \frac{pq}{\Gamma(p)} t_{n+1}^{p+q-1}\left\{B(1,q,p) - B\left(\frac{t_n}{t_{n+1}}, q, p\right)\right\},$$
$$y_{n+1} = (1-p)qt_{n+1}^{q-1}f_2(t_{n+1}, x_{n+1}^u, y_{n+1}^u, z_{n+1}^u)$$
$$+ \frac{pq}{\Gamma(p)}\sum_{j=0}^{n-1} f_2\left(t_j + \frac{h}{2}, \frac{x_j + x_{j+1}}{2}, \frac{y_j + y_{j+1}}{2}, \frac{z_j + z_{j+1}}{2}\right)$$
$$\times t_{n+1}^{p+q-1}\left[B\left(\frac{t_{j+1}}{t_{n+1}}, q, p\right) - B\left(\frac{t_j}{t_{n+1}}, q, p\right)\right]$$
$$+ \frac{pq}{\Gamma(p)} t_{n+1}^{p+q-1}\left\{B(1,q,p) - B\left(\frac{t_n}{t_{n+1}}, q, p\right)\right\},$$
$$z_{n+1} = (1-p)qt_{n+1}^{q-1}f_3(t_{n+1}, x_{n+1}^u, y_{n+1}^u, z_{n+1}^u)$$
$$+ \frac{pq}{\Gamma(p)}\sum_{j=0}^{n-1} f_3\left(t_j + \frac{h}{2}, \frac{x_j + x_{j+1}}{2}, \frac{y_j + y_{j+1}}{2}, \frac{z_j + z_{j+1}}{2}\right)$$
$$\times t_{n+1}^{p+q-1}\left[B\left(\frac{t_{j+1}}{t_{n+1}}, q, p\right) - B\left(\frac{t_j}{t_{n+1}}, q, p\right)\right]$$
$$+ \frac{pq}{\Gamma(p)} t_{n+1}^{p+q-1}\left\{B(1,q,p) - B\left(\frac{t_n}{t_{n+1}}, q, p\right)\right\},$$

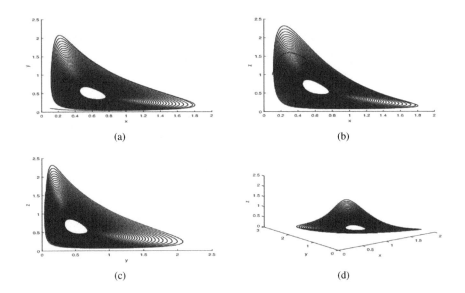

Figure 18.5 Model simulation of fractal-fractional Atangana-Baleanu model when $p = 1$ and $q = 1$.

where

$$x_{n+1}^u = \frac{q}{\Gamma(p)} \sum_{j=0}^{n} f_1(t_j, x_j, y_j, z_j) t_{n+1}^{p+q-1} \left[B\left(\frac{t_{j+1}}{t_{n+1}}, q, p\right) - B\left(\frac{t_j}{t_{n+1}}, q, p\right) \right],$$

$$y_{n+1}^u = \frac{q}{\Gamma(p)} \sum_{j=0}^{n} f_2(t_j, x_j, y_j, z_j) t_{n+1}^{p+q-1} \left[B\left(\frac{t_{j+1}}{t_{n+1}}, q, p\right) - B\left(\frac{t_j}{t_{n+1}}, q, p\right) \right],$$

$$z_{n+1}^u = \frac{q}{\Gamma(p)} \sum_{j=0}^{n} f_3(t_j, x_j, y_j, z_j) t_{n+1}^{p+q-1} \left[B\left(\frac{t_{j+1}}{t_{n+1}}, q, p\right) - B\left(\frac{t_j}{t_{n+1}}, q, p\right) \right].$$

For some fractal-fractional order values, we have the graphical results in Figures 18.5 and 18.6 using the middle-point interpolations for Mittag-Leffler case.

18.5 CONCLUSION

We presented in the given chapter a chaotic system using fractal-fractional operators. The fractal-fractional operators were proposed in sense of different kernels, such as power law, exponential law, and the Mittag-Leffler law kernel. We examined the existence and uniqueness of the model and then presented a discussion on the equilibrium points. Later, we implemented three different operators and presented a numerical scheme for each case. With some selective fractal and fractional order values, we obtained numerical results graphically.

A Chaotic Cancer Model in Fractal-Fractional Operators

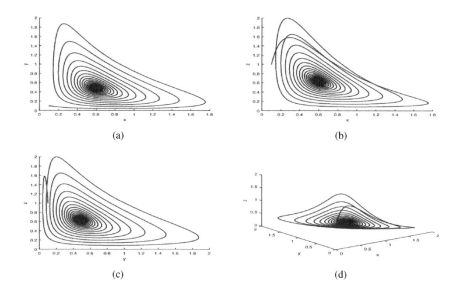

Figure 18.6 Model simulation of fractal-fractional Atangana-Baleanu model when $p = 0.99$ and $q = 0.96$.

19 A Multiple Chaotic Attractor Model under Fractal-Fractional Operators

19.1 INTRODUCTION

Multiple chaotic systems' chaos synchronisation has received a lot of attention recently in the field of non-linear research [142, 149, 185, 230]. It is in favor of its prospective uses in covert signaling, multilateral communications, and numerous other engineering disciplines. By utilising some sophisticated control strategies, many types of synchronisation of numerous chaotic systems have been suggested. On the basis of feedback control techniques, L and Liu tackled the perfect synchronisation of N-coupled chaotic systems with chain and ring connections [149]. The work of Tang and Fang was expanded to include several fractional-order chaotic systems. Chen et al. studied perfect synchronisation, anti-synchronisation, and hybrid synchronisation among various chaotic systems using the direct design method. Jiang et al. and Sun et al., respectively, looked at the generalised combination synchronisation of several chaotic systems [50, 232]. Based on adaptive approaches, Sun et al. discussed compound synchronisation between four memristor chaotic oscillator systems. A new transmission synchronisation technique of multiple chaotic systems was proposed in for a unique ring connection topology of multiple chaotic systems. Three alternative fractional-order chaotic systems were proposed for synchronisation by Xi and colleagues using adaptive function projective combination [49, 50, 220, 273]. It should be noted that the majority of the conclusions discussed above only address the synchronisation of multiple systems based on exact knowledge of the system parameters, and the effects of such systems' unknown parameters are not taken into account. In fact, it is difficult to precisely determine the values of the system parameters a priori, and unknown parameters inherently affect chaotic systems. Due to its advantages on observed swift and spectacular advancements leading to global stability and tracking outcomes for non-linear systems, the adaptive control approach is an effective means to estimate the unknown parameters in another study area [220, 228]. Numerous significant results have been given, and it has been used successfully to synchronise chaotic systems with unknowable parameters. As an illustration, Park investigated the adaptive synchronisation of a unified chaotic system with an unknown parameter [184]. In this chapter, we'll take a look at a well-known chaotic model that has been shown to match the characteristics

DOI: 10.1201/9781003359258-19

mentioned here and expand it to fit the fractal-fractional derivative framework. We consider in the present chapter a multiple chaotic attractor in different fractal-fractional operators. We first give the model and discuss their existence and uniqueness. We discuss the equilibrium points of the model and present their related results. We then use the concept of middle-point interpolations, and present numerical schemes for fractal-fractional system using different kernels. For each kernel, we present some graphical results for illustration.

19.2 MODEL DESCRIPTIONS

In the present section we consider the multiattractor chaotic model taken from ([119]), which is given by the following equations:

$$\begin{cases} \frac{dx(t)}{dt} = ax - yz, \\ \frac{dy(t)}{dt} = -by + xz, \\ \frac{dz(t)}{dt} = -cz + xyz + k, \end{cases} \tag{19.1}$$

where $x(t)$, $y(t)$, and $z(t)$ are the state variables. This system is chaotic for the given values of the parameters $a = 4$, $b = 9$, $c = 4$, $k = 4$ with the initial conditions on state variables $x(0) = 0.01$, $y(0) = 0.1$, and $z(0) = 1$. The fractal-fractional representation of the model (19.1) is given by

$$\begin{cases} {}^{FF}D_{0,t}^{p,\,q}\left(x(t)\right) = ax - yz, \\ {}^{FF}D_{0,t}^{p,\,q}\left(y(t)\right) = -by + xz, \\ {}^{FF}D_{0,t}^{p,\,q}\left(z(t)\right) = -cz + xyz + k, \end{cases} \tag{19.2}$$

where p represents the fractional orders while q is the fractal order.

19.3 EXISTENCE AND UNIQUENESS

We present in this section the conditions under which the system (19.2) admits a unique solution. We prove that if the solutions are bounded, then also the functions f_x, f_y, and f_z are bounded, respectively.

Proof 3 *Assuming that x, y, and z are bounded, we can find M_x, M_y, and M_z positive constants such that $\|x\|_\infty$, $\|y\|_\infty$, and $\|z\|_\infty$ verify $\|x\|_\infty \leq M_x$, $\|y\|_\infty \leq M_y$, and $\|z\|_\infty \leq M_z$.*

$$\begin{aligned} |f_x(t,x,y,z)| &= |ax - yz|, \\ &\leq |a||x| + |y||z|, \\ &\leq a \sup_{t \in D_x} |x(t)| + \sup_{t \in D_y} |y(t)| \sup_{t \in D_z} |z(t)|, \\ &\leq aM_x + M_y M_z < \infty, \\ |f_y(t,x,y,z)| &\leq bM_y + M_x M_z < \infty, \end{aligned}$$

A Multiple Chaotic Attractor Model under Fractal-Fractional Operators

$$|f_z(t,x,y,z)| \leq cM_z + M_xM_yM_z + k < \infty.$$

We have also

$$|f_x(t,x,y,z)|^2 = |ax - yz|^2,$$
$$\& \leq 2a^2|x|^2 + 2|y|^2|z|^2,$$
$$\leq 2a^2|x|^2 + 2M_y^2M_z^2,$$
$$\leq 2M_y^2M_z^2\left(1 + \frac{a^2}{M_y^2M_z^2}|x|^2\right),$$

if $\frac{a^2}{M_y^2M_z^2} < 1$, then

$$|f_x(t,x,y,z)|^2 \leq 2M_y^2M_z^2(1+|x|^2),$$
$$|f_y(t,x,y,z)|^2 \leq 2M_x^2M_z^2(1+|y|^2),$$

if $\frac{b^2}{M_x^2M_z^2} < 1$.

$$|f_z(t,x,y,z)|^2 = |-cz + xyz + k|^2,$$
$$\leq 3c^2|z|^2 + 3|x|^2|y|^2|z|^2 + 3k^2,$$
$$\leq 3c^2|z|^2 + 3M_x^2M_y^2|z|^2 + 3k^2,$$
$$\leq 3k^2\left(1 + \frac{c^2 + M_x^2M_y^2}{k^2}|z|^2\right),$$

if $\frac{c^2 + M_x^2M_y^2}{k^2} < 1$, then

$$|f_z(t,x,y,z)|^2 \leq 3k^2(1+|z|^2).$$

Therefore, if

$$\max\left\{\frac{c^2 + M_x^2M_y^2}{k^2}, \frac{a^2}{M_y^2M_z^2}, \frac{b^2}{M_x^2M_z^2}\right\} < 1,$$

then the system satisfies the linear growth condition. On the other hand,

$$|f_x(t,x_1,y,z) - f_x(t,x_2,y,z)| = a|x_1 - x_2| \leq \frac{3}{2}a|x_1 - x_2|,$$
$$|f_y(t,x,y_1,z) - f_y(t,x,y_2,z)| = b|y_1 - y_2| \leq \frac{3}{2}b|y_1 - y_2|.$$

$$\begin{aligned}|f_z(t,x,y,z_1)-f_x(t,x,y,z_2)| &= |c(z_1-z_2)+xy(z_1-z_2)|,\\ &\le c|z_1-z_2|+|x||y||z_1-z_2|,\\ &\le c|z_1-z_2|+M_xM_y|z_1-z_2|,\\ &\le (c+M_xM_y)|z_1-z_2|.\end{aligned}$$

Therefore, even by taking the square the system satisfies the Lipschitz conditions. Conclusion under the conditions of linear growth, the system admits a unique system of solutions.

19.3.1 EQUILIBRIA AND THEIR STABILITY

The present section determines the equilibria associated with the model (19.2) and their stability. We have the equilibrium points for the model (19.2) given by

$$E_0 = \left(0,0,\frac{k}{c}\right),$$
$$E_1 = \left(\sqrt{c\sqrt{ab}-k/a},\sqrt{c\sqrt{ab}-k/b},\sqrt{ab}\right),$$
$$E_2 = \left(-\sqrt{c\sqrt{ab}-k/a},-\sqrt{c\sqrt{ab}-k/b},\sqrt{ab}\right).$$

Theorem 2 *If $a,b,k > 0$ and c satisfies the condition given below,*

$$\frac{k}{\sqrt{ab}} < c < \frac{2k[(a^2+b^2)\sqrt{ab}+k(b-a)]}{\sqrt{ab}[(a+b)^2\sqrt{ab}+k(b-a)]},$$

then the points E_1 and E_2 of the model (19.2) are locally asymptotically stable.

Proof 4 *We have the Jacobian matrix of the model (19.2) given by*

$$J = \begin{pmatrix} a & -z & -y \\ z & -b & x \\ yz & xz & xy-c \end{pmatrix}.$$

The characteristics equation associated with J is given by

$$\lambda^3 + a_1\lambda^2 + a_2\lambda + a_3 = 0,$$

where

$$a_1 = \left(b-a+\frac{k}{\sqrt{ab}}\right),$$
$$a_2 = (a-b)\left(c-\frac{2k}{\sqrt{ab}}\right),$$

A Multiple Chaotic Attractor Model under Fractal-Fractional Operators 293

$$a_3 = 4ab\left(c - \frac{2k}{\sqrt{ab}}\right).$$

The Routh-Hurtwiz conditions should meet if

$$c < c_0 = \frac{2k[(a^2+b^2)\sqrt{ab}+k(b-a)]}{\sqrt{ab}[(a+b)^2\sqrt{ab}+k(b-a)]}.$$

If $c_0 < \frac{2k}{\sqrt{ab}}$, then E_1 and E_2 become asymptotically stable.

The stabling of the equilibrium point E_0 can be obtained by obtaining the Jacobian matrix, and then its chrematistics equation becomes

$$(\lambda+c)[c^2\lambda^2+(b-a)c^2\lambda+k^2-abc^2]=0. \tag{19.3}$$

If $c < \frac{2k}{\sqrt{ab}}$, then equation (24.1) provides eigenvalues with negative real parts for which the system (19.2) becomes locally asymptotically stable.

19.4 NUMERICAL PROCEDURE FOR THE CHAOTIC MODEL

This section presents the numerical schemes for the solution of fractal-fractional-order model based on the middle-point interpolation method. We consider three different kernels, such as power, exponential, and Mittag-Leffler, and present scheme for each case in the following:

19.4.1 NUMERICAL SCHEME FOR POWER LAW CASE

In this subsection, we consider the fractal-fractional Cauchy problem for the power law case,

$$_0^{FFP}D_t^{p,q}\mathbf{g}(t) = f(t,\mathbf{g}(t)), \text{ if } t > 0,$$

$$\mathbf{g}(0) = \mathbf{g}_0, \text{ if } t = 0,$$

where $\mathbf{g}(t) = (x(t),y(t),z(t))$ defines the equations of the model. We can write further

$$_0^{RL}D_t^p\mathbf{g}(t) = qt^{q-1}f(t,\mathbf{g}(t)), \text{ if } t > 0,$$

$$\mathbf{g}(0) = \mathbf{g}_0, \text{ if } t = 0.$$

Further, we write

$$\mathbf{g}(t) = \frac{q}{\Gamma(p)}\int_0^t \tau^{q-1}f(\tau,\mathbf{g}(\tau))(t-\tau)^{p-1}d\tau,$$

$$\mathbf{g}(0) = \mathbf{g}_0.$$

At $t = t_{n+1}$, we have

$$\mathbf{g}(t_{n+1}) = \frac{q}{\Gamma(p)} \int_0^{t_{n+1}} \tau^{q-1} f(\tau, \mathbf{g}(\tau))(t_{n+1} - \tau)^{p-1} d\tau,$$

$$\mathbf{g}(t_{n+1}) = \frac{q}{\Gamma(p)} \sum_{j=0}^{n} \int_{t_j}^{t_{j+1}} \tau^{q-1} f(\tau, \mathbf{g}(\tau))(t_{n+1} - \tau)^{p-1} d\tau.$$

Within $[t_j, t_{j+1}]$, we approximate the function $f(\tau, \mathbf{g}(\tau))$ using the middle-point method. Thus,

$$\mathbf{g}_{n+1} = \frac{q}{\Gamma(p)} \sum_{j=0}^{n} \int_{t_j}^{t_{j+1}} (t_{n+1} - \tau)^{p-1} \tau^{q-1} f\left(t_j + \frac{\Delta t}{2}, \frac{\mathbf{g}_j + \mathbf{g}_{j+1}}{2}\right) d\tau.$$

Further, we have

$$\mathbf{g}_{n+1} = \frac{q}{\Gamma(p)} \sum_{j=0}^{n} f\left(t_j + \frac{\Delta t}{2}, \frac{\mathbf{g}_j + \mathbf{g}_{j+1}}{2}\right) \int_{t_j}^{t_{j+1}} \tau^{q-1} (t_{n+1} - \tau)^{p-1} d\tau,$$

noting that

$$\int_{t_j}^{t_{j+1}} \tau^{q-1}(t_{n+1} - \tau)^{p-1} d\tau = \int_0^{t_{j+1}} \tau^{q-1}(t_{n+1} - \tau)^{p-1} d\tau - \int_0^{t_j} \tau^{q-1}(t_{n+1} - \tau)^{p-1} d\tau,$$

where

$$\int_0^{t_{j+1}} \tau^{q-1}(t_{n+1} - \tau)^{p-1} d\tau = t_{n+1}^{p+q-1} B\left(\frac{t_{j+1}}{t_{n+1}}, q, p\right),$$

$$\int_0^{t_j} \tau^{q-1}(t_{n+1} - \tau)^{p-1} d\tau = t_{n+1}^{p+q-1} B\left(\frac{t_j}{t_{n+1}}, q, p\right).$$

Therefore,

$$\mathbf{g}_{n+1} = \frac{q}{\Gamma(p)} t_{n+1}^{p+q-1} \sum_{j=0}^{n-1} f\left(t_j + \frac{\Delta t}{2}, \frac{\mathbf{g}_j + \mathbf{g}_{j+1}}{2}\right) \left\{ B\left(\frac{t_{j+1}}{t_{n+1}}, q, p\right) - B\left(\frac{t_j}{t_{n+1}}, q, p\right) \right\}$$

$$+ \frac{q}{\Gamma(p)} t_{n+1}^{p+q-1} f\left(t_n + \frac{\Delta t}{2}, \frac{\mathbf{g}_n + \mathbf{g}_{n+1}^p}{2}\right) \left\{ B(1, q, p) - B\left(\frac{t_n}{t_{n+1}}, q, p\right) \right\}.$$

Now, for our system (19.2), we can write the scheme finally for the case of power law kernel, given by

$$x_{n+1} = \frac{q}{\Gamma(p)} t_{n+1}^{p+q-1} \sum_{j=0}^{n-1} f_1\left(t_j + \frac{\Delta t}{2}, \frac{x_j + x_{j+1}}{2}, \frac{y_j + y_{j+1}}{2}, \frac{z_j + z_{j+1}}{2}\right) \widehat{B_1}$$

$$+\frac{q}{\Gamma(p)}t_{n+1}^{p+q-1}f_1\left(t_n+\frac{\Delta t}{2}, \frac{x_n+x_{n+1}^u}{2}, \frac{y_n+y_{n+1}^u}{2}, \frac{z_n+z_{n+1}^u}{2}\right)\widehat{B}_2,$$

$$y_{n+1} = \frac{q}{\Gamma(p)}t_{n+1}^{p+q-1}\sum_{j=0}^{n-1}f_2\left(t_j+\frac{\Delta t}{2}, \frac{x_j+x_{j+1}}{2}, \frac{y_j+y_{j+1}}{2}, \frac{z_j+z_{j+1}}{2}\right)\widehat{B}_1$$

$$+\frac{q}{\Gamma(p)}t_{n+1}^{p+q-1}f_2\left(t_n+\frac{\Delta t}{2}, \frac{x_n+x_{n+1}^u}{2}, \frac{y_n+y_{n+1}^u}{2}, \frac{z_n+z_{n+1}^u}{2}\right)\widehat{B}_2,$$

$$z_{n+1} = \frac{q}{\Gamma(p)}t_{n+1}^{p+q-1}\sum_{j=0}^{n-1}f_3\left(t_j+\frac{\Delta t}{2}, \frac{x_j+x_{j+1}}{2}, \frac{y_j+y_{j+1}}{2}, \frac{z_j+z_{j+1}}{2}\right)\widehat{B}_1$$

$$+\frac{q}{\Gamma(p)}t_{n+1}^{p+q-1}f_3\left(t_n+\frac{\Delta t}{2}, \frac{x_n+x_{n+1}^u}{2}, \frac{y_n+y_{n+1}^u}{2}, \frac{z_n+z_{n+1}^u}{2}\right)\widehat{B}_2,$$

where

$$f_1(t,x,y,z) = ax - yz,$$
$$f_2(t,x,y,z) = -by + xz,$$
$$f_3(t,x,y,z) = -cz + xyz + k$$

and

$$\widehat{B}_1 = \left\{B\left(\frac{t_{j+1}}{t_{n+1}}, q, p\right) - B\left(\frac{t_j}{t_{n+1}}, q, p\right)\right\},$$
$$\widehat{B}_2 = \left\{B(1, q, p) - B\left(\frac{t_n}{t_{n+1}}, q, p\right)\right\}.$$

Utilising the above scheme, we have the graphical results for the chaotic model in Caputo derivative using middle-point interpolations in Figures 19.1 and 19.3.

19.4.2 NUMERICAL SCHEME FOR EXPONENTIAL CASE

We consider in this subsection the following fractal-fractional Cauchy problem:

$$_0^{FFE}D_t^{p,q}\mathbf{g}(t) = f(t,\mathbf{g}(t)), \text{ if } t > 0,$$
$$\mathbf{g}(0) = \mathbf{g}_0, \text{ if } t = 0.$$

We can write further

$$_0^{CFR}D_t^p\mathbf{g}(t) = qt^{q-1}f(t,\mathbf{g}(t)), \text{ if } t > 0,$$
$$\mathbf{g}(0) = \mathbf{g}_0, \text{ if } t = 0.$$

We have further the following:

$$\mathbf{g}(t) = qt^{q-1}(1-p)f(t,\mathbf{g}(t))t^{q-1} + pq\int_0^t \tau^{q-1}f(\tau,\mathbf{g}(\tau))d\tau, \text{ if } t > 0,$$
$$\mathbf{g}(0) = \mathbf{g}_0, \text{ if } t = 0.$$

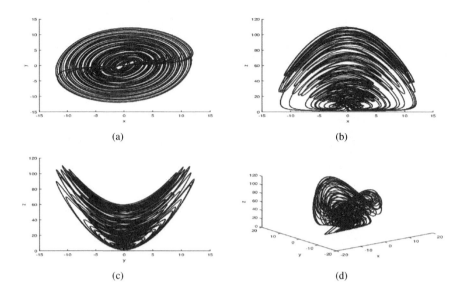

Figure 19.1 Model simulation of fractal-fractional Caputo model when $p = 1$ and $q = 1$.

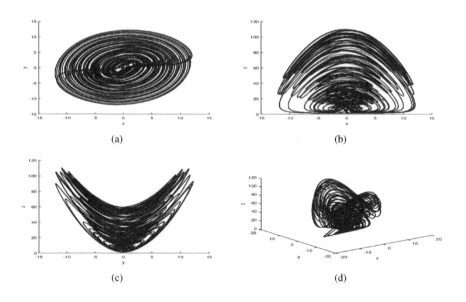

Figure 19.2 Model simulation of fractal-fractional Caputo model when $p = 0.98$ and $q = 1$.

A Multiple Chaotic Attractor Model under Fractal-Fractional Operators

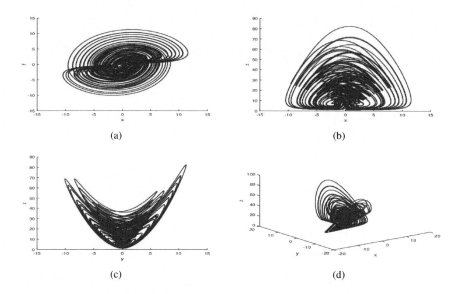

Figure 19.3 Model simulation of fractal-fractional Caputo model when $p = 0.96$ and $q = 0.92$.

At $t = t_{n+1}$ and $t = t_n$, we have

$$\mathbf{g}(t_{n+1}) = (1-p)qt_{n+1}^{q-1}f(t_{n+1}, \mathbf{g}_{n+1}^p) + pq\int_0^{t_{n+1}} \tau^{q-1}f(\tau, \mathbf{g}(\tau))d\tau,$$

and

$$\mathbf{g}(t_n) = (1-p)qt_n^{q-1}f(t_n, \mathbf{g}_n) + pq\int_0^{t_n} \tau^{q-1}f(\tau, \mathbf{g}(\tau))d\tau.$$

Therefore,

$$\mathbf{g}(t_{n+1}) = \mathbf{g}_n + (1-p)q\left(t_{n+1}^{q-1}f(t_{n+1}, \mathbf{g}_{n+1}^u) - t_n^{q-1}f(t_n, \mathbf{g}_n)\right)$$
$$+ pq\int_{t_n}^{t_{n+1}} \tau^{q-1}f(\tau, \mathbf{g}(\tau))d\tau,$$

$$\mathbf{g}_{n+1} = \mathbf{g}_n + (1-p)q\left[t_{n+1}^{q-1}f(t_{n+1}, \mathbf{g}_{n+1}^u) - t_n^{q-1}f(t_n, \mathbf{g}_n)\right]$$
$$+ p(\Delta t)^q f\left(t_n + \frac{\Delta t}{2}, \mathbf{g}_n + (\Delta t)^q f(t_n, \mathbf{g}_n)\{(n+1)^q - n^q\}\right),$$

where

$$\mathbf{g}_{n+1}^u = \mathbf{g}_n + p(\Delta t)^q f\left(t_n + \frac{\Delta t}{2}, \mathbf{g}_n + (\Delta t)^q f(t_n, \mathbf{g}_n)\{(n+1)^q - n^q\}\right).$$

Finally, the scheme for our proposed model can be shown by the following form:

$$\begin{aligned}
x_{n+1} &= x_n + (1-p)q\left[t_{n+1}^{q-1}f_1(t_{n+1},x_{n+1}^u,y_{n+1}^u,z_{n+1}^u) - t_n^{q-1}f_1(t_n,x_n,y_n,z_n)\right] \\
&+ p(\Delta t)^q f_1\left(t_n + \frac{\Delta t}{2}, x_n + (\Delta t)^q f_1(t_n,x_n,y_n,z_n)\{(n+1)^q - n^q\},\right. \\
&\quad y_n + (\Delta t)^q f_1(t_n,x_n,y_n,z_n)\{(n+1)^q - n^q\}, z_n + (\Delta t)^q f_1(t_n,x_n,y_n,z_n) \\
&\quad \left.\times \{(n+1)^q - n^q\}\right),
\end{aligned}$$

$$\begin{aligned}
y_{n+1} &= x_n + (1-p)q\left[t_{n+1}^{q-1}f_2(t_{n+1},x_{n+1}^u,y_{n+1}^u,z_{n+1}^u) - t_n^{q-1}f_2(t_n,x_n,y_n,z_n)\right] \\
&+ p(\Delta t)^q f_2\left(t_n + \frac{\Delta t}{2}, x_n + (\Delta t)^q f_2(t_n,x_n,y_n,z_n)\{(n+1)^q - n^q\},\right. \\
&\quad y_n + (\Delta t)^q f_2(t_n,x_n,y_n,z_n)\{(n+1)^q - n^q\}, z_n + (\Delta t)^q f_2(t_n,x_n,y_n,z_n) \\
&\quad \left.\times \{(n+1)^q - n^q\}\right),
\end{aligned}$$

$$\begin{aligned}
z_{n+1} &= z_n + (1-p)q\left[t_{n+1}^{q-1}f_3(t_{n+1},x_{n+1}^u,y_{n+1}^u,z_{n+1}^u) - t_n^{q-1}f_3(t_n,x_n,y_n,z_n)\right] \\
&+ p(\Delta t)^q f_3\left(t_n + \frac{\Delta t}{2}, x_n + (\Delta t)^q f_3(t_n,x_n,y_n,z_n)\{(n+1)^q - n^q\},\right. \\
&\quad y_n + (\Delta t)^q f_3(t_n,x_n,y_n,z_n)\{(n+1)^q - n^q\}, z_n + (\Delta t)^q f_3(t_n,x_n,y_n,z_n) \\
&\quad \left.\times \{(n+1)^q - n^q\}\right),
\end{aligned}$$

where

$$\begin{aligned}
x_{n+1}^u &= x_n + p(\Delta t)^q f_1\left(t_n + \frac{\Delta t}{2}, x_n + (\Delta t)^q f_1(t_n,x_n,y_n,z_n)\{(n+1)^q - n^q\},\right. \\
&\quad y_n + (\Delta t)^q f_1(t_n,x_n,y_n,z_n)\{(n+1)^q - n^q\}, z_n + (\Delta t)^q f_1(t_n,x_n,y_n,z_n) \\
&\quad \left.\times \{(n+1)^q - n^q\}\right),
\end{aligned}$$

$$\begin{aligned}
y_{n+1}^u &= y_n + p(\Delta t)^q f_2\left(t_n + \frac{\Delta t}{2}, x_n + (\Delta t)^q f_2(t_n,x_n,y_n,z_n)\{(n+1)^q - n^q\},\right. \\
&\quad y_n + (\Delta t)^q f_2(t_n,x_n,y_n,z_n)\{(n+1)^q - n^q\}, z_n + (\Delta t)^q f_2(t_n,x_n,y_n,z_n) \\
&\quad \left.\times \{(n+1)^q - n^q\}\right),
\end{aligned}$$

$$\begin{aligned}
z_{n+1}^u &= z_n + p(\Delta t)^q f_3\left(t_n + \frac{\Delta t}{2}, x_n + (\Delta t)^q f_3(t_n,x_n,y_n,z_n)\{(n+1)^q - n^q\},\right. \\
&\quad y_n + (\Delta t)^q f_3(t_n,x_n,y_n,z_n)\{(n+1)^q - n^q\}, z_n + (\Delta t)^q f_3(t_n,x_n,y_n,z_n) \\
&\quad \left.\times \{(n+1)^q - n^q\}\right).
\end{aligned}$$

A Multiple Chaotic Attractor Model under Fractal-Fractional Operators

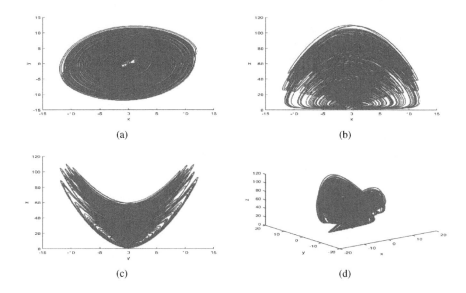

Figure 19.4 Model simulation of fractal-fractional Caputo-Fabrizio model when $p = 1$ and $q = 1$.

Using the numerical scheme presented above for the case of exponential kernel using middle-point interpolation, we achieve the graphical results in Figures 19.4 and 19.6.

19.4.3 NUMERICAL SCHEME FOR THE MITTAG-LEFFLER CASE

This subsection presents the numerical scheme for Mittag-Leffler law using middle-point interpolation. We consider the following fractal-fractional Cauchy problem for the Mittag-Leffler case:

$$_0^{FFM}D_t^{p,q}\mathbf{g}(t) = f(t,\mathbf{g}(t)), \text{ if } t > 0,$$

$$\mathbf{g}(0) = \mathbf{g}_0, \text{ if } t = 0.$$

We can write further

$$_0^{ABR}D_t^p\mathbf{g}(t) = qt^{q-1}f(t,\mathbf{g}(t)), \text{ if } t > 0,$$

$$\mathbf{g}(0) = \mathbf{g}_0, \text{ if } t = 0.$$

Further, we write

$$\mathbf{g}(t) = (1-p)qt^{q-1}f(t,\mathbf{g}(t)) + \frac{pq}{\Gamma(p)}\int_0^t \tau^{q-1}f(\tau,\mathbf{g}(\tau))(t-\tau)^{p-1}d\tau,$$

$$\mathbf{g}(0) = \mathbf{g}_0.$$

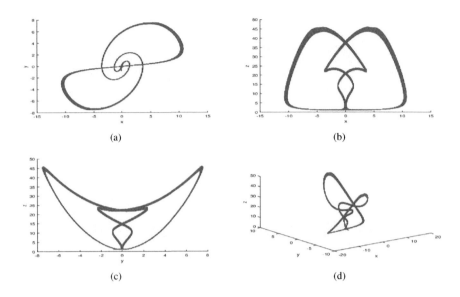

Figure 19.5 Model simulation of fractal-fractional Caputo-Fabrizio model when $p = 0.99$ and $q = 1$.

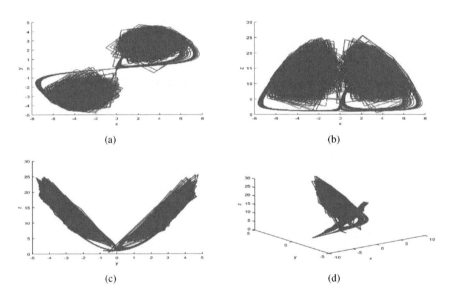

Figure 19.6 Model simulation of fractal-fractional Caputo-Fabrizio model when $p = 0.94$ and $q = 0.99$.

For $t = t_{n+1}$, we have

$$\mathbf{g}(t_{n+1}) = (1-p)qt_{n+1}^{q-1} f(t_{n+1}, \mathbf{g}^u(t_{n+1}))$$
$$+ \frac{pq}{\Gamma(p)} \sum_{j=0}^{n} \int_{t_j}^{t_{j+1}} \tau^{q-1} f(\tau, \mathbf{g}(\tau))(t_{n+1} - \tau)^{p-1} d\tau,$$

whereas

$$\mathbf{g}^u(t_{n+1}) = \frac{q}{\Gamma(p)} \int_0^{t_{n+1}} \tau^{q-1}(t_{n+1} - \tau)^{p-1} f(\tau, \mathbf{g}(\tau)) d\tau,$$
$$= \frac{q}{\Gamma(p)} \sum_{j=0}^{n} \int_{t_j}^{t_{j+1}} \tau^{q-1}(t_{n+1} - \tau)^{p-1} f(\tau, \mathbf{g}(\tau)) d\tau.$$

Hence, we use the Euler approximations,

$$\mathbf{g}_{n+1}^u = \frac{q}{\Gamma(p)} \sum_{j=0}^{n} \int_{t_j}^{t_{j+1}} f(t_j, \mathbf{g}_j)(t_{n+1} - \tau)^{p-1} \tau^{q-1} d\tau,$$
$$= \frac{q}{\Gamma(p)} \sum_{j=0}^{n} f(t_j, \mathbf{g}_j) t_{n+1}^{p+q-1} \left[B\left(\frac{t_{j+1}}{t_{n+1}}, q, p\right) - B\left(\frac{t_j}{t_{n+1}}, q, p\right) \right].$$

Therefore, the numerical scheme is given as

$$\mathbf{g}_{n+1} = \frac{q}{\Gamma(p)} t_{n+1}^{p+q-1} \sum_{j=0}^{n-1} f\left(t_j + \frac{\Delta t}{2}, \frac{\mathbf{g}_j + \mathbf{g}_{j+1}}{2}\right) \left[B\left(\frac{t_{j+1}}{t_{n+1}}, q, p\right) - B\left(\frac{t_j}{t_{n+1}}, q, p\right) \right]$$
$$+ \frac{q}{\Gamma(p)} t_{n+1}^{p+q-1} f\left(t_n + \frac{\Delta t}{2}, \frac{\mathbf{g}_n}{2} + \frac{\mathbf{g}_{n+1}^u}{2}\right) \left\{ B(1, q, p) - B\left(\frac{t_n}{t_{n+1}}, q, p\right) \right\}$$

and

$$\mathbf{g}_{(t_{n+1})} \approx \mathbf{g}_{n+1} = (1-p)qt_{n+1}^{q-1} f(t_{n+1}, \mathbf{g}_{n+1}^u)$$
$$+ \frac{pq}{\Gamma(p)} \sum_{j=0}^{n-1} f\left(t_j + \frac{h}{2}, \frac{\mathbf{g}_j + \mathbf{g}_{j+1}}{2}\right) t_{n+1}^{p+q-1}$$
$$\times \left[B\left(\frac{t_{j+1}}{t_{n+1}}, q, p\right) - B\left(\frac{t_j}{t_{n+1}}, q, p\right) \right]$$
$$+ \frac{pq}{\Gamma(p)} t_{n+1}^{p+q-1} \left\{ B(1, q, p) - B\left(\frac{t_n}{t_{n+1}}, q, p\right) \right\},$$

where,

$$\mathbf{g}_{n+1}^u = \frac{q}{\Gamma(p)} \sum_{j=0}^{n} f(t_j, \mathbf{g}_j) t_{n+1}^{p+q-1} \left[B\left(\frac{t_{j+1}}{t_{n+1}}, q, p\right) - B\left(\frac{t_j}{t_{n+1}}, q, p\right) \right].$$

We now write the scheme for our system given by

$$x_{n+1} = (1-p)qt_{n+1}^{q-1}f_1(t_{n+1}, x_{n+1}^u, y_{n+1}^u, z_{n+1}^u)$$
$$+ \frac{pq}{\Gamma(p)} \sum_{j=0}^{n-1} f_1\left(t_j + \frac{h}{2}, \frac{x_j+x_{j+1}}{2}, \frac{y_j+y_{j+1}}{2}, \frac{z_j+z_{j+1}}{2}\right) t_{n+1}^{p+q-1}$$
$$\times \left[B\left(\frac{t_{j+1}}{t_{n+1}}, q, p\right) - B\left(\frac{t_j}{t_{n+1}}, q, p\right)\right]$$
$$+ \frac{pq}{\Gamma(p)} t_{n+1}^{p+q-1} \left\{B(1, q, p) - B\left(\frac{t_n}{t_{n+1}}, q, p\right)\right\},$$

$$y_{n+1} = (1-p)qt_{n+1}^{q-1}f_2(t_{n+1}, x_{n+1}^u, y_{n+1}^u, z_{n+1}^u)$$
$$+ \frac{pq}{\Gamma(p)} \sum_{j=0}^{n-1} f_2\left(t_j + \frac{h}{2}, \frac{x_j+x_{j+1}}{2}, \frac{y_j+y_{j+1}}{2}, \frac{z_j+z_{j+1}}{2}\right) t_{n+1}^{p+q-1}$$
$$\times \left[B\left(\frac{t_{j+1}}{t_{n+1}}, q, p\right) - B\left(\frac{t_j}{t_{n+1}}, q, p\right)\right]$$
$$+ \frac{pq}{\Gamma(p)} t_{n+1}^{p+q-1} \left\{B(1, q, p) - B\left(\frac{t_n}{t_{n+1}}, q, p\right)\right\},$$

$$z_{n+1} = (1-p)qt_{n+1}^{q-1}f_3(t_{n+1}, x_{n+1}^u, y_{n+1}^u, z_{n+1}^u)$$
$$+ \frac{pq}{\Gamma(p)} \sum_{j=0}^{n-1} f_3\left(t_j + \frac{h}{2}, \frac{x_j+x_{j+1}}{2}, \frac{y_j+y_{j+1}}{2}, \frac{z_j+z_{j+1}}{2}\right) t_{n+1}^{p+q-1}$$
$$\times \left[B\left(\frac{t_{j+1}}{t_{n+1}}, q, p\right) - B\left(\frac{t_j}{t_{n+1}}, q, p\right)\right]$$
$$+ \frac{pq}{\Gamma(p)} t_{n+1}^{p+q-1} \left\{B(1, q, p) - B\left(\frac{t_n}{t_{n+1}}, q, p\right)\right\},$$

where

$$x_{n+1}^u = \frac{q}{\Gamma(p)} \sum_{j=0}^{n} f_1(t_j, x_j, y_j, z_j) t_{n+1}^{p+q-1} \left[B\left(\frac{t_{j+1}}{t_{n+1}}, q, p\right) - B\left(\frac{t_j}{t_{n+1}}, q, p\right)\right],$$
$$y_{n+1}^u = \frac{q}{\Gamma(p)} \sum_{j=0}^{n} f_2(t_j, x_j, y_j, z_j) t_{n+1}^{p+q-1} \left[B\left(\frac{t_{j+1}}{t_{n+1}}, q, p\right) - B\left(\frac{t_j}{t_{n+1}}, q, p\right)\right],$$
$$z_{n+1}^u = \frac{q}{\Gamma(p)} \sum_{j=0}^{n} f_3(t_j, x_j, y_j, z_j) t_{n+1}^{p+q-1} \left[B\left(\frac{t_{j+1}}{t_{n+1}}, q, p\right) - B\left(\frac{t_j}{t_{n+1}}, q, p\right)\right].$$

Figures 19.7–19.8 are obtained for the fractal-fractional-order model based on the new idea of middle-point interpolation in the sense of Atangana-Baleanu derivative.

19.5 CONCLUSION

The presented work explored the dynamics of a chaotic system under fractal-fractional operators in the sense of power law, exponential law, and the Mittag-Leffler law. Initially, we presented the model descriptions, and then we proved the existence

A Multiple Chaotic Attractor Model under Fractal-Fractional Operators   303

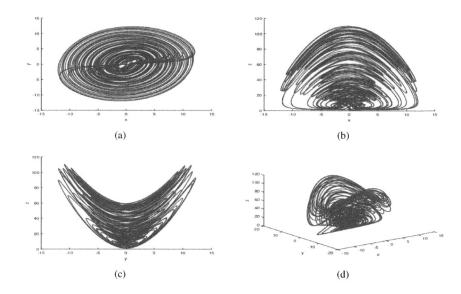

Figure 19.7 Model simulation of fractal-fractional Atangana-Baleanu model when $p = 1$ and $q = 1$.

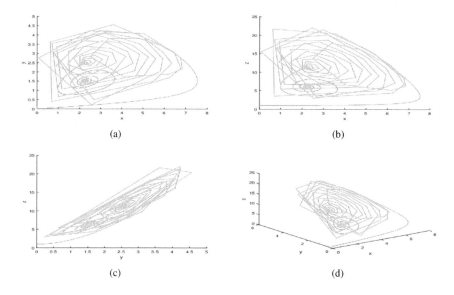

Figure 19.8 Model simulation of fractal-fractional Atangana-Baleanu model when $p = 0.94$ and $q = 0.99$.

and uniqueness. Further, a discussion has been made on the equilibrium points and their related results. Moreover, we utilised the concept of middle-point interpolation formula and obtained numerical schemes for each. Some graphical results for each were presented for the validations of the schemes.

20 The Dynamics of Multiple Chaotic Attractor with Fractal-Fractional Operators

20.1 INTRODUCTION

Due to many applications of chaotic models in science and engineering, it is of interest of the researchers and scientists to study further more advances in the applications of chaotic models. Recently, a very rapid increase in the publications of chaotic models is observed in literature; see [54, 254, 255, 277]. In practical purpose, the applications of chaotic models can be seen in image water marking, chaotic communications, autonomous mobile robots, etc. Due to these applications, some chaotic models are considered in [54, 254, 255]. The literature of chaos suggests that mostly the models have unstable equilibrium points [239, 254], while some systems have self-cited attractors [61, 218].

The fractional order systems that related to chaotic system are reported in [267–269]. The advancement in the fractional calculus by defining new operators the subject of fractional calculus becomes the important research area among the researchers these days. It is well known that these days mostly, the real-life problems are particularly gaining much attentions in formulations with the operators of Caputo, CF, and AB. The deficiency of the Caputo is to some extent recovered in CF, but still it is not the correct operator that understands the dynamics of a practical phenomenon well, so the definition of AB operators arose and the deficiency of Caputo and CF has been recovered. The applications of such operators and their uses in many realistic problems can be seen in [15, 18, 21, 23, 78, 85, 94, 95, 103–105, 109–111, 169, 194, 242, 243, 262]. In particular the application of the fractional derivatives can be observed in [11, 12, 20, 55, 180] and much more. In the present chapter, we focus on considering a chaotic model in studying its dynamics with more advanced operators known as fractal-fractional operators. We consider in this chapter a fractal-fractional-order chaotic system under power, exponential and Mittag-Leffler operators. Using the power law, exponential law, and the Mittag-Leffler law, we give the formulation of fractal-fractional-order system of ordinary differential equation. Initially, we study the model equilibrium points and discuss their results. The existence and uniqueness for the model is given. Further, we use the middle-point interpolation method to derive the numerical schemes for power law kernel, exponential kernel, and the Mittag-Leffler kernel. The scheme for each case of fractal-fractional differential

20.2 MODEL DESCRIPTIONS

We consider the following chaotic model governed by three non-linear differential equations in the following form:

$$\begin{cases} \frac{dx(t)}{dt} = y - A_1, \\ \frac{dy(t)}{dt} = -\beta xz + A_2, \\ \frac{dz(t)}{dt} = xy - z + y^2 - yz. \end{cases} \quad (20.1)$$

In model (20.1), the state variables are given by $x(t)$, $y(t)$, and $z(t)$ and A_1, A_2, and β are the parameter. The model (20.1) behaves chaotically for the set of parameter values: $A_1 = 0.95$, $\beta = 1.45$, $A_2 = 0.01$ and for the initial conditions of the state variables: $x(0) = 1$, $y(0) = 1$, and $z(0) = 1$. Using the fractal-fractional ordinary differential equation structure, we can write the model (20.1) in the below form:

$$\begin{cases} {}^{FF}D_{0,t}^{p,\,q}\bigl(x(t)\bigr) = y - A_1, \\ {}^{FF}D_{0,t}^{p,\,q}\bigl(y(t)\bigr) = -\beta xz + A_2, \\ {}^{FF}D_{0,t}^{p,\,q}\bigl(z(t)\bigr) = xy - z + y^2 - yz, \end{cases} \quad (20.2)$$

where p and q denote, respectively, the fractional and the fractal order. For onward analysis, we consider the model (20.2).

20.3 EXISTENCE AND UNIQUENESS OF THE MODEL

In this section, we study the existence and puniness of the model (20.2). It is assumed that for all $t \in [0, T]$, the functions $x(t)$, $y(t)$, and $z(t)$ are bounded. Further, we write the model (20.2) in the form:

$$f_x(t,x,y,z) = y - A_1,$$

$$f_y(t,x,y,z) = -\beta xz + A_2,$$

$$f_z(t,x,y,z) = xy - z + y^2 - yz.$$

We prove that f_x, f_y, and f_z satisfy the linear growth and Lipschitz conditions. We first show that the functions satisfy the linear growth property.

$$\|f_x(t,x,y,z)\| = \|y - A_1\|,$$
$$\leq \|y\|_\infty + |A|,$$
$$\leq M_y + |A| \leq \infty.$$

$$|f_x(t,x,y,z)|^2 = |y - A_1|^2,$$

$$\leq |y^2| + 2A_1|y| + A_1^2,$$
$$\leq |y^2| + 2A_1|y|^2 + A_1^2,$$
$$\leq |A_1^2|(1 + \frac{1+2A_1}{A_1^2}|y|^2).$$

If $\frac{2A_1}{A_1^2} = \frac{1}{A_1^2} + \frac{2}{A_1} < 1$, then

$$|f_x(t,x,y,z)|^2 \leq A_1^2(1+|y|^2),$$
$$\leq A_1^2(1+|y|_\infty^2)(1+|x|^2),$$
$$\leq K_x(1+|x|^2),$$

where $K_x = A^2(1+|y|_\infty^2)$.

$$|f_y(t,x,y,z)|^2 = |-\beta xz + A_2|^2,$$
$$= |-\beta xz + A_2|^2,$$
$$\leq 2\beta^2|x|^2|z|^2 + 2A_2^2,$$
$$\leq (2\beta^2 M_x^2 x M_z^2 + 2A_2)(1+|y|^2),$$
$$\leq K_y(1+|y|^2),$$

where $K_y = 2\beta^2 M_x^2 M_z^2 + 2A_2$.

$$|f_z(t,x,y,z)|^2 = 3|x|^2|y|^2 + 3|y|^2 + 3|1+y|^2|z|^2,$$
$$\leq 3M_x^2 M_y^2 + 3M_y^2 + 3(2+2M_x^2)|z|^2,$$
$$\leq 3(M_x^2 M_y^2 + M_y^2)(1 + \frac{2+2M_x^2}{M_x^2 M_y^2 + M_y^2}|z|),$$

if $\frac{2+2M_x^2}{M_x^2 M_y^2 + M_y^2} < 1$, then

$$|f_z(t,x,y,z)|^2 \leq K_z(1+|z|^2),$$

where $K_z = 3(M_x^2 M_y^2 + M_y^2)$. Therefore,

$$\max\left\{\frac{1}{A_1^2} + \frac{2}{A_1}, \frac{2+2M_x^2}{M_x^2 M_y^2 + M_y^2}\right\} < 1,$$

then the system admits a linear growth property. Next, we show the Lipschitz conditions:

$$|f_x(t,x_1,y,z) - f_x(t,x_2,y,z)|^2 = 0 \leq \overline{K}_x |x_1 - x_2|^2,$$

$$|f_y(t,x,y_1,z) - f_y(t,x,y_2,z)|^2 = 0 \leq \overline{K}_y |y_1 - y_2|^2.$$

$$|f_z(t,x,y,z_1) - f_z(t,x,y,z_2)|^2 = |(z_1 - z_2)(-1 - y)|,$$
$$\leq (1 + 2M_y + M_y^2)|z_1 - z_2|^2,$$
$$\leq \overline{K}_z |z_1 - z_2|^2,$$

where $\overline{K}_z = (1 + 2M_y + M_y^2)$. Conclusion is if the following inequality is satisfied

$$\max\left\{\frac{1}{A_1^2} + \frac{2}{A_1}, \frac{2 + 2M_x^2}{M_x^2 M_y^2 + M_y^2}\right\} < 1,$$

then the system admits linear growth and Lipschitz condition. Thus, we have a unique solution.

EQUILIBRIUM POINTS AND THEIR ANALYSIS

The possible equilibrium points of the model (20.2) shall be determined by equating the rate of change of the system (20.2) equal to zero,

$$\begin{cases} {}^{FF}D_{0,t}^{p,q}\left(x(t)\right) = 0, \\ {}^{FF}D_{0,t}^{p,q}\left(y(t)\right) = 0, \\ {}^{FF}D_{0,t}^{p,q}\left(z(t)\right) = 0. \end{cases} \quad (20.3)$$

Solving (20.3), we shall get the two possible equilibrium points, say $E_1 = (x_1^*, y_1^*, z_1^*)$ and $E_2 = (x_2^*, y_2^*, z_2^*)$, where

$$x_1^* = -\frac{\sqrt{A_1^3\beta + 4(A_1+1)A_2}}{2\sqrt{A_1}\sqrt{\beta}} - \frac{A_1}{2}, \quad y_1^* = A_1,$$

$$z_1^* = \frac{A_1^2\beta - \sqrt{A_1}\sqrt{\beta}\sqrt{A_1^3\beta + 4A_2A_1 + 4A_2}}{2(A_1\beta + \beta)},$$

and

$$x_2^* = \frac{\sqrt{A_1^3\beta + 4(A_1+1)A_2}}{2\sqrt{A_1}\sqrt{\beta}} - \frac{A_1}{2}, \quad y_2^* = A_1,$$

$$z_2^* = \frac{A_1^2\beta + \sqrt{A_1}\sqrt{\beta}\sqrt{A_1^3\beta + 4A_2A_1 + 4A_2}}{2(A_1\beta + \beta)}.$$

In order to discuss the asymptotic stability of the above system at the two given equilibrium points, we first need to obtain the Jacobian matrix, which is obtained as given by

$$J = \begin{pmatrix} 0 & 1 & 0 \\ -\beta z^* & 0 & -\beta x^* \\ y^* & x^* + 2y^* - z^* & -y^* - 1 \end{pmatrix}. \tag{20.4}$$

At the equilibrium point E_1, we have the eigenvalues, when using the values of the model parameters, A_1, A_2, and β, $\lambda_1 = -2.27342$, $\lambda_2 = 0.948828$, and $\lambda_3 = -0.625405$. This shows that E_1 is unstable. Further, we find the asymptotic stability of the system at E_2. After the evaluation of the Jacobian matrix at E_2, we obtain $\lambda_1 = -1.94113$, $\lambda_2 = -0.00443393 + 0.833644I$, and $\lambda_3 = -0.00443393 - 0.833644I$. All the eigenvalues contain negative real parts, so the system is stable locally asymptotically at E_2.

20.4 NUMERICAL PROCEDURE FOR THE CHAOTIC MODEL

The chaotic model can be studied numerically in order to determin their graphical behaviour, whether it is chaotic or not. In this regard, numerical approaches are widely used for this purpose. In the present section, we aim to develop a new numerical scheme based on the middle-point interpolation method for the numerical solution of the chaotic model using power, exponential and the Mittag-Leffler kernel. In the following, we shall show the details of the schemes for each kernel.

20.4.1 NUMERICAL SCHEME FOR POWER LAW CASE

In this subsection, we consider the fractal-fractional Cauchy problem for the power law case,

$$\substack{FFP\\0}D_t^{p,q}\mathbf{g}(t) = f(t,\mathbf{g}(t)), \text{ if } t > 0,$$

$$\mathbf{g}(0) = \mathbf{g}_0, \text{ if } t = 0,$$

where $\mathbf{g}(t) = (x(t), y(t), z(t))$ defines the equations of the model. We can write further

$$\substack{RL\\0}D_t^p\mathbf{g}(t) = qt^{q-1}f(t,\mathbf{g}(t)), \text{ if } t > 0,$$

$$\mathbf{g}(0) = \mathbf{g}_0, \text{ if } t = 0.$$

Further, we write

$$\mathbf{g}(t) = \frac{q}{\Gamma(p)}\int_0^t \tau^{q-1}f(\tau,\mathbf{g}(\tau))(t-\tau)^{p-1}d\tau,$$

$$\mathbf{g}(0) = \mathbf{g}_0.$$

At $t = t_{n+1}$, we have

$$\mathbf{g}(t_{n+1}) = \frac{q}{\Gamma(p)} \int_0^{t_{n+1}} \tau^{q-1} f(\tau, \mathbf{g}(\tau)) (t_{n+1} - \tau)^{p-1} d\tau,$$

$$\mathbf{g}(t_{n+1}) = \frac{q}{\Gamma(p)} \sum_{j=0}^{n} \int_{t_j}^{t_{j+1}} \tau^{q-1} f(\tau, \mathbf{g}(\tau)) (t_{n+1} - \tau)^{p-1} d\tau.$$

Within $[t_j, t_{j+1}]$, we approximate the function $f(\tau, \mathbf{g}(\tau))$ using the middle-point method. Thus,

$$\mathbf{g}_{n+1} = \frac{q}{\Gamma(p)} \sum_{j=0}^{n} \int_{t_j}^{t_{j+1}} (t_{n+1} - \tau)^{p-1} \tau^{q-1} f\left(t_j + \frac{\Delta t}{2}, \frac{\mathbf{g}_j + \mathbf{g}_{j+1}}{2}\right) d\tau.$$

Further, we have

$$\mathbf{g}_{n+1} = \frac{q}{\Gamma(p)} \sum_{j=0}^{n} f\left(t_j + \frac{\Delta t}{2}, \frac{\mathbf{g}_j + \mathbf{g}_{j+1}}{2}\right) \int_{t_j}^{t_{j+1}} \tau^{q-1} (t_{n+1} - \tau)^{p-1} d\tau,$$

noting that

$$\int_{t_j}^{t_{j+1}} \tau^{q-1} (t_{n+1} - \tau)^{p-1} d\tau = \int_0^{t_{j+1}} \tau^{q-1} (t_{n+1} - \tau)^{p-1} d\tau$$
$$- \int_0^{t_j} \tau^{q-1} (t_{n+1} - \tau)^{p-1} d\tau,$$

where

$$\int_0^{t_{j+1}} \tau^{q-1} (t_{n+1} - \tau)^{p-1} d\tau = t_{n+1}^{p+q-1} B\left(\frac{t_{j+1}}{t_{n+1}}, q, p\right),$$

$$\int_0^{t_j} \tau^{q-1} (t_{n+1} - \tau)^{p-1} d\tau = t_{n+1}^{p+q-1} B\left(\frac{t_j}{t_{n+1}}, q, p\right).$$

Therefore,

$$\mathbf{g}_{n+1} = \frac{q}{\Gamma(p)} t_{n+1}^{p+q-1} \sum_{j=0}^{n-1} f\left(t_j + \frac{\Delta t}{2}, \frac{\mathbf{g}_j + \mathbf{g}_{j+1}}{2}\right) \left\{ B\left(\frac{t_{j+1}}{t_{n+1}}, q, p\right) - B\left(\frac{t_j}{t_{n+1}}, q, p\right) \right\}$$
$$+ \frac{q}{\Gamma(p)} t_{n+1}^{p+q-1} f\left(t_n + \frac{\Delta t}{2}, \frac{\mathbf{g}_n + \mathbf{g}_{n+1}^p}{2}\right) \left\{ B(1, q, p) - B\left(\frac{t_n}{t_{n+1}}, q, p\right) \right\}.$$

Now, for our system (20.2), we can write the scheme finally for the case of power law kernel, given by

$$x_{n+1} = \frac{q}{\Gamma(p)} t_{n+1}^{p+q-1} \sum_{j=0}^{n-1} f_1\left(t_j + \frac{\Delta t}{2}, \frac{x_j + x_{j+1}}{2}, \frac{y_j + y_{j+1}}{2}, \frac{z_j + z_{j+1}}{2}\right) \widehat{B_1}$$

The Dynamics of Multiple Chaotic Attractor with Fractal-Fractional Operators 311

$$y_{n+1} = \begin{aligned} & +\frac{q}{\Gamma(p)}t_{n+1}^{p+q-1}f_1\left(t_n+\frac{\Delta t}{2},\frac{x_n+x_{n+1}^u}{2},\frac{y_n+y_{n+1}^u}{2},\frac{z_n+z_{n+1}^u}{2}\right)\widehat{B}_2, \\ & \frac{q}{\Gamma(p)}t_{n+1}^{p+q-1}\sum_{j=0}^{n-1}f_2\left(t_j+\frac{\Delta t}{2},\frac{x_j+x_{j+1}}{2},\frac{y_j+y_{j+1}}{2},\frac{z_j+z_{j+1}}{2}\right)\widehat{B}_1 \\ & +\frac{q}{\Gamma(p)}t_{n+1}^{p+q-1}f_2\left(t_n+\frac{\Delta t}{2},\frac{x_n+x_{n+1}^u}{2},\frac{y_n+y_{n+1}^u}{2},\frac{z_n+z_{n+1}^u}{2}\right)\widehat{B}_2, \end{aligned}$$

$$z_{n+1} = \begin{aligned} & \frac{q}{\Gamma(p)}t_{n+1}^{p+q-1}\sum_{j=0}^{n-1}f_3\left(t_j+\frac{\Delta t}{2},\frac{x_j+x_{j+1}}{2},\frac{y_j+y_{j+1}}{2},\frac{z_j+z_{j+1}}{2}\right)\widehat{B}_1 \\ & +\frac{q}{\Gamma(p)}t_{n+1}^{p+q-1}f_3\left(t_n+\frac{\Delta t}{2},\frac{x_n+x_{n+1}^u}{2},\frac{y_n+y_{n+1}^u}{2},\frac{z_n+z_{n+1}^u}{2}\right)\widehat{B}_2, \end{aligned}$$

where

$$f_1(t,x,y,z) = y - A_1,$$
$$f_2(t,x,y,z) = -\beta xz + A_2,$$
$$f_3(t,x,y,z) = xy - z + y^2 - yz,$$

and

$$\widehat{B}_1 = \left\{B\left(\frac{t_{j+1}}{t_{n+1}},q,p\right) - B\left(\frac{t_j}{t_{n+1}},q,p\right)\right\},$$
$$\widehat{B}_2 = \left\{B(1,q,p) - B\left(\frac{t_n}{t_{n+1}},q,p\right)\right\}.$$

For the above scheme, we have the graphical results in Figures 20.1–20.3 by using different fractal and fractional order values.

20.4.2 NUMERICAL SCHEME FOR EXPONENTIAL CASE

We consider in this subsection the following fractal-fractional Cauchy problem:

$$_0^{FFE}D_t^{p,q}\mathbf{g}(t) = f(t,\mathbf{g}(t)), \text{ if } t > 0,$$
$$\mathbf{g}(0) = \mathbf{g}_0, \text{ if } t = 0.$$

We can write further

$$_0^{CFR}D_t^p\mathbf{g}(t) = qt^{q-1}f(t,\mathbf{g}(t)), \text{ if } t > 0,$$
$$\mathbf{g}(0) = \mathbf{g}_0, \text{ if } t = 0.$$

We have further the following

$$\mathbf{g}(t) = qt^{q-1}(1-p)f(t,\mathbf{g}(t))t^{q-1} + pq\int_0^t \tau^{q-1}f(\tau,\mathbf{g}(\tau))d\tau, \text{ if } t > 0,$$
$$\mathbf{g}(0) = \mathbf{g}_0, \text{ if } t = 0.$$

312 Numerical Methods for Fractal-Fractional Differential

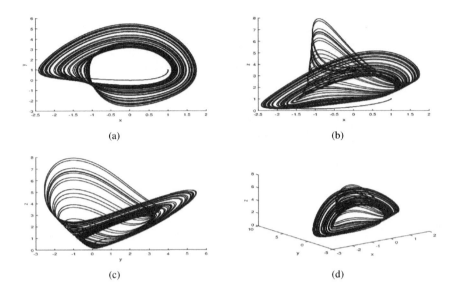

Figure 20.1 Model simulation of fractal-fractional Caputo model when $p = 1$ and $q = 1$.

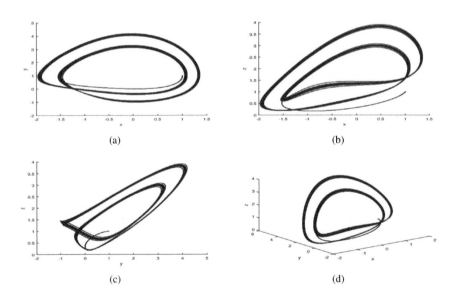

Figure 20.2 Model simulation of fractal-fractional Caputo model when $p = 0.98$ and $q = 1$.

The Dynamics of Multiple Chaotic Attractor with Fractal-Fractional Operators

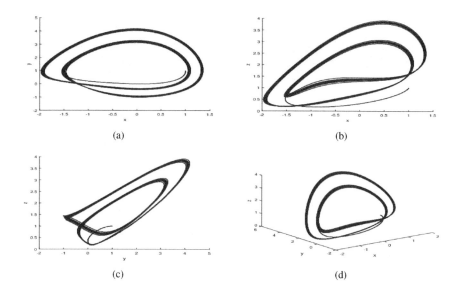

Figure 20.3 Model simulation of fractal-fractional Caputo model when $p = 0.98$ and $q = 0.98$.

At $t = t_{n+1}$ and $t = t_n$, we have

$$\mathbf{g}(t_{n+1}) = (1-p)qt_{n+1}^{q-1}f(t_{n+1}, \mathbf{g}_{n+1}^p) + pq \int_0^{t_{n+1}} \tau^{q-1} f(\tau, \mathbf{g}(\tau)) d\tau,$$

and

$$\mathbf{g}(t_n) = (1-p)qt_n^{q-1}f(t_n, \mathbf{g}_n) + pq \int_0^{t_n} \tau^{q-1} f(\tau, \mathbf{g}(\tau)) d\tau.$$

Therefore,

$$\mathbf{g}(t_{n+1}) = \mathbf{g}_n + (1-p)q \left(t_{n+1}^{q-1} f(t_{n+1}, \mathbf{g}_{n+1}^u) - t_n^{q-1} f(t_n, \mathbf{g}_n) \right)$$
$$+ pq \int_{t_n}^{t_{n+1}} \tau^{q-1} f(\tau, \mathbf{g}(\tau)) d\tau,$$

$$\mathbf{g}_{n+1} = \mathbf{g}_n + (1-p)q \left[t_{n+1}^{q-1} f(t_{n+1}, \mathbf{g}_{n+1}^u) - t_n^{q-1} f(t_n, \mathbf{g}_n) \right]$$
$$+ p(\Delta t)^q f \left(t_n + \frac{\Delta t}{2}, \mathbf{g}_n + (\Delta t)^q f(t_n, \mathbf{g}_n) \{(n+1)^q - n^q\} \right),$$

where

$$\mathbf{g}_{n+1}^u = \mathbf{g}_n + p(\Delta t)^q f \left(t_n + \frac{\Delta t}{2}, \mathbf{g}_n + (\Delta t)^q f(t_n, \mathbf{g}_n) \{(n+1)^q - n^q\} \right).$$

Finally, the scheme for our proposed model can be shown by the following form:

$$\begin{aligned}
x_{n+1} = &\ x_n + (1-p)q\left[t_{n+1}^{q-1}f_1(t_{n+1},x_{n+1}^u,y_{n+1}^u,z_{n+1}^u) - t_n^{q-1}f_1(t_n,x_n,y_n,z_n)\right] \\
&+ p(\Delta t)^q f_1\left(t_n + \frac{\Delta t}{2}, x_n + (\Delta t)^q f_1(t_n,x_n,y_n,z_n)\{(n+1)^q - n^q\},\right. \\
&\ y_n + (\Delta t)^q f_1(t_n,x_n,y_n,z_n)\{(n+1)^q - n^q\}, z_n + (\Delta t)^q f_1(t_n,x_n,y_n,z_n) \\
&\ \left.\{(n+1)^q - n^q\}\right),
\end{aligned}$$

$$\begin{aligned}
y_{n+1} = &\ x_n + (1-p)q\left[t_{n+1}^{q-1}f_2(t_{n+1},x_{n+1}^u,y_{n+1}^u,z_{n+1}^u) - t_n^{q-1}f_2(t_n,x_n,y_n,z_n)\right] \\
&+ p(\Delta t)^q f_2\left(t_n + \frac{\Delta t}{2}, x_n + (\Delta t)^q f_2(t_n,x_n,y_n,z_n)\{(n+1)^q - n^q\},\right. \\
&\ y_n + (\Delta t)^q f_2(t_n,x_n,y_n,z_n)\{(n+1)^q - n^q\}, z_n + (\Delta t)^q f_2(t_n,x_n,y_n,z_n) \\
&\ \left.\{(n+1)^q - n^q\}\right),
\end{aligned}$$

$$\begin{aligned}
z_{n+1} = &\ z_n + (1-p)q\left[t_{n+1}^{q-1}f_3(t_{n+1},x_{n+1}^u,y_{n+1}^u,z_{n+1}^u) - t_n^{q-1}f_3(t_n,x_n,y_n,z_n)\right] \\
&+ p(\Delta t)^q f_3\left(t_n + \frac{\Delta t}{2}, x_n + (\Delta t)^q f_3(t_n,x_n,y_n,z_n)\{(n+1)^q - n^q\},\right. \\
&\ y_n + (\Delta t)^q f_3(t_n,x_n,y_n,z_n)\{(n+1)^q - n^q\}, z_n + (\Delta t)^q f_3(t_n,x_n,y_n,z_n) \\
&\ \left.\{(n+1)^q - n^q\}\right),
\end{aligned}$$

where

$$\begin{aligned}
x_{n+1}^u = &\ x_n + p(\Delta t)^q f_1\left(t_n + \frac{\Delta t}{2}, x_n + (\Delta t)^q f_1(t_n,x_n,y_n,z_n)\{(n+1)^q - n^q\},\right. \\
&\ y_n + (\Delta t)^q f_1(t_n,x_n,y_n,z_n)\{(n+1)^q - n^q\}, z_n + (\Delta t)^q f_1(t_n,x_n,y_n,z_n) \\
&\ \left.\{(n+1)^q - n^q\}\right),
\end{aligned}$$

$$\begin{aligned}
y_{n+1}^u = &\ y_n + p(\Delta t)^q f_2\left(t_n + \frac{\Delta t}{2}, x_n + (\Delta t)^q f_2(t_n,x_n,y_n,z_n)\{(n+1)^q - n^q\},\right. \\
&\ y_n + (\Delta t)^q f_2(t_n,x_n,y_n,z_n)\{(n+1)^q - n^q\}, z_n + (\Delta t)^q f_2(t_n,x_n,y_n,z_n) \\
&\ \left.\{(n+1)^q - n^q\}\right),
\end{aligned}$$

$$\begin{aligned}
z_{n+1}^u = &\ z_n + p(\Delta t)^q f_3\left(t_n + \frac{\Delta t}{2}, x_n + (\Delta t)^q f_3(t_n,x_n,y_n,z_n)\{(n+1)^q - n^q\},\right. \\
&\ y_n + (\Delta t)^q f_3(t_n,x_n,y_n,z_n)\{(n+1)^q - n^q\}, z_n + (\Delta t)^q f_3(t_n,x_n,y_n,z_n)
\end{aligned}$$

The Dynamics of Multiple Chaotic Attractor with Fractal-Fractional Operators 315

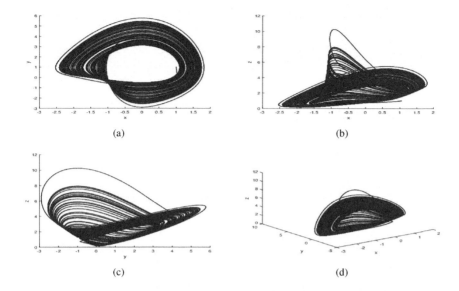

Figure 20.4 Model simulation of fractal-fractional Caputo-Fabrizio model when $p = 1$ and $q = 1$.

$$\{(n+1)^q - n^q\}\Big).$$

We use the obtained scheme given above for exponential kernel, and the results are presented graphically in Figures 20.4 and 20.5. We used various values of the fractal and fractional order parameters and presented the graphical results.

20.4.3 NUMERICAL SCHEME FOR THE MITTAG-LEFFLER CASE

This subsection presents the numerical scheme for Mittag-Leffler law using middle-point interpolation. We consider the following fractal-fractional Cauchy problem for the Mittag-Leffler case:

$$_0^{FFM}D_t^{p,q}\mathbf{g}(t) = f(t,\mathbf{g}(t)), \text{ if } t > 0,$$

$$\mathbf{g}(0) = \mathbf{g}_0, \text{ if } t = 0.$$

We can write further

$$_0^{ABR}D_t^p\mathbf{g}(t) = qt^{q-1}f(t,\mathbf{g}(t)), \text{ if } t > 0,$$

$$\mathbf{g}(0) = \mathbf{g}_0, \text{ if } t = 0.$$

Further, we write

$$\mathbf{g}(t) = (1-p)qt^{q-1}f(t,\mathbf{g}(t)) + \frac{pq}{\Gamma(p)}\int_0^t \tau^{q-1}f(\tau,\mathbf{g}(\tau))(t-\tau)^{p-1}d\tau,$$

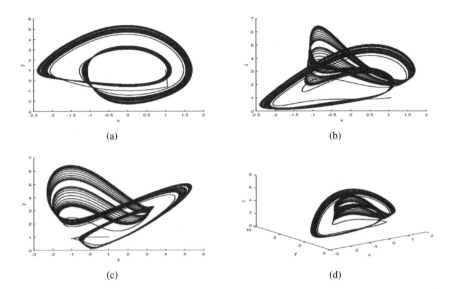

Figure 20.5 Model simulation of fractal-fractional Caputo-Fabrizio model when $p = 0.99$ and $q = 0.9$.

$$\mathbf{g}(0) = \mathbf{g}_0.$$

For $t = t_{n+1}$, we have

$$\mathbf{g}(t_{n+1}) = (1-p)qt_{n+1}^{q-1}f(t_{n+1},\mathbf{g}^u(t_{n+1}))$$
$$+ \frac{pq}{\Gamma(p)}\sum_{j=0}^{n}\int_{t_j}^{t_{j+1}} \tau^{q-1}f(\tau,\mathbf{g}(\tau))(t_{n+1}-\tau)^{p-1}d\tau,$$

whereas

$$\mathbf{g}^u(t_{n+1}) = \frac{q}{\Gamma(p)}\int_0^{t_{n+1}} \tau^{q-1}(t_{n+1}-\tau)^{p-1}f(\tau,\mathbf{g}(\tau))d\tau,$$
$$= \frac{q}{\Gamma(p)}\sum_{j=0}^{n}\int_{t_j}^{t_{j+1}} \tau^{q-1}(t_{n+1}-\tau)^{p-1}f(\tau,\mathbf{g}(\tau))d\tau.$$

Hence, we use the Euler approximations,

$$\mathbf{g}_{n+1}^u = \frac{q}{\Gamma(p)}\sum_{j=0}^{n}\int_{t_j}^{t_{j+1}} f(t_j,\mathbf{g}_j)(t_{n+1}-\tau)^{p-1}\tau^{q-1}d\tau,$$
$$= \frac{q}{\Gamma(p)}\sum_{j=0}^{n} f(t_j,\mathbf{g}_j)t_{n+1}^{p+q-1}\left[B\left(\frac{t_{j+1}}{t_{n+1}},q,p\right) - B\left(\frac{t_j}{t_{n+1}},q,p\right)\right].$$

Therefore, the numerical scheme is given as

$$\mathbf{g}_{n+1} = \frac{q}{\Gamma(p)} t_{n+1}^{p+q-1} \sum_{j=0}^{n-1} f\left(t_j + \frac{\Delta t}{2}, \frac{\mathbf{g}_j + \mathbf{g}_{j+1}}{2}\right) \left[B\left(\frac{t_{j+1}}{t_{n+1}}, q, p\right) - B\left(\frac{t_j}{t_{n+1}}, q, p\right)\right]$$

$$+ \frac{q}{\Gamma(p)} t_{n+1}^{p+q-1} f\left(t_n + \frac{\Delta t}{2}, \frac{\mathbf{g}_n}{2} + \frac{\mathbf{g}_{n+1}^u}{2}\right) \left\{B(1, q, p) - B\left(\frac{t_n}{t_{n+1}}, q, p\right)\right\},$$

and

$$\mathbf{g}_{(t_{n+1})} \approx \mathbf{g}_{n+1} = (1-p) q t_{n+1}^{q-1} f(t_{n+1}, \mathbf{g}_{n+1}^u)$$

$$+ \frac{pq}{\Gamma(p)} \sum_{j=0}^{n-1} f\left(t_j + \frac{h}{2}, \frac{\mathbf{g}_j + \mathbf{g}_{j+1}}{2}\right) t_{n+1}^{p+q-1}$$

$$\times \left[B\left(\frac{t_{j+1}}{t_{n+1}}, q, p\right) - B\left(\frac{t_j}{t_{n+1}}, q, p\right)\right]$$

$$+ \frac{pq}{\Gamma(p)} t_{n+1}^{p+q-1} \left\{B(1, q, p) - B\left(\frac{t_n}{t_{n+1}}, q, p\right)\right\},$$

where,

$$\mathbf{g}_{n+1}^u = \frac{q}{\Gamma(p)} \sum_{j=0}^{n} f(t_j, \mathbf{g}_j) t_{n+1}^{p+q-1} \left[B\left(\frac{t_{j+1}}{t_{n+1}}, q, p\right) - B\left(\frac{t_j}{t_{n+1}}, q, p\right)\right].$$

We now write the scheme for our system given by

$$x_{n+1} = (1-p) q t_{n+1}^{q-1} f_1(t_{n+1}, x_{n+1}^u, y_{n+1}^u, z_{n+1}^u)$$

$$+ \frac{pq}{\Gamma(p)} \sum_{j=0}^{n-1} f_1\left(t_j + \frac{h}{2}, \frac{x_j + x_{j+1}}{2}, \frac{y_j + y_{j+1}}{2}, \frac{z_j + z_{j+1}}{2}\right) t_{n+1}^{p+q-1}$$

$$\times \left[B\left(\frac{t_{j+1}}{t_{n+1}}, q, p\right) - B\left(\frac{t_j}{t_{n+1}}, q, p\right)\right]$$

$$+ \frac{pq}{\Gamma(p)} t_{n+1}^{p+q-1} \left\{B(1, q, p) - B\left(\frac{t_n}{t_{n+1}}, q, p\right)\right\},$$

$$y_{n+1} = (1-p) q t_{n+1}^{q-1} f_2(t_{n+1}, x_{n+1}^u, y_{n+1}^u, z_{n+1}^u)$$

$$+ \frac{pq}{\Gamma(p)} \sum_{j=0}^{n-1} f_2\left(t_j + \frac{h}{2}, \frac{x_j + x_{j+1}}{2}, \frac{y_j + y_{j+1}}{2}, \frac{z_j + z_{j+1}}{2}\right) t_{n+1}^{p+q-1}$$

$$\times \left[B\left(\frac{t_{j+1}}{t_{n+1}}, q, p\right) - B\left(\frac{t_j}{t_{n+1}}, q, p\right)\right]$$

$$+ \frac{pq}{\Gamma(p)} t_{n+1}^{p+q-1} \left\{B(1, q, p) - B\left(\frac{t_n}{t_{n+1}}, q, p\right)\right\},$$

$$z_{n+1} = (1-p) q t_{n+1}^{q-1} f_3(t_{n+1}, x_{n+1}^u, y_{n+1}^u, z_{n+1}^u)$$

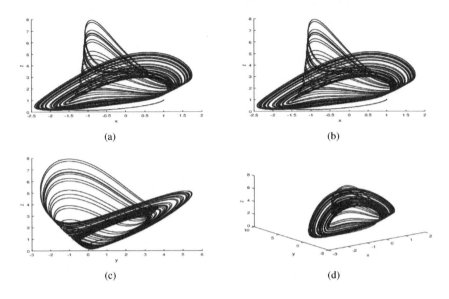

Figure 20.6 Model simulation of fractal-fractional Atangana-Baleanu model when $p = 1$ and $q = 1$.

$$+ \frac{pq}{\Gamma(p)} \sum_{j=0}^{n-1} f_3\left(t_j + \frac{h}{2}, \frac{x_j + x_{j+1}}{2}, \frac{y_j + y_{j+1}}{2}, \frac{z_j + z_{j+1}}{2}\right) t_{n+1}^{p+q-1}$$
$$\times \left[B\left(\frac{t_{j+1}}{t_{n+1}}, q, p\right) - B\left(\frac{t_j}{t_{n+1}}, q, p\right)\right]$$
$$+ \frac{pq}{\Gamma(p)} t_{n+1}^{p+q-1} \left\{B(1, q, p) - B\left(\frac{t_n}{t_{n+1}}, q, p\right)\right\},$$

where

$$x_{n+1}^u = \frac{q}{\Gamma(p)} \sum_{j=0}^{n} f_1(t_j, x_j, y_j, z_j) t_{n+1}^{p+q-1} \left[B\left(\frac{t_{j+1}}{t_{n+1}}, q, p\right) - B\left(\frac{t_j}{t_{n+1}}, q, p\right)\right],$$
$$y_{n+1}^u = \frac{q}{\Gamma(p)} \sum_{j=0}^{n} f_2(t_j, x_j, y_j, z_j) t_{n+1}^{p+q-1} \left[B\left(\frac{t_{j+1}}{t_{n+1}}, q, p\right) - B\left(\frac{t_j}{t_{n+1}}, q, p\right)\right],$$
$$z_{n+1}^u = \frac{q}{\Gamma(p)} \sum_{j=0}^{n} f_3(t_j, x_j, y_j, z_j) t_{n+1}^{p+q-1} \left[B\left(\frac{t_{j+1}}{t_{n+1}}, q, p\right) - B\left(\frac{t_j}{t_{n+1}}, q, p\right)\right].$$

The numerical results for middle-point interpolation based scheme for fractal-fractional system in the sense of Mittag-Leffler kernel, we the simulation results given in Figures 20.6–20.8, for various values of fractal and fractional order parameters.

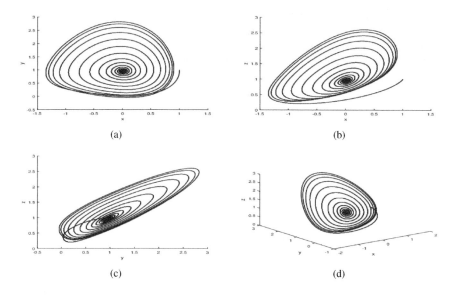

Figure 20.7 Model simulation of fractal-fractional Atangana-Baleanu model when $p = 0.98$ and $q = 1$.

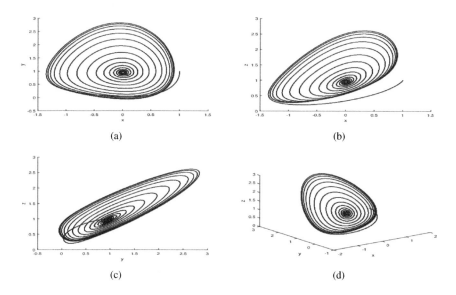

Figure 20.8 Model simulation of fractal-fractional Atangana-Baleanu model when $p = 0.98$ and $q = 0.98$.

20.5 CONCLUSION

We considered a fractal-fractional order chaotic system in various operators. The model equilibrium points are obtained, and their stability analysis has been discussed. We have proven the existence and uniqueness of the solution of the fractal-fractional order system. Then, we considered each case of fractal-fractional system in the sense of power law, exponential law, and Mittag-Leffler law and obtained numerical schemes based on middle-point interpolations. The numerical results are obtained graphically for each case using fractal and fractional order parameter values.

21 Dynamics of 3D Chaotic Systems with Fractal-Fractional Operators

21.1 INTRODUCTION

The concept of chaos theory emerged from the realisation that deterministic physical systems, despite appearing to be simple, can exhibit unpredictable behaviour. There are three possible outcomes for non-linear systems: equilibrium, steady oscillation, and chaotic change. For a deterministic equation to depict non-random chaos, these three regimes—known as the logistic map—can be described by straightforward mathematical equations. Mandelbrot, however, used fractal geometry, a novel type of geometry, to model chaotic systems. On all scales, fractals exhibit irregularity, but to the same degree. The propensity for repeating self-similarity in ferns, cauliflowers, and broccoli is an example of this in nature. Fractals can exhibit randomness or nonlinearity in addition to linear self-similarity. In order to model coastlines, mountains, and cloud patterns, random fractals have been used in advanced computer graphics. Non-linear, deterministic chaos can be seen in the equations used to model turbulence in liquids and weather patterns. They have been applied to simulate the unexpected elements of human behaviour in social and economic systems, as well as the morphology of cities. To evaluate changes in the state of such systems, complex metrics have been devised, and some of these metrics can now be calculated using computer software. The literature deals the chaotic systems with some specific equilibria. Chaotic systems considered in the literature without equilibrium points are studied in [199, 266], those with equilibria and stable are discussed in [116, 263], with equilibria of curves see, [32, 196], with surfaces of equilibria [92, 93]. The chaotic models studied with non-hyperbolic equilibria are given in [270]. The chaotic model with symmetry feature is considered in [129], and amplitude control is in [133, 134]. The generation of the multiwing chaotic models is studied in [278] and multiscroll attractors in [171, 237]. In non-linear dynamics the multiscroll attractors are the interested research area. The articles with megastability, extreme multistability, and with memristive elements of chaotic models are studied, respectively, in [27, 29, 222, 231], and [30, 31].

The chaotic models considered a lot of attention in integer order study as described above. But the fractional calculus is also gaining much attention from the researchers since the last two decades and particularly since the last years with

DOI: 10.1201/9781003359258-21

the definitions of the new operators. It is well-known that the fractional order models are considered important because of the orders, which allows one to have more information at each value of the fractional order. Some important works related to fractional-order chaotic and non-chaotic models considered in literature are [17, 19, 95, 105, 109, 110, 194, 212] and much more. The mathematical models that first address with the new fractional operators and their applications to the real data are considered in [19, 139, 261]. In the present chapter, we investigate the dynamics of a chaotic system using different fractal-fractional operators. The fractal-fractional operators considered in this chapter are the Caputo, Caputo-Fabrizio, and Atangana-Baleanu. We give details of the model equilibrium points and show the existence and uniqueness of the model solutions. We present the numerical approaches based on Euler approach and obtain the method for the case of power law, exponential law, and the Mittag-Leffler law. We have many graphical results for the case of fractal-fractional model using different fractal-fractional orders.

21.2 MODEL DESCRIPTIONS AND THEIR ANALYSIS

We consider the following chaotic model governed by the system of non-linear differential equations:

$$\begin{cases} \frac{dx(t)}{dt} = z, \\ \frac{dy(t)}{dt} = -z(ay+bxz), \\ \frac{dz(t)}{dt} = x+y^2 - r. \end{cases} \quad (21.1)$$

In system (21.1), $x(t)$, $y(t)$, and $z(t)$ denote the state variables and a, b, and r are the parameters. The system becomes chaotic for the numerical values of these parameters, $a = 12, b = 0.4$, and $r = 1$ together with the initial conditions of the state variables $x(0) = 0$, $y(0) = 0$, and $z(0) = 0$. In order to formulate the integer model (21.1) in the form of fractal-fractional form, we write the system (21.1) in the following form:

$$\begin{cases} {}^{FF}D_{0,t}^{p,q}\left(x(t)\right) = z, \\ {}^{FF}D_{0,t}^{p,q}\left(y(t)\right) = -z(ay+bxz), \\ {}^{FF}D_{0,t}^{p,q}\left(z(t)\right) = x+y^2 - r, \end{cases} \quad (21.2)$$

where p represents the fractional order and q is the fractal order.

21.3 EXISTENCE AND UNIQUENESS

In this section, we study the existence and puniness of the model (21.2). It is assumed that for all $t \in [0, T]$, the functions $x(t)$, $y(t)$, and $z(t)$ are bounded. Further, we write the model (21.2) in the form:

$$f_x(t, x, y, z) = z,$$

$$f_y(t, x, y, z) = -z(ay+bxz),$$

$$f_z(t,x,y,z) = x+y^2-r.$$

We prove that $f_x, f_y,$ and f_z satisfy the linear growth and Lipschitz conditions. We first show that the functions satisfy the linear growth property.

$$|f_x(t,x,y,z)|^2 = |z|^2,$$
$$\leq M_z^2(1+|x|^2),$$
$$|f_y(t,x,y,z)|^2 = |z|^2 a^2 |y|^2,$$
$$\leq M_z^2(1+|y|^2),$$

$$|f_z(t,x,y,z)|^2 = 0 \leq (1+|z|^2),$$

so the system admits a linear growth property. Next, we show the Lipschitz conditions:

$$|f_x(t,x_1,y,z) - f_x(t,x_2,y,z)|^2 = 0 \leq |x_1-x_2|^2,$$
$$|f_y(t,x,y_1,z) - f_y(t,x,y_2,z)|^2 = |z|^2 a^2 |y_1-y_2|^2,$$
$$\leq M_z^2 a^2 |y_1-y_2|^2,$$
$$|f_z(t,x,y,z_1) - f_z(t,x,y,z_2)|^2 = = 0 \leq |z_1-z_2|^2.$$

Thus, the system admits linear growth and Lipschitz condition. So, we have a unique solution.

21.3.1 EQUILIBRIUM POINTS AND THEIR ANALYSIS

We can obtain the equilibrium points of the above system (21.2) by equating the right side equal to zero and get the following equilibrium points, $E_1 = (r-y^2, -\sqrt{r-x}, 0)$ and $E_2 = (r-y^2, \sqrt{r-x}, 0)$. In order to discuss the local asymptotic stability at the given equilibrium points, we need to obtain first the Jacobian matrix of the system (21.2), which is given by

$$J = \begin{pmatrix} 0 & 0 & 1 \\ -bz^{*2} & -az^* & -ay^* - 2bx^*z^* \\ 1 & 2y^* & 0 \end{pmatrix}.$$

The Jacobian matrix J at the equilibrium point E_1 gives the eigenvalues $\lambda_1 = 0$, $\lambda_2 = -\sqrt{24x^* - 23}$, and $\lambda_3 = \sqrt{24x^* - 23}$, and for the equilibrium point E_2, we have the eigenvalues $\lambda_1 = 0, \lambda_2 = -\sqrt{24x^* - 23}$, and $\lambda_3 = \sqrt{24x^* - 23}$. It can be observed that for some positive value of x^*, one can get a real value for $\lambda_{2,3}$, but it will be opposite, and hence both the equilibrium points are unstable.

21.4 NUMERICAL PROCEDURE FOR THE CHAOTIC MODEL USING EULER-BASED METHOD

This section explains the procedure for the numerical solution of fractal-fractional order system in different fractal-fractional operators. The new scheme will be considered, which is based on the Euler method for the solution of fractal-fractional operators in the sense of power law, exponential law, and the Mittag-Leffler law. In the following subsection, each scheme will be obtained with numerical results.

21.4.1 EULER-BASED NUMERICAL SCHEME FOR FF-CAPUTO OPERATOR

We present a numerical scheme for system (21.2), which is based on the Euler method for the numerical solution of fractal-fractional differential equation in the sense of Caputo derivative. In order to get the numerical scheme with Euler based, we need to consider the Cauchy problem below,

$$_{0}^{FFP}D_{t}^{p,q}g(t) = f(t,g(t)), \text{ if } t > 0,$$

$$g(0) = g_0, \text{ if } t = 0. \qquad (21.3)$$

$$_{0}^{RL}D_{t}^{p}g(t) = qt^{q-1}f(t,g(t)) \text{ if } t > 0$$

$$g(0) = g_0 \text{ if } t = 0.$$

We write

$$g(t) = \frac{q}{\Gamma(p)} \int_0^t \tau^{q-1} f(\tau, g(\tau))(t-\tau)^{p-1} d\tau, \; g(0) = g_0.$$

When $t = t_{n+1}$, we get the following:

$$g(t_{n+1}) = \frac{q}{\Gamma(p)} \int_0^{t_{n+1}} \tau^{q-1} f(\tau, g(\tau))(t_{n+1}-\tau)^{p-1} d\tau.$$

Finally, we have the scheme after simplifying the results,

$$g(t_{n+1}) = \frac{q}{\Gamma(p)} \sum_{j=0}^{n} \int_{t_j}^{t_{j+1}} \tau^{q-1} f(\tau, g(\tau))(t_{n+1}-\tau)^{p-1} d\tau,$$

$$\approx \frac{q}{\Gamma(p)} \sum_{j=0}^{n} f(t_j, g_j) \int_{t_j}^{t_{j+1}} \tau^{q-1}(t_{n+1}-\tau)^{p-1} d\tau,$$

$$\approx \frac{q}{\Gamma(p)} \sum_{j=0}^{n} f(t_j, g_j) t_{n+1}^{p+q-1} \left[B\left(\frac{t_{j+1}}{t_{n+1}}, q, p\right) - B\left(\frac{t_j}{t_{n+1}}, q, p\right) \right]. \quad (21.4)$$

Applying the scheme above (21.4) to our model, then we have the following representations:

$$x_{n+1} = \frac{q}{\Gamma(p)} \sum_{j=0}^{n} f_1(t_j, x_j, y_j, z_j) t_{n+1}^{p+q-1} \left[B\left(\frac{t_{j+1}}{t_{n+1}}, q, p\right) - B\left(\frac{t_j}{t_{n+1}}, q, p\right) \right],$$

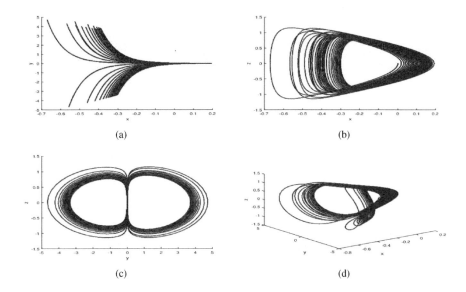

Figure 21.1 Model simulation of fractal-fractional Caputo model when $p = 1$ and $q = 1$.

$$y_{n+1} = \frac{q}{\Gamma(p)} \sum_{j=0}^{n} f_2(t_j, x_j, y_j, z_j) t_{n+1}^{p+q-1} \left[B\left(\frac{t_{j+1}}{t_{n+1}}, q, p\right) - B\left(\frac{t_j}{t_{n+1}}, q, p\right) \right],$$

$$z_{n+1} = \frac{q}{\Gamma(p)} \sum_{j=0}^{n} f_3(t_j, x_j, y_j, z_j) t_{n+1}^{p+q-1} \left[B\left(\frac{t_{j+1}}{t_{n+1}}, q, p\right) \right.$$
$$\left. - B\left(\frac{t_j}{t_{n+1}}, q, p\right) \right]. \tag{21.5}$$

The graphical solution using the scheme (21.5) has been obtained for the model (21.2) shown in Figures 21.1–21.3 for some arbitrary values of the fractal and fractional orders.

21.4.2 EULER-BASED NUMERICAL SCHEME FOR FF-CF OPERATOR

We give the details of the numerical scheme based on Euler method for FF-Caputo-Fabrizio derivative. We write the model in the form of Caputo-Fabrizio fractal-fractional operator in the following:

$$^{CF}D_{0,t}^{p}\left(x(t)\right) = qt^{q-1}f_1(x,y,z,t),$$
$$^{CF}D_{0,t}^{p}\left(y(t)\right) = qt^{q-1}f_2(x,y,z,t),$$
$$^{CF}D_{0,t}^{p}\left(z(t)\right) = qt^{q-1}f_3(x,y,z,t). \tag{21.6}$$

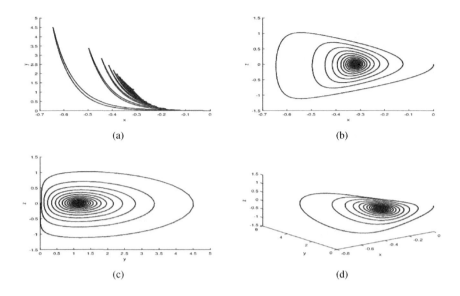

Figure 21.2 Model simulation of fractal-fractional Caputo model when $p = 0.98$ and $q = 0.97$.

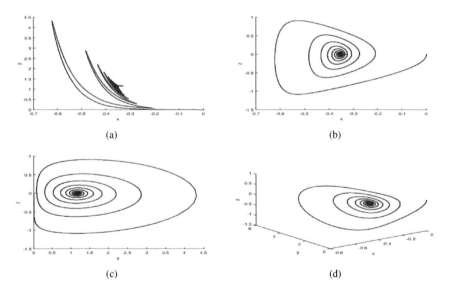

Figure 21.3 Model simulation of fractal-fractional Caputo model when $p = 0.96$ and $q = 0.94$.

In order to get the numerical scheme for the exponential kernel, we write the following differential equation in general form:

$$_{0}^{FFF}D_{t}^{p,q}g(t) = f(t,g(t)), \text{ if } t > 0,$$

Dynamics of 3D Chaotic Systems with Fractal-Fractional Operators

$$g(0) = g_0, \text{ if } t = 0. \tag{21.7}$$

And also, we write

$$_0^{CFR}D_t^p g(t) = qt^{q-1}f(t,g(t)) \text{ if } t > 0$$

$$g(0) = g_0 \text{ if } t = 0.$$

Then, it can be written as

$$g(t) = (1-p)qt^{q-1}f(t,g(t))$$
$$+ pq \int_0^t \tau^{q-1} f(\tau, g(\tau))d\tau, \text{ if } t > 0,$$

$$g(0) = g_0, \text{ if } t = 0.$$

For $t = t_{n+1}$, we have

$$g(t_{n+1}) = qt_{n+1}^{q-1} f(t_{n+1}, g(t_{n+1}))$$
$$+ pq \int_0^{t_{n+1}} \tau^{q-1} f(\tau, g(\tau))d\tau,$$
$$= qt_{n+1}^{q-1} f(t_{n+1}, g^u(t_{n+1}))$$
$$+ pq \sum_{j=0}^{n} \int_{t_j}^{t_{j+1}} f(\tau, g(\tau)) \tau^{q-1} d\tau.$$

Finally, we get

$$g_{n+1} = qt_{n+1}^{q-1} f(t_{n+1}, g^u(t_{n+1})) + p(\Delta t)^q \sum_{j=0}^{n} f(t_j, g_j) \left\{ (j+1)^q - j^q \right\}, \tag{21.8}$$

where

$$g^u(t_{n+1}) = g_n + p(\Delta t)^q \{(n+1)^q - n^q\} f(t_n, g_n).$$

The scheme (21.8) can be considered for the model (21.2), and hence the scheme for the system (21.2) is as follows:

$$x_{n+1} = qt_{n+1}^{q-1} f_1(t_{n+1}, x^u(t_{n+1}), y^u(t_{n+1}), z^u(t_{n+1}))$$
$$+ p(\Delta t)^q \sum_{j=0}^{n} f_1(t_j, x_j, y_j, z_j) \left\{ (j+1)^\beta - j^q \right\},$$

$$y_{n+1} = qt_{n+1}^{q-1} f_2(t_{n+1}, x^u(t_{n+1}), y^u(t_{n+1}), z^u(t_{n+1}))$$

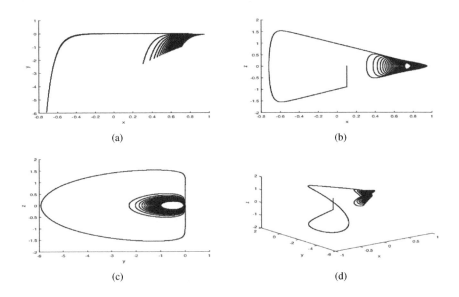

Figure 21.4 Model simulation of fractal-fractional Caputo-Fabrizio model when $p = 1$ and $q = 1$.

$$z_{n+1} = \begin{aligned} &+ p(\Delta t)^q \sum_{j=0}^{n} f_2(t_j, x_j, y_j, z_j) \left\{ (j+1)^q - j^q \right\}, \\ & qt_{n+1}^{q-1} f_3(t_{n+1}, x^u(t_{n+1}), y^u(t_{n+1}), z^u(t_{n+1})) \\ &+ p(\Delta t)^q \sum_{j=0}^{n} f_3(t_j, x_j, y_j, z_j) \left\{ (j+1)^q - j^q \right\}, \end{aligned} \quad (21.9)$$

where

$$x^u(t_{n+1}) = x_n + p(\Delta t)^q \{(n+1)^q - n^q\} f_1(t_n, x_n, y_n, z_n),$$
$$y^u(t_{n+1}) = y_n + p(\Delta t)^q \{(n+1)^q - n^q\} f_2(t_n, x_n, y_n, z_n),$$
$$z^u(t_{n+1}) = z_n + p(\Delta t)^q \{(n+1)^q - n^q\} f_3(t_n, x_n, y_n, z_n). \quad (21.10)$$

The numerical scheme presented above is used to obtain the graphical results for the fractal-fractional model in Caputo-Fabrizio sense and obtained have Figures 21.4–21.6.

Dynamics of 3D Chaotic Systems with Fractal-Fractional Operators

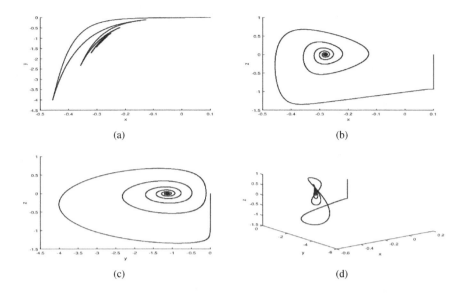

Figure 21.5 Model simulation of fractal-fractional Caputo-Fabrizio model when $p = 0.98$ and $q = 0.97$.

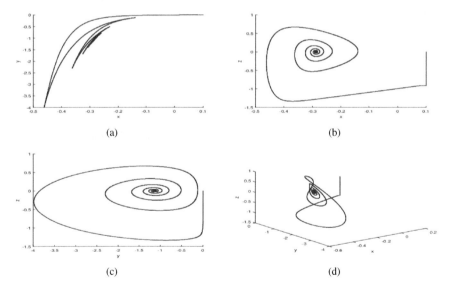

Figure 21.6 Model simulation of fractal-fractional Caputo-Fabrizio model when $p = 0.98$ and $q = 0.99$.

21.4.3 EULER-BASED NUMERICAL SCHEME FOR FF ATANGANA-BALEANU OPERATOR

In order to get a numerical scheme for the numerical solution of fractal-fractional model in the sense of Atangana-Baleanu operator using the new approach of Euler method, we first express the model (21.2) in the following form:

$$^{ABR}D^{\alpha}_{0,t}\left(x(t)\right) = qt^{q-1}f_1(x,y,z,t),$$
$$^{ABR}D^{\alpha}_{0,t}\left(y(t)\right) = qt^{q-1}f_2(x,y,z,t),$$
$$^{ABR}D^{\alpha}_{0,t}\left(z(t)\right) = qt^{q-1}f_3(x,y,z,t). \quad (21.11)$$

Then, we consider a fractal-fractional Cauchy problem with the generalised Mittag-Leffler function given by

$$^{FFM}_0 D^{p,q}_t g(t) = f(t,g(t)), \text{ if } t > 0,$$
$$g(0) = g_0, \text{ if } t = 0. \quad (21.12)$$

Then, we write in the form:

$$^{ABR}_0 D^{p}_t g(t) = qt^{q-1}f(t,g(t)) \text{ if } t > 0$$
$$g(0) = g_0 \text{ if } t = 0.$$

Simplifying the results, and finally, we have the scheme

$$g_{n+1} = qt^{q-1}_{n+1} f(t_{n+1}, g^{u}_{n+1})(1-p) + \frac{pq}{\Gamma(p)}(\Delta t)^{p+q-1}(n+1)^{p+q}$$
$$\times \sum_{j=0}^{n} f(t_j, g_j)\left[B\left(\frac{t_{j+1}}{t_{n+1}}, q, p\right) - B\left(\frac{t_j}{t_{n+1}}, q, p\right)\right], \quad (21.13)$$

where

$$g^{u}_{n+1} = \frac{pq}{\Gamma(p)}(\Delta t)^{p+q-1}(n+1)^{p+\beta-1} \sum_{j=0}^{n} f(t_j, g_j)$$
$$\times \left[B\left(\frac{t_{j+1}}{t_{n+1}}, q, p\right) - B\left(\frac{t_j}{t_{n+1}}, q, p\right)\right].$$

We consider the scheme (21.13) and apply to our proposed system; we have

$$x_{n+1} = qt^{q-1}_{n+1} f_1(t_{n+1}, x^u_{n+1}, y^u_{n+1}, z^u_{n+1})(1-p)$$
$$+ \frac{pq}{\Gamma(p)}(\Delta t)^{p+q-1}(n+1)^{p+q}$$
$$\times \sum_{j=0}^{n} f_1(t_j, x_j, y_j, z_j)\left[B\left(\frac{t_{j+1}}{t_{n+1}}, q, p\right) - B\left(\frac{t_j}{t_{n+1}}, q, p\right)\right],$$
$$y_{n+1} = qt^{q-1}_{n+1} f_1(t_{n+1}, x^u_{n+1}, y^u_{n+1}, z^u_{n+1})(1-p)$$
$$+ \frac{pq}{\Gamma(p)}(\Delta t)^{p+q-1}(n+1)^{p+q}$$

$$\times \sum_{j=0}^{n} f_2(t_j, x_j, y_j, z_j) \left[B\left(\frac{t_{j+1}}{t_{n+1}}, q, p\right) - B\left(\frac{t_j}{t_{n+1}}, q, p\right) \right],$$

$$z_{n+1} = qt_{n+1}^{q-1} f_3(t_{n+1}, x_{n+1}^u, y_{n+1}^u, z_{n+1}^u)(1-p)$$

$$+ \frac{pq}{\Gamma(p)} (\Delta t)^{p+q-1} (n+1)^{p+q}$$

$$\times \sum_{j=0}^{n} f_1(t_j, x_j, y_j, z_j) \left[B\left(\frac{t_{j+1}}{t_{n+1}}, q, p\right) - B\left(\frac{t_j}{t_{n+1}}, q, p\right) \right], \quad (21.14)$$

where

$$x_{n+1}^u = \frac{pq}{\Gamma(p)} (\Delta t)^{p+q-1} (n+1)^{p+q-1} \sum_{j=0}^{n} f_1(t_j, x_j, y_j, z_j)$$

$$\times \left[B\left(\frac{t_{j+1}}{t_{n+1}}, q, p\right) - B\left(\frac{t_j}{t_{n+1}}, q, p\right) \right],$$

$$x_{n+1}^u = \frac{pq}{\Gamma(p)} (\Delta t)^{p+q-1} (n+1)^{p+q-1} \sum_{j=0}^{n} f_2(t_j, x_j, y_j, z_j)$$

$$\times \left[B\left(\frac{t_{j+1}}{t_{n+1}}, q, p\right) - B\left(\frac{t_j}{t_{n+1}}, q, p\right) \right],$$

$$z_{n+1}^u = \frac{pq}{\Gamma(p)} (\Delta t)^{p+q-1} (n+1)^{p+q-1} \sum_{j=0}^{n} f_3(t_j, x_j, y_j, z_j)$$

$$\times \left[B\left(\frac{t_{j+1}}{t_{n+1}}, q, p\right) - B\left(\frac{t_j}{t_{n+1}}, q, p\right) \right].$$

The Euler-based fractal-fractional scheme (21.14) for the numerical solution of the model has been used to get the graphical results in Figures 21.7–21.9. Different numerical values of fractal and factional orders are considered, and the graphical results are obtained.

21.5 CONCLUSION

In this chapter, we investigated the dynamics of new chaotic system under different fractal-fractional operators using the power law kernel, exponential kernel, and the Mittag-Leffler kernel. We presented the algorithm which is recently obtained in the present book. For each fractal-fractional order system, we presented the scheme and obtained the simulation results for various values of fractal-fractional orders.

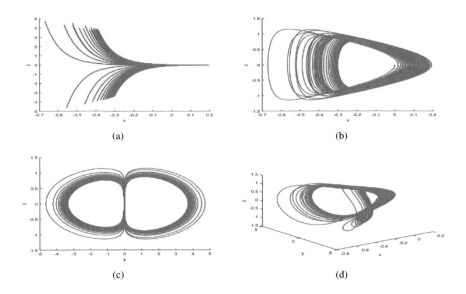

Figure 21.7 Model simulation of fractal-fractional Atangana-Baleanu model when $p = 1$ and $q = 1$.

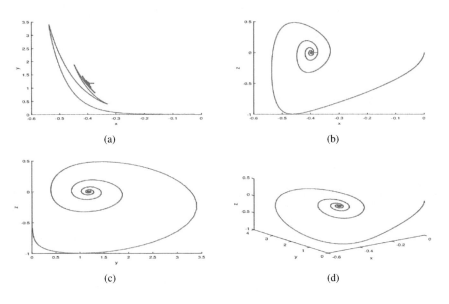

Figure 21.8 Model simulation of fractal-fractional Atangana-Baleanu model when $p = 0.98$ and $q = 0.96$.

Dynamics of 3D Chaotic Systems with Fractal-Fractional Operators

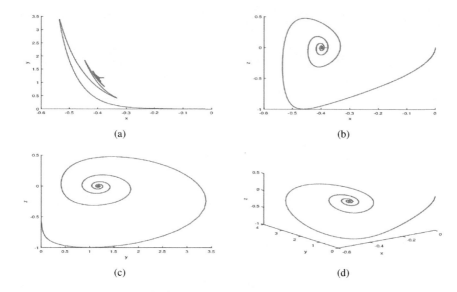

Figure 21.9 Model simulation of fractal-fractional Atangana-Baleanu model when $p = 0.98$ and $q = 0.99$.

22 The Hidden Attractors Model with Fractal-Fractional Operators

22.1 INTRODUCTION

One can see the abrupt transition to unanticipated (unwanted or unknown) attractors in multistable systems, especially when there are attractors with very small basins or previously unknown attractors. Catastrophic occurrences such as rapid climate change, deadly epidemics, financial crises, and malfunctions of commercial equipment can result from such a transition. The dramatic example of a catastrophe brought on by a quick switch to an undesirable attractor is the YF-22 Boeing crash in April 1992. In general, one must first identify all coexisting attractors before using the proper governing scheme to keep the system on the target attractor. In the vicinity of unstable fixed points, the majority of the well-known instances of chaotic and regular attractors, including those of van der Pol, Belousov-Zhabotinsky, Lorenz, Rossler, Chua, and many others, can be found. One can start with the beginning conditions in a close vicinity of the unstable fixed point on the unstable manifold and see how it is attracted. These attractors are known as the self-excited attractors and can be easily localised numerically using the conventional computational approach. Ueda quantitatively created the classic illustration of a self-excited chaotic attractor in a Duffing system in 1961, albeit it didn't become widely known until later. The behaviour in such mathematical models is known as chaotic solution. Researchers always look to have such dynamical system problems, which have such properties described above that have many applications in science and engineering and other social sciences. After the discovery of the Lorenz system in 1963 [150], various numerical and experimental results are obtained for the chaotic system in different areas of science and engineering. In the subject of chemistry the chaotic phenomenon has been investigated in [203]. In classical mechanics, such as the fundamental models of classical mechanics, the example of pendulum with periodic excitations [216], the mathematical models describe the fluid dynamics [71] and the hydrodynamics [86]. In brain activity the unique features of chaotic signals confirm the existence of the chaotic behaviour [24] or electrocardiogram [198] data. In biology, the chaotic behaviour can be observed as the emergence of a population of different species [163]; in economics the commodity exchange can exhibit the chaotic behaviour [164] and the models associated with the ecological phenomenon also have the chaotic behaviour [82]. All these examples are associated with the dynamical systems which depend on time. It is also known that the chaotic behaviour exists in the electrical circuits [137] and also in electronic circuits. We present in this chapter a dynamical chaotic model with fractal-fractional operators. We consider a chaotic system and determine the suitable values of the initial conditions for which the model becomes chaotic. Then, we present the possible fixed points of the model. Further, we formulated the model in each operator and design a numerical procedure for their solution. We consider a new numerical approach using the Euler-based method for the case of fractal-fractional differential equations in the sense of power law, exponential law, and the Mittag-Leffler law. Using the newly developed scheme, we present the numerical results for each model and discuss it in detail.

22.2 MODEL AND ITS ANALYSIS

The chaotic model we consider here is given by the following equations [265]:

$$\begin{cases} \frac{dx(t)}{dt} = y, \\ \frac{dy(t)}{dt} = -x + yz, \\ \frac{dz(t)}{dt} = -x - 15xy - xz, \end{cases} \tag{22.1}$$

where $x(t)$, $y(t)$, and $z(t)$ are the state variables. The initial values for the model parameters are $x(0) = 0.01$, $y(0) = 0.05$, and $z(0) = 0.5$. The fractal-fractional representation of the model (22.1) is given by

$$\begin{cases} {}^{FF}D_{0,t}^{p,q}\left(x(t)\right) = y, \\ {}^{FF}D_{0,t}^{p,q}\left(y(t)\right) = -x + yz, \\ {}^{FF}D_{0,t}^{p,q}\left(z(t)\right) = -x - 15xy - xz, \end{cases} \tag{22.2}$$

where p represents the fractional order, while q is the fractal order.

22.3 EXISTENCE AND UNIQUENESS

In this section, we study the existence and puniness of the model (22.2). It is assumed that for all $t \in [0, T]$, the functions $x(t)$, $y(t)$, and $z(t)$ are bounded. Further, we write the model (22.2) in the form:

$$f_x(t, x, y, z) = y,$$

$$f_y(t, x, y, z) = -x + yz,$$

$$f_z(t, x, y, z) = -x - 15xy - xz.$$

We prove that f_x, f_y, and f_z satisfy the linear growth and Lipschitz conditions. We first show that the functions satisfy the linear growth property.

$$|f_x(t, x, y, z)|^2 = |z|^2 \le M_z^2(1 + |x|^2),$$

The Hidden Attractors Model with Fractal-Fractional Operators 337

$$|f_y(t,x,y,z)|^2 = |-x+yz|^2 \leq 2|x|^2 + 2|z|^2|y|^2,$$
$$\leq 2M_x^2 + 2M_y^2|y|^2,$$
$$\leq 2M_x^2\left(1 + \frac{M_z^2}{M_y^2}|y|^2\right)$$

if $\frac{M_z^2}{M_y^2} < 1$, then

$$|f_y(t,x,y,z)|^2 \leq 2M_x^2(1+|y|^2).$$

$$|f_z(t,x,y,z)|^2 = |-x - 15xy - xz|^2,$$
$$\leq 3|x|^2 + 45|x|^2|y|^2 + 3|x|^2|z|^2,$$
$$\leq 3M_x^2 + 45M_x^2M_y^2 + 3M_x^2|z|^2,$$
$$\leq 3(M_x^2 + 15M_x^2M_y^2)\left(1 + \frac{M_x^2}{3(M_x^2 + 15M_x^2M_y^2)}|z|^2\right),$$
$$\leq (M_x^2 + 15M_x^2M_y^2)(1+|z|^2),$$

if $\frac{M_x^2}{3(M_x^2+15M_x^2M_y^2)} < 1$, then

$$|f_z(t,x,y,z)|^2 \leq (M_x^2 + 15M_x^2M_y^2)(1+|z|^2),$$
$$\leq K_z(1+|z|^2),$$

where $K_z = (M_x^2 + 15M_x^2M_y^2)$. The linear growth property is satisfied by the system. Now, we show the Lipschitz conditions:

$$|f_x(t,x_1,y,z) - f_x(t,x_2,y,z)|^2 = 0 \leq |x_1 - x_2|^2,$$
$$|f_y(t,x,y_1,z) - f_y(t,x,y_2,z)|^2 = M_y|y_1 - y_2|^2,$$
$$|f_z(t,x,y,z_1) - f_z(t,x,y,z_2)|^2 = |x|^2|z_1 - z_2|^2 \leq M_x^2|z_1 - z_2|^2.$$

If $\frac{M_z}{M_x} < 1$, then the system satisfies the linear growth and the Lipschitz conditions. So, the system admits a unique system of solutions.

22.3.1 EQUILIBRIUM POINTS AND THEIR ANALYSIS

For the fractal-fractional model (22.2), we obtain the equilibrium points by setting

$$\begin{cases} 0 = y, \\ 0 = -x + yz, \\ 0 = -x - 15xy - xz; \end{cases}$$

we observe that there is no equilibrium point for this model.

22.4 NUMERICAL PROCEDURE FOR THE CHAOTIC MODEL

The aim of the present section is to present the novel procedure for the solution of the chaotic problem in different fractal-fractional operators. We consider three different newly defined operators known as fractal-fractional Caputo, fractal-fractional Caputo-Fabrizio, and the fractal-fractional Atangana-Baleanu.

22.4.1 NUMERICAL SCHEME WITH EULER FOR FF-CAPUTO OPERATOR

This subsection determines the numerical procedure for the solution of the fractal-fractional model given by (22.2) in the sense of fractal-fractional-Caputo operator. In order to get the numerical scheme with Euler, we consider the Cauchy problem given by

$$\begin{aligned} {}_0^{FFP}D_t^{p,q}g(t) &= f(t,g(t)), \text{ if } t > 0, \\ g(0) &= g_0, \text{ if } t = 0. \\ {}_0^{RL}D_t^p g(t) &= qt^{q-1}f(t,g(t)) \text{ if } t > 0 \\ g(0) &= g_0 \text{ if } t = 0. \end{aligned} \quad (22.3)$$

We write

$$g(t) = \frac{q}{\Gamma(p)} \int_0^t \tau^{q-1} f(\tau, g(\tau))(t-\tau)^{p-1} d\tau, \ g(0) = g_0.$$

When $t = t_{n+1}$, we get the following:

$$g(t_{n+1}) = \frac{q}{\Gamma(p)} \int_0^{t_{n+1}} \tau^{q-1} f(\tau, g(\tau))(t_{n+1}-\tau)^{p-1} d\tau,$$

and finally, we write

$$\begin{aligned} g(t_{n+1}) &= \frac{q}{\Gamma(p)} \sum_{j=0}^{n} \int_{t_j}^{t_{j+1}} \tau^{q-1} f(\tau, g(\tau))(t_{n+1}-\tau)^{p-1} d\tau, \\ &\approx \frac{q}{\Gamma(p)} \sum_{j=0}^{n} f(t_j, g_j) \int_{t_j}^{t_{j+1}} \tau^{q-1} (t_{n+1}-\tau)^{p-1} d\tau, \\ &\approx \frac{q}{\Gamma(p)} \sum_{j=0}^{n} f(t_j, g_j) t_{n+1}^{p+q-1} \left[B\left(\frac{t_{j+1}}{t_{n+1}}, q, p\right) \right. \\ &\quad \left. - B\left(\frac{t_j}{t_{n+1}}, q, p\right) \right]. \end{aligned} \quad (22.4)$$

Applying the scheme above (22.4) to our model, we have the following representations:

$$x_{n+1} = \frac{q}{\Gamma(p)} \sum_{j=0}^{n} f_1(t_j, x_j, y_j, z_j) t_{n+1}^{p+q-1} \left[B\left(\frac{t_{j+1}}{t_{n+1}}, q, p\right) - B\left(\frac{t_j}{t_{n+1}}, q, p\right) \right],$$

The Hidden Attractors Model with Fractal-Fractional Operators

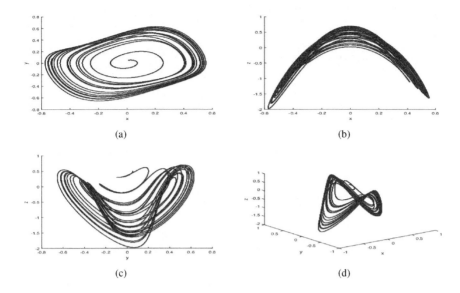

Figure 22.1 Model simulation of fractal-fractional Caputo model when $p = 1$ and $q = 1$.

$$y_{n+1} = \frac{q}{\Gamma(p)} \sum_{j=0}^{n} f_2(t_j, x_j, y_j, z_j) t_{n+1}^{p+q-1} \left[B\left(\frac{t_{j+1}}{t_{n+1}}, q, p\right) - B\left(\frac{t_j}{t_{n+1}}, q, p\right) \right],$$

$$z_{n+1} = \frac{q}{\Gamma(p)} \sum_{j=0}^{n} f_3(t_j, x_j, y_j, z_j) t_{n+1}^{p+q-1} \left[B\left(\frac{t_{j+1}}{t_{n+1}}, q, p\right) \right.$$

$$\left. - B\left(\frac{t_j}{t_{n+1}}, q, p\right) \right]. \tag{22.5}$$

We used the numerical procedure described above for the fractal-fractional model in the sense of Caputo derivative and obtained the graphical results for some selected values of the fractal and fractional order parameters; see Figures 22.1 to 22.4.

22.4.2 NUMERICAL SCHEME WITH EULER FF CAPUTO-FABRIZIO OPERATOR

The aim of this subsection is to present a numerical procedure for the solution of the model in fractal-fractional Caputo-Fabrizio sense. We first made the following representation:

$$^{CF}D_{0,t}^{p}\big(x(t)\big) = qt^{q-1}f_1(x,y,z,t),$$

$$^{CF}D_{0,t}^{p}\big(y(t)\big) = qt^{q-1}f_2(x,y,z,t),$$

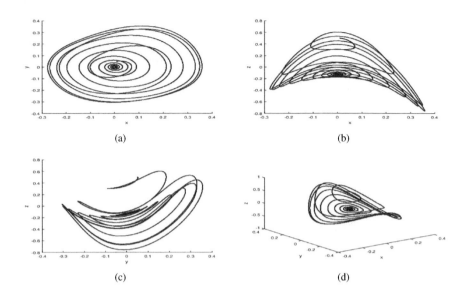

Figure 22.2 Model simulation of fractal-fractional Caputo model when $p = 0.99$ and $q = 0.98$.

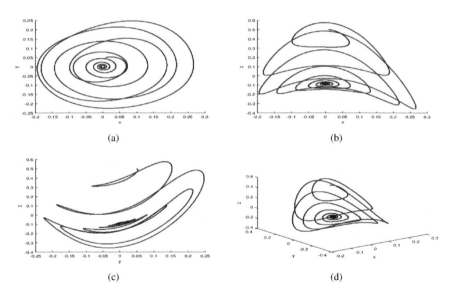

Figure 22.3 Model simulation of fractal-fractional Caputo model when $p = 0.97$ and $q = 0.9$.

$$^{CF}D_{0,t}^{p}\Big(z(t)\Big) \;=\; qt^{q-1}f_3(x,y,z,t).$$

The Hidden Attractors Model with Fractal-Fractional Operators

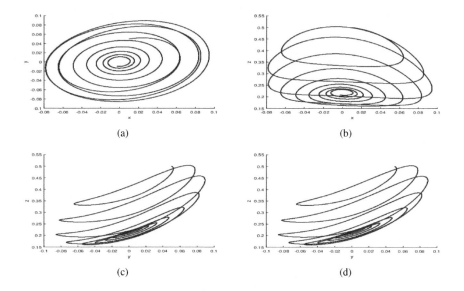

Figure 22.4 Model simulation of fractal-fractional Caputo model when $p = 0.9$ and $q = 0.9$.

In order to get the numerical scheme for exponential kernel, we write the following differential equation in general form:

$$_{0}^{FFF}D_{t}^{p,q}g(t) = f(t,g(t)), \text{ if } t > 0,$$

$$g(0) = g_0, \text{ if } t = 0. \quad (22.6)$$

And also, we write

$$_{0}^{CFR}D_{t}^{p}g(t) = qt^{q-1}f(t,g(t)) \text{ if } t > 0$$

$$g(0) = g_0 \text{ if } t = 0.$$

Then, it can be written as

$$g(t) = (1-p)qt^{q-1}f(t,g(t))$$

$$+ pq\int_{0}^{t} \tau^{q-1}f(\tau,g(\tau))d\tau, \text{ if } t > 0,$$

$$g(0) = g_0, \text{ if } t = 0.$$

For $t = t_{n+1}$, we have

$$g(t_{n+1}) = qt_{n+1}^{q-1}f(t_{n+1},g(t_{n+1}))$$

$$+pq\int_0^{t_{n+1}} \tau^{q-1}f(\tau,g(\tau))d\tau,$$

$$= qt_{n+1}^{q-1}f(t_{n+1},g^u(t_{n+1}))$$

$$+pq\sum_{j=0}^{n}\int_{t_j}^{t_{j+1}} f(\tau,g(\tau))\tau^{q-1}d\tau.$$

Finally, we get

$$g_{n+1} = qt_{n+1}^{q-1}f(t_{n+1},g^u(t_{n+1}))$$

$$+p(\Delta t)^q \sum_{j=0}^{n} f(t_j,g_j)\left\{(j+1)^q - j^q\right\}, \qquad (22.7)$$

where

$$g^u(t_{n+1}) = g_n + p(\Delta t)^q\{(n+1)^q - n^q\}f(t_n,g_n).$$

The scheme (22.7) can be considered for the model (22.2), and hence the scheme for the system (22.2) is as follows:

$$x_{n+1} = qt_{n+1}^{q-1}f_1(t_{n+1},x^u(t_{n+1}),y^u(t_{n+1}),z^u(t_{n+1}))$$

$$+p(\Delta t)^q \sum_{j=0}^{n} f_1(t_j,x_j,y_j,z_j)\left\{(j+1)^\beta - j^q\right\},$$

$$y_{n+1} = qt_{n+1}^{q-1}f_2(t_{n+1},x^u(t_{n+1}),y^u(t_{n+1}),z^u(t_{n+1}))$$

$$+p(\Delta t)^q \sum_{j=0}^{n} f_2(t_j,x_j,y_j,z_j)\left\{(j+1)^q - j^q\right\},$$

$$z_{n+1} = qt_{n+1}^{q-1}f_3(t_{n+1},x^u(t_{n+1}),y^u(t_{n+1}),z^u(t_{n+1}))$$

$$+p(\Delta t)^q \sum_{j=0}^{n} f_3(t_j,x_j,y_j,z_j)\left\{(j+1)^q - j^q\right\}, \qquad (22.8)$$

where

$$x^u(t_{n+1}) = x_n + p(\Delta t)^q\{(n+1)^q - n^q\}f_1(t_n,x_n,y_n,z_n),$$

$$y^u(t_{n+1}) = y_n + p(\Delta t)^q\{(n+1)^q - n^q\}f_2(t_n,x_n,y_n,z_n),$$

$$z^u(t_{n+1}) = z_n + p(\Delta t)^q\{(n+1)^q - n^q\}f_3(t_n,x_n,y_n,z_n). \qquad (22.9)$$

The graphical results shown in Figures 22.6 to 22.7 represent the behaviour of the fractal-fractional model in the sense of Caputo-Fabrizio derivative when selecting different values of the fractal and fractional order considering the scheme presented above.

Next, we present the numerical procedure for the fractal-fractional in the sense of Atangana-Baleanu derivative in the following subsection.

The Hidden Attractors Model with Fractal-Fractional Operators

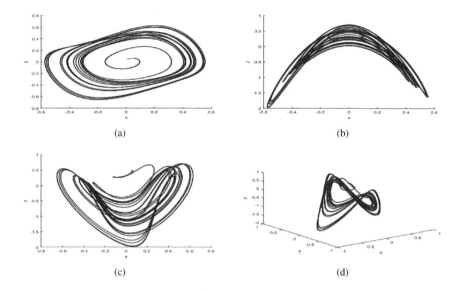

Figure 22.5 Model simulation of fractal-fractional Caputo-Fabrizio model when $p = 1$ and $q = 1$.

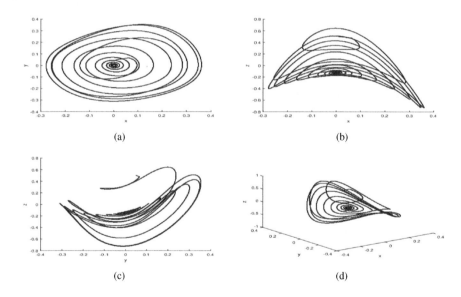

Figure 22.6 Model simulation of fractal-fractional Caputo-Fabrizio model when $p = 0.98$ and $q = 0.96$.

22.4.3 NUMERICAL SCHEME WITH EULER FF ATANGANA-BALEANU

In order to have a numerical procedure for the solution of fractal-fractional model in the sense of Atangana-Baleanu operator, we express first the model (22.2) in the

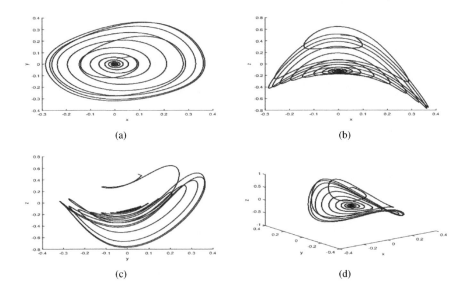

Figure 22.7 Model simulation of fractal-fractional Caputo-Fabrizio model when $p = 0.98$ and $q = 0.94$.

following form:

$$^{ABR}D_{0,t}^{p}\left(x(t)\right) = qt^{q-1}f_1(x,y,z,t),$$

$$^{ABR}D_{0,t}^{p}\left(y(t)\right) = qt^{q-1}f_2(x,y,z,t),$$

$$^{ABR}D_{0,t}^{p}\left(z(t)\right) = qt^{q-1}f_3(x,y,z,t).$$

Then, we consider a fractal-fractional Cauchy problem with the generalised Mittag-Leffler function given by

$$_{0}^{FFM}D_t^{p,q}g(t) = f(t,g(t)), \text{ if } t > 0,$$

$$g(0) = g_0, \text{ if } t = 0. \qquad (22.10)$$

Then, we write in the form:

$$_{0}^{ABR}D_t^{p}g(t) = qt^{q-1}f(t,g(t)) \text{ if } t > 0$$

$$g(0) = g_0 \text{ if } t = 0.$$

Simplifying the results, and finally, we have the scheme

$$g_{n+1} = qt_{n+1}^{q-1}f(t_{n+1},g_{n+1}^u)(1-p) + \frac{pq}{\Gamma(p)}(\Delta t)^{p+q-1}(n+1)^{p+q}$$

$$\times \sum_{j=0}^{n} f(t_j, g_j) \left[B\left(\frac{t_{j+1}}{t_{n+1}}, q, p\right) - B\left(\frac{t_j}{t_{n+1}}, q, p\right) \right], \quad (22.11)$$

where

$$\begin{aligned}
g_{n+1}^u &= \frac{pq}{\Gamma(p)} (\Delta t)^{p+q-1} (n+1)^{p+\beta-1} \sum_{j=0}^{n} f(t_j, g_j) \\
&\quad \times \left[B\left(\frac{t_{j+1}}{t_{n+1}}, q, p\right) - B\left(\frac{t_j}{t_{n+1}}, q, p\right) \right].
\end{aligned}$$

We consider the scheme (22.11) and apply to our proposed system; we have

$$\begin{aligned}
x_{n+1} &= q t_{n+1}^{q-1} f_1(t_{n+1}, x_{n+1}^u, y_{n+1}^u, z_{n+1}^u)(1-p) \\
&\quad + \frac{pq}{\Gamma(p)} (\Delta t)^{p+q-1} (n+1)^{p+q} \\
&\quad \times \sum_{j=0}^{n} f_1(t_j, x_j, y_j, z_j) \left[B\left(\frac{t_{j+1}}{t_{n+1}}, q, p\right) - B\left(\frac{t_j}{t_{n+1}}, q, p\right) \right], \\
y_{n+1} &= q t_{n+1}^{q-1} f_1(t_{n+1}, x_{n+1}^u, y_{n+1}^u, z_{n+1}^u)(1-p) \\
&\quad + \frac{pq}{\Gamma(p)} (\Delta t)^{p+q-1} (n+1)^{p+q} \\
&\quad \times \sum_{j=0}^{n} f_2(t_j, x_j, y_j, z_j) \left[B\left(\frac{t_{j+1}}{t_{n+1}}, q, p\right) - B\left(\frac{t_j}{t_{n+1}}, q, p\right) \right], \\
z_{n+1} &= q t_{n+1}^{q-1} f_3(t_{n+1}, x_{n+1}^u, y_{n+1}^u, z_{n+1}^u)(1-p) \\
&\quad + \frac{pq}{\Gamma(p)} (\Delta t)^{p+q-1} (n+1)^{p+q} \\
&\quad \times \sum_{j=0}^{n} f_1(t_j, x_j, y_j, z_j) \left[B\left(\frac{t_{j+1}}{t_{n+1}}, q, p\right) - B\left(\frac{t_j}{t_{n+1}}, q, p\right) \right], \quad (22.12)
\end{aligned}$$

where

$$\begin{aligned}
x_{n+1}^u &= \frac{pq}{\Gamma(p)} (\Delta t)^{p+q-1} (n+1)^{p+q-1} \sum_{j=0}^{n} f_1(t_j, x_j, y_j, z_j) \\
&\quad \times \left[B\left(\frac{t_{j+1}}{t_{n+1}}, q, p\right) - B\left(\frac{t_j}{t_{n+1}}, q, p\right) \right], \\
x_{n+1}^u &= \frac{pq}{\Gamma(p)} (\Delta t)^{p+q-1} (n+1)^{p+q-1} \sum_{j=0}^{n} f_2(t_j, x_j, y_j, z_j) \\
&\quad \times \left[B\left(\frac{t_{j+1}}{t_{n+1}}, q, p\right) - B\left(\frac{t_j}{t_{n+1}}, q, p\right) \right], \\
z_{n+1}^u &= \frac{pq}{\Gamma(p)} (\Delta t)^{p+q-1} (n+1)^{p+q-1} \sum_{j=0}^{n} f_3(t_j, x_j, y_j, z_j)
\end{aligned}$$

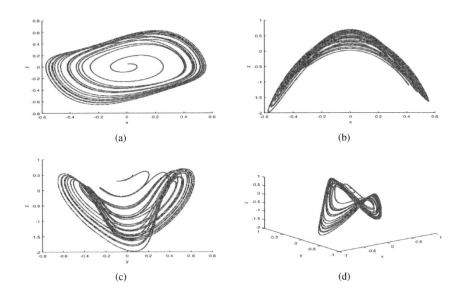

Figure 22.8 Model simulation of fractal-fractional Atangana-Baleanu model when $p = 1$ and $q = 1$.

$$\times \left[B\left(\frac{t_{j+1}}{t_{n+1}}, q, p\right) - B\left(\frac{t_j}{t_{n+1}}, q, p\right) \right]. \tag{22.13}$$

We obtained the graphical results shown in Figures 22.8 to 22.10 for the fractal-fractional model in the sense of Atangana-Baleanu derivative presented above for some selected values of the fractal and fractional order parameters by using the scheme described above.

22.5 CONCLUSION

We presented a dynamical chaotic system in the framework of fractal-fractional operators. Various fractal-fractional operators are used to obtain the results. A new numerical scheme for the case of power law kernel, exponential kernel, and the Mittag-Leffler kernel is obtained. Graphical results for each case with different fractal and fractional orders are obtained and discussed.

The Hidden Attractors Model with Fractal-Fractional Operators

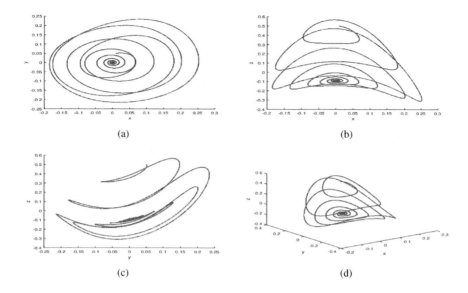

Figure 22.9 Model simulation of fractal-fractional Atangana-Baleanu model when $p = 0.98$ and $q = 0.97$.

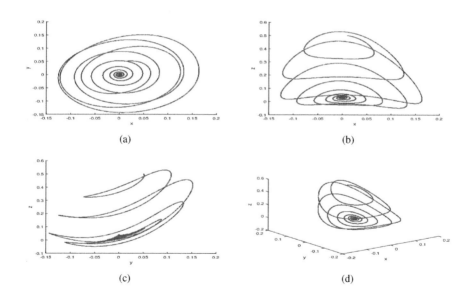

Figure 22.10 Model simulation of fractal-fractional Atangana-Baleanu model when $p = 0.96$ and $q = 0.96$.

23 An SIR Epidemic Model with Fractal-Fractional Derivative

23.1 INTRODUCTION

In mathematical biology or disease epidemic modelling, the most important concept was the development of the SIR model. The SIR model stands for S—susceptible, IV-infected, and R—recovered or removed. This concept was introduced in 1927. Later many developments have been made by extending the work. The SIR modelling approach is now used by the researchers in many fields of science and engineering. With the concept of ordinary differential equations, the SIR-type models are numerous in literature; see, for example, [87, 136, 258] and much more. The SIR-type models extended to fractional differential equations, and later the concept of the new introduced derivative; the work is extended further. With the advancement in the field of fractional calculus, the modelling of an SIR type has been extended to a more, general, and complex modelling; see [14, 68, 107, 108, 241] and much more for the understanding of science and engineering problems. With the development of the fractal-fractional operators, the idea of modelling in fractional extended to fractal-fractional operators. Hence, some literature that used the concept of fractal-fractional operators is available in literature; see, for example, [8, 16, 19, 64, 106, 138, 140, 261] and much more.

In the present chapter, we consider an SIR-type simple model and explore its dynamics under the framework of fractal-fractional operators. We consider different fractal-fractional operators such as the power law, exponential law, and the Mittag-Leffler law. With the use of these laws, we also develop a new numerical solution for the fractal-fractional differential equations using Euler approach. This chapter investigates the dynamics of an SIR epidemic model in the framework of fractal-fractional derivative. Initially, we consider the model in integer case and then by applying the fractal-fractional operators. We consider the stability of the equilibrium points, existence, and uniqueness of solutions of the model. Giving the novel numerical approaches using the Euler approach for the models in fractal-fractional differential equations. Many graphical results for each fractal-fractional case are given for the demonstration of the approach.

23.2 MODEL FORMULATION

We consider here to contract a simple mathematical model of the type of an SIR model. The SIR model usually consists of the three components, that is the healthy or susceptible individuals, $S(t)$, the infected individuals, $I(t)$, and the individuals recovered or removed from the infection, $R(t)$. Usually, one can write the total population of the humans, population, denoted by $N(t)$, with $N(t) = S(t) + I(t) + R(t)$. The population of the healthy individuals is increased by birth of the newly born individuals, say Λ, while this population can be reduced with the natural death of the individuals, μ. The diseases contact of the healthy with the infected individuals also decreases the population of the healthy individuals, βSI, where β is the contact rate. The infected individuals are increased by the contact rate βSI, and it is decreased by the disease death and natural mortality rates. Further, the recovery of the infected individuals also decreases the population of the infected population. The recovered individuals only depend on the recovery of the infected individuals, which increases its populations, while the natural mortality rate decreases its populations. With the above discussion, we can write the evolutionary dynamics of an SIR model through the following system of non-linear differential equations:

$$\begin{cases} \frac{dS(t)}{dt} = \Lambda - \beta S(t)I(t) - \mu S(t), \\ \frac{dI(t)}{dt} = \beta S(t)I(t) - (\mu + d + \gamma)I(t), \\ \frac{dR(t)}{dt} = \gamma I(t) - \mu R(t). \end{cases} \qquad (23.1)$$

The initial conditions for the model (23.1) are

$$S(0) = S_0 \geq 0,\ I(0) = I_0 \geq 0,\ \text{and}\ R(0) = R_0 \geq 0. \qquad (23.2)$$

In the model (23.1) in terms of the fractal-fractional representation, we give

$$\begin{cases} {}^{FF}D_{0,t}^{p,\,q}\left(S(t)\right) = \Lambda - \beta S(t)I(t) - \mu S(t), \\ {}^{FF}D_{0,t}^{p,\,q}\left(I(t)\right) = \beta S(t)I(t) - (\mu + d + \gamma)I(t), \\ {}^{FF}D_{0,t}^{p,\,q}\left(R(t)\right) = \gamma I(t) - \mu R(t), \end{cases} \qquad (23.3)$$

where p and q, respectively, defines the fractional and the fractal order. In the section below, we show the positivity of the system (23.3).

23.3 POSITIVITY OF THE MODEL

We show that the system admits a system of positive solutions. In order to show the positivity of the system (23.3). We know that all the model parameters are positive. We assume that $S(t)I(t)$ is positive

$$\begin{aligned} I'(t) &= \beta S(t)I(t) - (\mu + d + \gamma)I(t), \\ &\geq (\mu + d + \gamma)I(t), \\ \frac{I'(t)}{I(t)} &\geq -(\mu + d + \gamma), \end{aligned}$$

$$I(t) \geq I(0)\exp\left(-(\mu+d+\gamma)t\right) \geq 0. \quad (23.4)$$

Therefore, for every $t \geq 0$, $I(t) \geq 0$.

$$R'(t) = \gamma I(t) - \mu R(t) \geq -\mu R(t),$$

$$\frac{R'(t)}{R(t)} \geq -\mu,$$

$$R(t) \geq R(0)\exp\left(-\mu t\right). \quad (23.5)$$

$$S'(t) = \Lambda - \beta S(t)I(t) - \mu S(t),$$

$$\geq -\beta S(t)I(t) - \mu S(t),$$

$$\frac{S'(t)}{S(t)} \geq -\beta I(t) - \mu S(t),$$

$$\log(S(t)) \geq -\beta \int_0^t I(t) - \mu t,$$

$$S(t) \geq S(0)\exp\left(-\beta \int_0^t I(\tau)d\tau - \mu t\right).$$

Therefore,

$$S(t) \geq S(0)\exp\left(-\beta \int_0^t I(\tau)d\tau - \mu t\right).$$

Therefore, if $S(t)I(t) > 0$, we have for every $t \geq 0$

$$I(t) \geq I(0)\exp\left(-(\mu+d+\gamma)t\right),$$

$$R(t) \geq R(0)\exp\left(-\mu t\right),$$

$$S(t) \geq S(0)\exp\left(-\beta \int_0^t I(\tau)d\tau - \mu t\right).$$

Thus, under this condition, the solutions are positive. If $S(t)I(t) < 0$, then

$$S'(t) = \Lambda - \beta S(t)I(t) - \mu S(t) \geq -\mu S(t).$$

Since $I(t)S(t < 0)$, then $-\beta S(t)I(t) > 0$.

$$\frac{S'(t)}{S(t)} > -\mu S(t),$$

which gives

$$S(t) \geq S(0)\exp\left(-\mu S(t)\right).$$

$$I(t) = \beta S(t)I(t) - (\mu + d + \gamma)I(t).$$

Since $S(t) > 0$ and $S(t)I(t) < 0$, it implies $I(t) < 0$, which is not for what we shall expect. Therefore, $S(t)I(t) > 0$.

23.4 EXISTENCE AND UNIQUENESS

This section considers the existence and uniqueness of the fractal-fractional model (23.3). We assume that for all $t \in [0, T]$, the functions $S(t)$, $I(t)$, and $R(t)$ are bounded. We need to write first the model equations in the form:

$$\begin{aligned} S'(t) &= f_S(t, S, I, R) = \Lambda - \beta SI - \mu S, \\ I'(t) &= f_I(t, S, I, R) = \beta SI - (\mu + d + \gamma)I, \\ R'(t) &= f_R(t, S, I, R) = \gamma I - \mu R, \end{aligned} \quad (23.6)$$

and

$$\|S\| \leq M_S, \quad \|I\| \leq M_I, \quad \text{and} \quad \|R\| \leq M_R.$$

We shall prove that f_1, f_2, and f_3 satisfy both the linear growth and the Lipschitz conditions. Let's start with the linear growth,

$$\begin{aligned} \left| f_S(t, S, I, R) \right|^2 &= \left| \Lambda - \beta SI - \mu S \right|^2, \\ &< 2\Lambda^2 + 2|S|^2|I|^2 + 2\mu^2|S|^2, \\ &< 2\Lambda^2 + 2|S|^2 \sup_{t \in D_I} |I(t)|^2 + 2\mu^2|S|^2, \\ &< 2\Lambda^2 + 2|S|^2 M_I + 2\mu^2|S|^2, \\ &< 2\Lambda^2 \left(1 + \frac{M_I + \mu^2}{\Lambda^2} |S|^2 \right). \end{aligned}$$

If $\frac{M_I + \mu^2}{\Lambda^2} < 1$, then

$$\begin{aligned} \left| f_S(t, S, I, R) \right|^2 &< 2\Lambda^2 \left(1 + |S|^2 \right), \\ &< K_S \left(1 + |S|^2 \right), \end{aligned}$$

where $K_S = 2\Lambda^2$.

$$\left| f_I(t, S, I, R) \right|^2 = \left| \beta S(t)I(t) - (\mu + d + \gamma)I(t) \right|^2,$$

An SIR Epidemic Model with Fractal-Fractional Derivative

$$\begin{aligned}
&= \left|\beta S - (\mu + d + \gamma)\right|^2 |I(t)|^2, \\
&\leq \left(2\beta^2 |S|^2 + 2(\mu + d + \gamma)^2\right)|I(t)|^2, \\
&\leq \left(2\beta^2 M_S + 2(\mu + d + \gamma)^2\right)(1 + |I(t)|^2), \\
&\leq K_I(1 + |I(t)|^2),
\end{aligned}$$

where $K_I = 2\beta^2 M_S + 2(\mu + d + \gamma)^2$.

$$\begin{aligned}
\left|f_R(t, S, I, R)\right|^2 &= |\gamma I(t) - \mu R(t)|^2, \\
&\leq 2\gamma^2 |I(t)|^2 + 2\mu^2 |R(t)|, \\
&\leq 2\gamma^2 M_I + 2\mu^2 |R(t)|^2, \\
&\leq 2\gamma^2 M_I \left(1 + \frac{\mu^2 |R|^2}{\gamma^2 M_I}\right).
\end{aligned}$$

If $\frac{\mu^2}{\gamma^2 M_I} < 1$, then

$$\left|f_R(t, S, I, R)\right|^2 \leq K_R(1 + |R|^2),$$

where $K_R = 2\gamma^2 M_I$. Therefore if

$$\max\left\{\frac{\mu^2}{\gamma^2 M_I}, \frac{M_I + \mu^2}{\Lambda^2}\right\} < 1,$$

then the system satisfies the linear growth property. We now verify the Lipschitz conditions of the system,

$$\begin{aligned}
\left|f(t, S_1, I, R) - f(t, S_2, I, R)\right| &= \left|\beta I(S_1 - S_2) + \mu(S_1 - S_2)\right|, \\
&\leq (\beta M_I + \mu)\left|S_1 - S_2\right|, \\
&\leq K_S \left|S_1 - S_2\right|,
\end{aligned}$$

where $K_S = (\beta M_I + \mu)$.

$$\left|f(t, S, I_1, R) - f(t, S, I_2, R)\right| \leq K_I |I_1 - I_2|,$$

$$\left|f(t, S, I, R_1) - f(t, S, I, R_2)\right| \leq K_R |R_1 - R_2|,$$

where $K_I = (\beta M_S + \mu + d + \gamma)$ and $M_R = \mu$. The system satisfies the Lipschitz conditions, so the model admits a unique solution.

The equilibrium points related with the model associated with the model (23.3) can be determined by using the time rate of change equal to zero; we have the result in the following section.

23.4.1 EQUILIBRIUM POINTS AND THEIR ANALYSIS

This section focuses to determine the possible equilibrium points of the model and obtain the their stability analysis. Often, the models associated with humans, population give two equilibrium points, the disease free and the endemic. In the absence of the disease infection, we can have the disease-free equilibrium denoted by E_0, and it is given by

$$E_0 = (S^0, 0, 0) = \left(\frac{\Lambda}{\mu}, 0, 0\right). \tag{23.7}$$

The endemic equilibrium where the disease exists in the population can be denoted by $E_1 = (S^*, I^*, R^*)$ and is given by

$$S^* = \frac{\gamma + d + \mu}{\beta}, \quad I^* = \frac{\mu(\mathscr{R}_0 - 1)}{\beta}, \quad \text{and} \quad R^* = \frac{\gamma(\mathscr{R}_0 - 1)}{\beta}, \tag{23.8}$$

where

$$\mathscr{R}_0 = \frac{\beta \Lambda}{\mu(\gamma + d + \mu)}. \tag{23.9}$$

The disease will exist permanently in the population if $\mathscr{R}_0 > 1$. Hence, in the equilibria at E_1, we can see the existence of the endemic equilibria for $\mathscr{R}_0 > 1$. At E_0, we give the local asymptotic stability of the model (23.3) in the following theorems.

Theorem 3 *The equilibrium E_0 is locally asymptotically stable if $\mathscr{R}_0 < 1$.*

Proof 5 *We have the Jacobian matrix of the model (23.3) at E_0, given by*

$$J = \begin{pmatrix} -\mu & -\frac{\beta \Lambda}{\mu} & 0 \\ 0 & -(d+\gamma+\mu) + \frac{\beta \Lambda}{\mu} & 0 \\ 0 & \gamma & -\mu \end{pmatrix}. \tag{23.10}$$

The eigenvalues J give $\lambda_{1,2} = -\mu$ and $\lambda_3 = (\gamma + d + \mu)(\mathscr{R}_0 - 1) < 0$ if $\mathscr{R}_0 < 1$. It ensures that the model (23.3) at E_0 is locally asymptotically stable if $\mathscr{R}_0 < 1$.

Theorem 4 *If $\mathscr{R}_0 \leq 1$, then the model (23.3) is globally asymptotically stable.*

Proof 6 *We have the Lyapunov function*

$$L(t) = I. \tag{23.11}$$

The time differentiation of $L(t)$ along the solution of the model (23.3), we have

$$\begin{aligned} L'(t) &= I', \\ &= \beta SI - (\mu + d + \gamma)I, \end{aligned}$$

An SIR Epidemic Model with Fractal-Fractional Derivative

$$\begin{aligned}
&= (\beta S^0 - (\mu + d + \gamma))I, \\
&= (\mu + d + \gamma)\left(\frac{\beta S^0}{(\mu + d + \gamma)} - 1\right)I, \\
&= (\mu + d + \gamma)\left(\mathcal{R}_0 - 1\right)I. \quad (23.12)
\end{aligned}$$

If $\mathcal{R}_0 \leq 1$, then the model (23.3) is globally asymptotically stable at E_0.

23.4.2 GLOBAL STABILITY

This section invests in the global asymptotic stability of the model (23.3) at E_1. We have the following theorem.

Theorem 5 *If $\mathcal{R}_0 > 1$, then the model (23.3) at E_1 is locally asymptotically stable.*

Proof 7 *We have the Jacobian matrix at E_1, given by*

$$J_1 = \begin{pmatrix} -\beta I^* - \mu & -\beta S^* & 0 \\ \beta I^* & \beta S^* - (d+\gamma+\mu) & 0 \\ 0 & \gamma & -\mu \end{pmatrix}. \quad (23.13)$$

We have $\lambda_1 = -\mu < 0$, while the other two can be obtained through the following equation:

$$\lambda^2 + a_1 \lambda + a_2 = 0, \quad (23.14)$$

where

$$\begin{aligned}
a_1 &= (\gamma + d + \mu) + \beta I^* + \mu - \beta S^*, \\
&= \mu + \beta I^* = \mu \mathcal{R}_0, \\
a_2 &= \mu(\gamma + d + \mu) + (\gamma + d + \mu)\beta I^* - \beta \mu S^*, \\
&= (\gamma + d + \mu)\beta I^* = \mu(\gamma + d + \mu)(\mathcal{R}_0 - 1), \quad (23.15)
\end{aligned}$$

where $I^* = \frac{\mu(\mathcal{R}_0 - 1)}{\beta}$. We can see that a_1 and a_2 can be positive if $\mathcal{R}_0 > 1$, and it ensures that the model at E_1 is locally asymptotically stable.

Theorem 6 *The model at E_1 is globally asymptotically stable if $\mathcal{R}_0 > 1$.*

Proof 8 *We have the Lyapunov function*

$$L = S - S^* - S^* \log\left(\frac{S}{S^*}\right) + I - I^* - I^* \log\left(\frac{I}{I^*}\right). \quad (23.16)$$

Time differentiation along the solution of the model gives

$$L' = \left(1-\frac{S^*}{S}\right)S' + \left(1-\frac{I^*}{I}\right)I',$$
$$= \left(1-\frac{S^*}{S}\right)[\Lambda - \beta SI - \mu S] + \left(1-\frac{I^*}{I}\right)[\beta SI - (\mu+d+\gamma)I]. \quad (23.17)$$

Note that

$$\left(1-\frac{S^*}{S}\right)[\Lambda - \beta SI - \mu S] = \left(1-\frac{S^*}{S}\right)[\beta S^* I^* + \mu S^* - \beta SI - \mu S],$$
$$\leq -\mu S^*\left(\frac{S^*}{S} + \frac{S}{S^*} - 2\right)$$
$$+ \beta S^* I^*\left[1 - \frac{S^*}{S} - \frac{SI}{S^* I^*} + \frac{I}{I^*}\right]. \quad (23.18)$$

$$\left(1-\frac{I^*}{I}\right)[\beta SI - (\mu+d+\gamma)I] = \left(1-\frac{I^*}{I}\right)\left[\beta SI - \frac{\beta S^* I^*}{I^*}I\right],$$
$$= \beta S^* I^*\left(1 - \frac{I}{I^*} + \frac{SI}{S^* I^*} - \frac{S}{S^*}\right) \quad (23.19)$$

Using the above into (23.17), we have

$$L' = -(\mu + \beta I^*)S^*\left(\frac{S^*}{S} + \frac{S}{S^*} - 2\right) \leq 0. \quad (23.20)$$

Hence, the model at the endemic equilibrium E_1 is globally asymptotically stable.

23.5 NUMERICAL RESULTS AND THE SCHEMES

23.5.1 EULER SCHEME WITH POWER LAW CASE

We consider the Cauchy problem given by

$$\begin{aligned}{}^{FFP}_{0}D^{p,q}_t y(t) &= f(t, y(t)), \text{ if } t > 0,\\ y(0) &= y_0, \text{ if } t = 0. \end{aligned} \quad (23.21)$$

$$\begin{aligned}{}^{RL}_{0}D^p_t y(t) &= qt^{q-1}f(t, y(t)) \text{ if } t > 0\\ y(0) &= y_0 \text{ if } t = 0. \end{aligned} \quad (23.22)$$

We write

$$y(t) = \frac{q}{\Gamma(p)} \int_0^t \tau^{q-1} f(\tau, y(\tau))(t-\tau)^{p-1} d\tau, \ y(0) = y_0; \quad (23.23)$$

at $t = t_{n+1}$, we have

$$y(t_{n+1}) = \frac{q}{\Gamma(p)} \int_0^{t_{n+1}} \tau^{q-1} f(\tau, y(\tau))(t_{n+1} - \tau)^{p-1} d\tau, \qquad (23.24)$$

and finally, we write

$$\begin{aligned} y(t_{n+1}) &= \frac{q}{\Gamma(p)} \sum_{j=0}^{n} \int_{t_j}^{t_{j+1}} \tau^{q-1} f(\tau, y(\tau))(t_{n+1} - \tau)^{p-1} d\tau, \\ &\approx \frac{q}{\Gamma(p)} \sum_{j=0}^{n} f(t_j, y_j) \int_{t_j}^{t_{j+1}} \tau^{q-1}(t_{n+1} - \tau)^{p-1} d\tau, \\ &\approx \frac{q}{\Gamma(p)} \sum_{j=0}^{n} f(t_j, y_j) t_{n+1}^{p+q-1} \left[B\left(\frac{t_{j+1}}{t_{n+1}}, q, p\right) \right. \\ &\left. - B\left(\frac{t_j}{t_{n+1}}, q, p\right) \right]. \end{aligned} \qquad (23.25)$$

The above scheme can be applied to our system (23.3), and the desired scheme will be given as

$$\begin{aligned} S_{n+1} &= \frac{q}{\Gamma(p)} \sum_{j=0}^{n} f_1(t_j, S_j, I_j, R_j) t_{n+1}^{p+q-1} \left[B\left(\frac{t_{j+1}}{t_{n+1}}, q, p\right) - B\left(\frac{t_j}{t_{n+1}}, q, p\right) \right], \\ I_{n+1} &= \frac{q}{\Gamma(p)} \sum_{j=0}^{n} f_2(t_j, S_j, I_j, R_j) t_{n+1}^{p+q-1} \left[B\left(\frac{t_{j+1}}{t_{n+1}}, q, p\right) - B\left(\frac{t_j}{t_{n+1}}, q, p\right) \right], \\ R_{n+1} &= \frac{q}{\Gamma(p)} \sum_{j=0}^{n} f_3(t_j, S_j, I_j, R_j) t_{n+1}^{p+q-1} \left[B\left(\frac{t_{j+1}}{t_{n+1}}, q, p\right) \right. \\ &\left. - B\left(\frac{t_j}{t_{n+1}}, q, p\right) \right], \end{aligned} \qquad (23.26)$$

$$\begin{aligned} f_1(t, S, I, R) &= \Lambda - \beta S(t) I(t) - \mu S(t), \\ f_2(t, S, I, R) &= \beta S(t) I(t) - (\mu + d + \gamma) I(t), \\ f_3(t, S, I, R) &= \gamma I(t) - \mu R(t). \end{aligned} \qquad (23.27)$$

We use the new numerical scheme for the case of fractal-fractional power law and obtained the numerical results as given in Figures 23.1–23.3. The numerical values of the models parameters are $\Lambda = 0.3$, $\beta = 0.0004$, $\mu = 0.0004$, $d = 0.004$, and $\gamma = 0.035$. We consider the time unit per day. Figure 23.1 gives the model solution for the Caputo case when varying p while q is constant. Similarly, in Figures 23.2 and 23.3, we, respectively, vary q and keep p constant, and vary both p and q.

23.5.2 EULER SCHEME WITH EXPONENTIAL KERNEL

In this subsection, we consider the following Cauchy problem:

$$_0^{FFF} D_t^{p,q} y(t) = f(t, y(t)), \text{ if } t > 0,$$

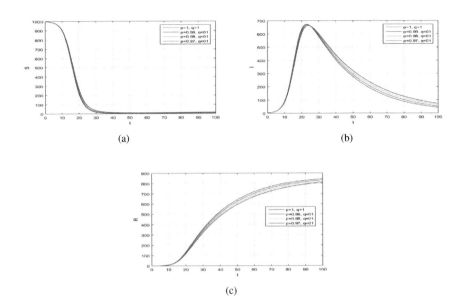

Figure 23.1 Simulation results for fractal-fractional model in Caputo case, when $p = 1, 0.99, 0.98,$ and 0.97 and $q = 1$.

$$y(0) = y_0, \text{ if } t = 0. \quad (23.28)$$

$${}_{0}^{CFR}D_t^p y(t) = qt^{q-1}f(t,y(t)) \text{ if } t > 0$$

$$y(0) = y_0 \text{ if } t = 0 \quad (23.29)$$

$$y(t) = (1-p)qt^{q-1}f(t,y(t))$$
$$+ pq \int_0^t \tau^{q-1} f(\tau, y(\tau)) d\tau, \text{ if } t > 0,$$
$$y(0) = y_0, \text{ if } t = 0; \quad (23.30)$$

at $t = t_{n+1}$, we have

$$y(t_{n+1}) = qt_{n+1}^{q-1} f(t_{n+1}, y(t_{n+1}))$$
$$+ pq \int_0^{t_{n+1}} \tau^{q-1} f(\tau, y(\tau)) d\tau,$$
$$= qt_{n+1}^{q-1} f(t_{n+1}, y^u(t_{n+1}))$$

An SIR Epidemic Model with Fractal-Fractional Derivative

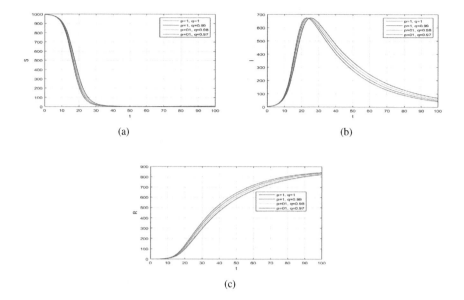

Figure 23.2 Simulation results for fractal-fractional model in Caputo case, when $q = 1, 0.99, 0.98$, and 0.97 and $p = 1$.

$$+pq \sum_{j=0}^{n} \int_{t_j}^{t_{j+1}} f(\tau, y(\tau)) \tau^{q-1} d\tau. \quad (23.31)$$

Further, we obtain

$$y_{n+1} = qt_{n+1}^{q-1} f(t_{n+1}, y^u(t_{n+1}))$$
$$+ p(\Delta t)^q \sum_{j=0}^{n} f(t_j, y_j) \{(j+1)^q - j^q\}, \quad (23.32)$$

where

$$y^u(t_{n+1}) = y_n + p(\Delta t)^q \{(n+1)^q - n^q\} f(t_n, y_n). \quad (23.33)$$

The scheme (23.32) can be used for the model (23.3). First, we write the model equations in the form given by

$$f_1(t, S, I, R) = \Lambda - \beta S(t)I(t) - \mu S(t),$$
$$f_2(t, S, I, R) = = \beta S(t)I(t) - (\mu + d + \gamma)I(t),$$
$$f_3(t, S, I, R) = = \gamma I(t) - \mu R(t), \quad (23.34)$$

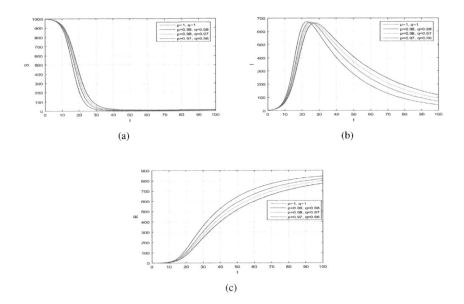

Figure 23.3 Simulation results for fractal-fractional model in Caputo case, when $p = 1, 0.99, 0.98,$ and 0.97 and $q = 1, 0.98, 0.97,$ and 0.96.

and hence the scheme for the system (23.3) is as follows:

$$S_{n+1} = qt_{n+1}^{q-1} f_1(t_{n+1}, S^u(t_{n+1}), I^u(t_{n+1}), R^u(t_{n+1}))$$
$$+ p(\Delta t)^q \sum_{j=0}^{n} f_1(t_j, S_j, I_j, R_j) \left\{ (j+1)^\beta - j^q \right\},$$
$$I_{n+1} = qt_{n+1}^{q-1} f_2(t_{n+1}, S^u(t_{n+1}), I^u(t_{n+1}), R^u(t_{n+1}))$$
$$+ p(\Delta t)^q \sum_{j=0}^{n} f_2(t_j, S_j, I_j, R_j) \left\{ (j+1)^q - j^q \right\},$$
$$R_{n+1} = qt_{n+1}^{q-1} f_3(t_{n+1}, S^u(t_{n+1}), I^u(t_{n+1}), R^u(t_{n+1}))$$
$$+ p(\Delta t)^q \sum_{j=0}^{n} f_3(t_j, S_j, I_j, R_j) \left\{ (j+1)^q - j^q \right\}, \quad (23.35)$$

where

$$S^u(t_{n+1}) = S_n + p(\Delta t)^q \{(n+1)^q - n^q\} f_1(t_n, S_n, I_n, R_n),$$
$$I^u(t_{n+1}) = I_n + p(\Delta t)^q \{(n+1)^q - n^q\} f_2(t_n, S_n, I_n, R_n),$$

An SIR Epidemic Model with Fractal-Fractional Derivative

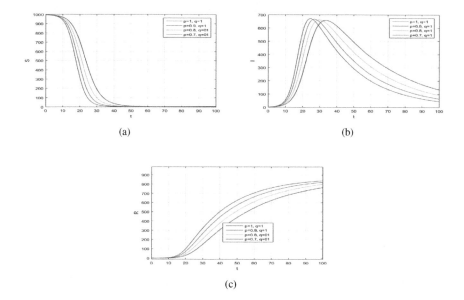

Figure 23.4 Simulation results for fractal-fractional model in Caputo-Fabrizio case, when $p = 1, 0.9, 0.8$, and 0.7 and $q = 1$.

$$R^u(t_{n+1}) = R_n + p(\Delta t)^q \{(n+1)^q - n^q\} f_3(t_n, S_n, I_n, R_n). \quad (23.36)$$

The model in CF-fractal-fractional differential equation under the same parameters and the initial conditions, we have Figures 23.4–23.6. Figures 23.4 and 23.5, respectively, show the model dynamics when p varies and q is constant, and when p is constant and q varies. Varying both p and q arbitrarily, we have Figure 23.6.

23.5.3 EULER SCHEME WITH MITTAG-LEFFLER KERNEL

In this section, a fractal-fractional Cauchy problem with the generalised Mittag-Leffler function is considered

$$\begin{aligned}
{}_0^{FFM}D_t^{p,q}y(t) &= f(t, y(t)), \text{ if } t > 0, \\
y(0) &= y_0, \text{ if } t = 0,
\end{aligned} \quad (23.37)$$

$$\begin{aligned}
{}_0^{ABR}D_t^p y(t) &= qt^{q-1} f(t, y(t)) \text{ if } t > 0 \\
y(0) &= y_0 \text{ if } t = 0.
\end{aligned} \quad (23.38)$$

$$y(t) = qt^{q-1} f(t, y(t))(1-p) + \frac{pq}{\Gamma(p)} \int_0^t$$

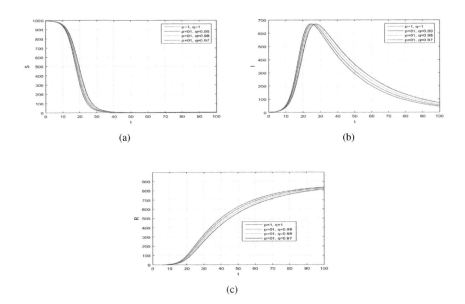

Figure 23.5 Simulation results for fractal-fractional model in Caputo-Fabrizio case, when $q = 1, 0.99, 0.98,$ and 0.97 and $p = 1$.

$$\times \tau^{\beta-1}(t-\tau)^{p-1} f(\tau, y(\tau)) d\tau, \text{ if } t > 0$$

$$y(0) = y_0 \text{ if } t = 0. \quad (23.39)$$

At $t = t_{n+1}$, we have

$$y(t_{n+1}) = q t_{n+1}^{q-1} f(t_{n+1}, y^u(t_{n+1}))(1-p)$$

$$+ \frac{pq}{\Gamma(p)} \int_0^{t_{n+1}} \tau^{q-1}(t_{n+1}-\tau)^{p-1} f(\tau, y(\tau)) d\tau. \quad (23.40)$$

Further, we write

$$y(t_{n+1}) = q t_{n+1}^{q-1} f(t_{n+1}, y^u(t_{n+1}))(1-p)$$

$$+ \frac{pq}{\Gamma(p)} \sum_{j=0}^{n} \int_{t_j}^{t_{j+1}} \tau^{q-1}(t_{n+1}-\tau)^{p-1} f(\tau, y(\tau)) d\tau. \quad (23.41)$$

Then, we have

$$y(t_{n+1}) \approx y_{n+1} = q t_{n+1}^{q-1} f(t_{n+1}, y_{n+1}^u)(1-p)$$

An SIR Epidemic Model with Fractal-Fractional Derivative

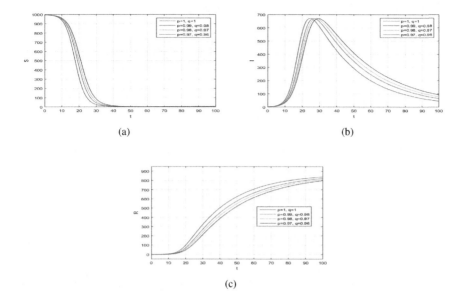

Figure 23.6 Simulation results for fractal-fractional model in Caputo-Fabrizio case, when $p = 1, 0.99, 0.98$, and 0.97 and $q = 1, 0.98, 0.97$, and 0.96.

$$+ \frac{pq}{\Gamma(p)} \sum_{j=0}^{n} \int_{t_j}^{t_{j+1}} \tau^{q-1} (t_{n+1} - \tau)^{p-1}$$
$$\times f(\tau, y(\tau)) d\tau. \tag{23.42}$$

We write as

$$y_{n+1} = q t_{n+1}^{q-1} f(t_{n+1}, y_{n+1}^u)(1-p)$$
$$+ \frac{pq}{\Gamma(p)} \sum_{j=0}^{n} f(t_j, y_j) \int_{t_j}^{t_{j+1}} (t_{n+1} - \tau)^{p-1} \tau^{q-1} d\tau. \tag{23.43}$$

Further, using the concept of generalised Beta-function, we write

$$y_{n+1} = q t_{n+1}^{q-1} f(t_{n+1}, y_{n+1}^u)(1-p) + \frac{pq}{\Gamma(p)} \sum_{j=0}^{n} f(t_j, y_j)$$
$$\times t_{n+1}^{p+q-1} \left[B\left(\frac{t_{j+1}}{t_{n+1}}, q, p\right) - B\left(\frac{t_j}{t_{n+1}}, q, p\right) \right]. \tag{23.44}$$

Further simplifying, we have

$$y_{n+1} = q t_{n+1}^{q-1} f(t_{n+1}, y_{n+1}^u)(1-p) + \frac{pq}{\Gamma(p)} (\Delta t)^{p+q-1} (n+1)^{p+q}$$

$$\times \sum_{j=0}^{n} f(t_j, y_j) \left[B\left(\frac{t_{j+1}}{t_{n+1}}, q, p\right) - B\left(\frac{t_j}{t_{n+1}}, q, p\right) \right], \quad (23.45)$$

where

$$\begin{aligned} y_{n+1}^u &= \frac{pq}{\Gamma(p)} (\Delta t)^{p+q-1} (n+1)^{p+\beta-1} \sum_{j=0}^{n} f(t_j, y_j) \\ &\quad \times \left[B\left(\frac{t_{j+1}}{t_{n+1}}, q, p\right) - B\left(\frac{t_j}{t_{n+1}}, q, p\right) \right]. \end{aligned} \quad (23.46)$$

Applying the scheme (23.47) to our system (23.3), given by

$$\begin{aligned}
S_{n+1} &= q t_{n+1}^{q-1} f_1(t_{n+1}, S_{n+1}^u, I_{n+1}^u, R_{n+1}^u)(1-p) \\
&\quad + \frac{pq}{\Gamma(p)} (\Delta t)^{p+q-1} (n+1)^{p+q} \\
&\quad \times \sum_{j=0}^{n} f_1(t_j, S_j, I_j, R_j) \left[B\left(\frac{t_{j+1}}{t_{n+1}}, q, p\right) - B\left(\frac{t_j}{t_{n+1}}, q, p\right) \right], \\
I_{n+1} &= q t_{n+1}^{q-1} f_1(t_{n+1}, S_{n+1}^u, I_{n+1}^u, R_{n+1}^u)(1-p) \\
&\quad + \frac{pq}{\Gamma(p)} (\Delta t)^{p+q-1} (n+1)^{p+q} \\
&\quad \times \sum_{j=0}^{n} f_2(t_j, S_j, I_j, R_j) \left[B\left(\frac{t_{j+1}}{t_{n+1}}, q, p\right) - B\left(\frac{t_j}{t_{n+1}}, q, p\right) \right], \\
R_{n+1} &= q t_{n+1}^{q-1} f_3(t_{n+1}, S_{n+1}^u, I_{n+1}^u, R_{n+1}^u)(1-p) \\
&\quad + \frac{pq}{\Gamma(p)} (\Delta t)^{p+q-1} (n+1)^{p+q} \\
&\quad \times \sum_{j=0}^{n} f_1(t_j, S_j, I_j, R_j) \left[B\left(\frac{t_{j+1}}{t_{n+1}}, q, p\right) - B\left(\frac{t_j}{t_{n+1}}, q, p\right) \right], \quad (23.47)
\end{aligned}$$

where

$$\begin{aligned}
S_{n+1}^u &= \frac{pq}{\Gamma(p)} (\Delta t)^{p+q-1} (n+1)^{p+q-1} \sum_{j=0}^{n} f_1(t_j, S_j, I_j, R_j) \\
&\quad \times \left[B\left(\frac{t_{j+1}}{t_{n+1}}, q, p\right) - B\left(\frac{t_j}{t_{n+1}}, q, p\right) \right], \\
I_{n+1}^u &= \frac{pq}{\Gamma(p)} (\Delta t)^{p+q-1} (n+1)^{p+q-1} \sum_{j=0}^{n} f_2(t_j, S_j, I_j, R_j) \\
&\quad \times \left[B\left(\frac{t_{j+1}}{t_{n+1}}, q, p\right) - B\left(\frac{t_j}{t_{n+1}}, q, p\right) \right], \\
R_{n+1}^u &= \frac{pq}{\Gamma(p)} (\Delta t)^{p+q-1} (n+1)^{p+q-1} \sum_{j=0}^{n} f_3(t_j, S_j, I_j, R_j)
\end{aligned}$$

An SIR Epidemic Model with Fractal-Fractional Derivative

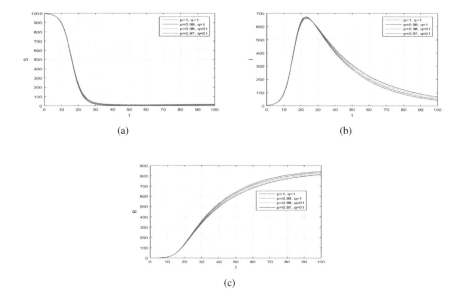

Figure 23.7 Simulation results for fractal-fractional model in Atangana-Baleanu case, when $p = 1, 0.99, 0.98,$ and 0.97 and $q = 1$.

$$\times \left[B\left(\frac{t_{j+1}}{t_{n+1}}, q, p\right) - B\left(\frac{t_j}{t_{n+1}}, q, p\right) \right]. \tag{23.48}$$

We solve the model in fractal-fractional Atangana-Baleanu case and obtained the required graphical results shown in Figures 23.7–23.9. In Figures 23.7 and 23.8, we give the solution behaviours of the model components when changing p and keeping q fixed, and changing q and keeping p fixed, respectively. While the behaviour of the solution is given in Figure 23.9, when changing both the orders p and q for various values.

23.6 CONCLUSION

We presented an SIR model with the applications of various fractal-fractional operators. The model dynamics for the fractal-fractional cases have been obtained and discussed. The local and global asymptotic stability of the model are obtained. The existence and uniqueness of the model solution are obtained and discussed. New fractal-fractional schemes for the solution of fractal-fractional differential equations are obtained. Numerical results for each case of fractal-fractional model have been obtained graphically.

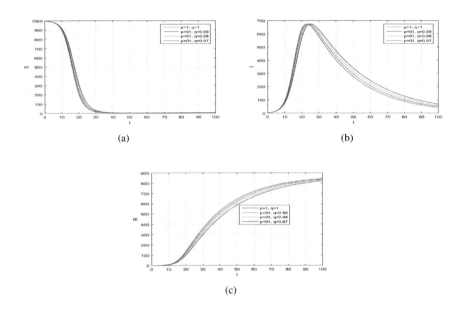

Figure 23.8 Simulation results for fractal-fractional model in Atangana-Baleanu case, when $q = 1, 0.99, 0.98,$ and 0.97 and $p = 1$.

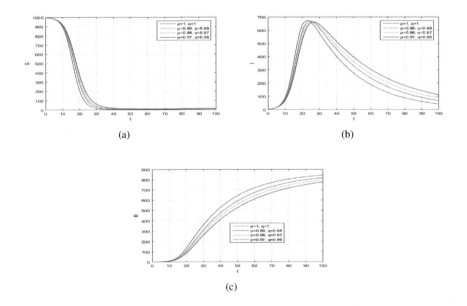

Figure 23.9 Simulation results for fractal-fractional model in Atangana-Baleanu case, when $p = 1, 0.99, 0.98,$ and 0.97 and $q = 1, 0.98, 0.97,$ and 0.96.

24 Application of Fractal-Fractional Operators to COVID-19 Infection

24.1 INTRODUCTION

The globe has faced and facing a new lethal disease known as COVID-19, which is mostly caused by the coronavirus. This virus was first detected in the Chinese city of Wuhan and quickly spread over the world. Because of the lockdown and control measures, the illness is nearly done in China, while the rest of the globe is fighting to reduce infection instances. It has been shown that patients who have been infected with coronavirus had respiratory illnesses and have recovered without any special therapy. People of advanced age, as well as those with underlying medical issues such as diabetes, cancer, chronic respiratory disorders, and cardiovascular disease, are more likely to acquire severe illnesses. The greatest strategy to limit the transmission of this virus and prevent it is to educate people about the COVID-19 virus, its causes, and its spread. To protect himself against this virus, he should wash his hands often or use an alcohol-based rub, and he should avoid touching his face. There is currently no particular therapy or vaccination for COVID-19; however, there are several ongoing trials to identify vaccines and potential therapeutic mechanisms, according to the WHO [179].

Because the coronavirus spreads quicker than the normal flu virus, the sickness spreads over the world and becomes an epidemic. Most nations have a considerable number of COVID-19 instances, but some, such as the United States, Italy, France, Iran, China, India, Spain, and Mexico, have a very high number of cases. According to the most current coronavirus update, the overall number of infected recorded cases is 22,423,016 with 787,909 fatalities, while the number of recovered cases is 15,907,858. The first coronavirus case in Saudi Arabia was reported on March 2, 2020 [1], when a Saudi citizen arrived from Iran via Bahrain. On March 14, a second example was recorded in which a person arrived from Iran via Bahrain as the companion of the first without indicating that he had visited Iran [2].

Many researchers produced mathematical models to better comprehend the coronavirus's dynamics and complexity. The authors in [272] investigated a mathematical model that addressed the dynamics of coronavirus in Canada. In [68], a mathematical model is developed to investigate data from China, Italy, and France. The authors of [107] developed a mathematical model in terms of fractional derivatives and investigated the dynamics of coronavirus using real data from China. In [14], novel mathematical research regarding the coronavirus and the influence of the lockdown on society is investigated. The authors employed real data from the United Kingdom, the United States, and Italy to do numerical analysis using a fractional mathematical model in [244]. In [238], the authors have studied the mathematical and computational study of the COVID-19 illness in Mexico. Non-pharmaceutical treatments' effects on the dynamics of new coronaviruses have been documented in [175]. In [241], the authors used optimum control techniques to investigate the dynamics of new coronaviruses. Using real data, the authors of [108] investigated the dynamical analysis of COVID-19 with quarantine and isolation. The authors in [177] investigated mathematical research on coronavirus utilising real cases from Nigeria. In [9], a mathematical model based on real statistics examples from Ghana was examined. The study in [282] investigated a mathematical model with singular and non-singular operators.

Fractional operators are commonly employed in scientific and engineering sectors due to the various applications of fractional order models. In modelling actual data, it has a good solution behaviour and is recognised as one of the finest applications. Furthermore, the crossover behaviour, memory, and hereditary features increase its strength. Some complicated models and their applications in science and engineering sectors include [14, 68, 107, 108, 241] and much more for the understanding of science and engineering problems. The concept of modelling in fractional was extended to fractal-fractional operators with the introduction of fractal-fractional operators. As a result, some literature that employed the notion of fractal-fractional operators is accessible in the literature, such as [8, 16, 19, 64, 106, 138, 140, 261] and much more. This chapter studies the dynamics of COVID-19 infection in the framework of the newly introduced fractal-fractional operators. The model will be designed in various fractal-fractional-fractional operators such as the power law kernel, exponential kernel, and the Mittag-Leffler kernel. Then, we consider the equilibrium points and obtain its stability results. We use the real data of COVID-19 cases in Saudi Arabia and parameterised the model. The real cases shall be used in numerical simulations with the use of the aforementioned fractal-fractional operator. The numerical simulations for each case of the fractal-fractional operators, we shall use and design the fractal-fractional Euler method.

24.2 MATHEMATICAL MODEL

This chapter considers the mathematical model given in [6] for the dynamics of the COVID-19 infection. The authors in [6] denoted the total population of humans by $N(t)$ and divided further into susceptible, $S(t)$, exposed $E(t)$, symptomatic $I(t)$, asymptomatic $A(t)$, and the recovered $R(t)$, so $N(t) = S(t) + E(t) + I(t) + A(t) + R(t)$. Using the concentration of the coronavirus in the environment, a new class, say $B(t)$, has been considered. With these details, the model has the following form:

$$\frac{dS}{dt} = \Lambda - (\theta_1 E + \theta_2 I + \theta_3 A + \theta_4 B)\frac{S}{N} - dS,$$

Application of Fractal-Fractional Operators to COVID-19 Infection   369

$$\frac{dE}{dt} = (\theta_1 E + \theta_2 I + \theta_3 A + \theta_4 B)\frac{S}{N} - (\varpi + d)E,$$
$$\frac{dI}{dt} = (1-\xi)\varpi E - (d + d_1 + \omega_1)I,$$
$$\frac{dA}{dt} = \xi\varpi E - (d + \omega_2)A,$$
$$\frac{dR}{dt} = \omega_1 I + \omega_2 A - dR,$$
$$\frac{dB}{dt} = \kappa_1 E + \kappa_2 I + \kappa_3 A - \delta B, \quad (24.1)$$

where the initial conditions are given by

$$S(0) = S_0 \geq 0, E(0) = E_0 \geq 0, I(0) = I_0 \geq 0, A(0) = A_0 \geq 0,$$
$$R(0) = R_0 \geq 0, B(0 = B_0 \geq 0.$$

The recruitment rate of the healthy people is given by Λ, while its disease natural death rate is d. The coefficients of contacts θ_i for $i = 1, 2, 3,$ and 4 are defined, respectively, the contact of the healthy with the exposed, susceptible, infected (showing symptoms), and asymptomatic (no disease symptoms). The exposed individuals develop symptoms and become infected at a rate $(1-\xi)\varpi$, where the asymptomatic infection occurs with a rate $\xi\varpi$. The death rate of symptomatic people is given by the rate d_1, while the recovery from infected and asymptomatic infected occur at rate ω_1 and ω_2, respectively. The parameters κ_i for $i = 1, 2,$ and 3 are the rate by which the exposed, infected, and asymptotically infected people contributed the virus to the environment reservoir, while δ is the removal rate of the virus from the environment reservoir. The biological feasible region for the given system is

$$\Omega = \left\{(S(t), E(t), I(t), A(t), R(t)) \in \mathbb{R}_+^5 : N(t) \leq \frac{\Lambda}{d}, B(t) \in \mathbb{R}_+ : \frac{\Lambda}{d}\frac{\kappa_1 + \kappa_2 + \kappa_3}{\delta}\right\}.$$

24.2.1 FRACTAL-FRACTIONAL ORDER COVID-19 MODEL

In this subsection, we apply the fractal-fractional operators to the model (24.1) and obtain the following generalised model:

$$^{FF}D_{0,t}^{p,q}\big(S(t)\big) = \Lambda - (\theta_1 E + \theta_2 I + \theta_3 A + \theta_4 B)\frac{S}{N} - dS,$$
$$^{FF}D_{0,t}^{p,q}\big(E(t)\big) = (\theta_1 E + \theta_2 I + \theta_3 A + \theta_4 B)\frac{S}{N} - (\varpi + d)E,$$
$$^{FF}D_{0,t}^{p,q}\big(I(t)\big) = (1-\xi)\varpi E - (d + d_1 + \omega_1)I,$$
$$^{FF}D_{0,t}^{p,q}\big(A(t)\big) = \xi\varpi E - (d + \omega_2)A,$$

$$^{FF}D_{0,t}^{p,q}\left(R(t)\right) = \omega_1 I + \omega_2 A - dR,$$

$$^{FF}D_{0,t}^{p,q}\left(B(t)\right) = \kappa_1 E + \kappa_2 I + \kappa_3 A - \delta B, \tag{24.2}$$

where p denotes the fractional order, while q is the fractal order. All the initial conditions for the fractal-fractional order system (24.2) are non-negative. The analysis of the model (24.2) and their related results are shown in the upcoming sections.

24.3 EXISTENCE AND UNIQUENESS

This section considers the existence and uniqueness of the fractal-fractional model (24.2). We assume that for all $t \in [0,T]$, the functions $S(t), E(t), I(t), A(t), R(t)$, and $B(t)$ are bounded. We need to write first the model equations in the form:

$$f_1(t,S,E,I,A,R,B) = \Lambda - (\theta_1 E + \theta_2 I + \theta_3 A + \theta_4 B)\frac{S}{N} - dS,$$

$$f_2(t,S,E,I,A,R,B) = (\theta_1 E + \theta_2 I + \theta_3 A + \theta_4 B)\frac{S}{N} - (\varpi + d)E,$$

$$f_3(t,S,E,I,A,R,B) = (1-\xi)\varpi E - (d + d_1 + \omega_1)I,$$

$$f_4(t,S,E,I,A,R,B) = \xi \varpi E - (d + \omega_2)A,$$

$$f_5(t,S,E,I,A,R,B) = \omega_1 I + \omega_2 A - dR,$$

$$f_6(t,S,E,I,A,R,B) = \kappa_1 E + \kappa_2 I + \kappa_3 A - \delta B, \tag{24.3}$$

and

$$\|S\| \leq M_S, \ \|E\| \leq M_E, \ \|I\| \leq M_I, \ \|A\| \leq M_A, \ \|R\| \leq M_R, \ \|B\| \leq M_B.$$

We shall prove that $f_1,...,f_6$ satisfy both the linear growth and the Lipschitz conditions. We start with the linear growth

$$\begin{aligned}\left|f_1(t,S,...,B)\right|^2 &\leq 3\Lambda^2 + 3|\theta_1 E + \theta_2 I + \theta_3 A + \theta_4 B|^2 \frac{|S|^2}{|N|^2} + 2d^2|S|^2,\\ &\leq 3\Lambda^2 + 12\left\{\frac{\theta_1^2|E| + \theta_2^2|I| + \theta_3^2|A| + \theta_4^2|B|}{|N|^2}\right\}|S|^2 + 3d^2|S|^2,\\ &\leq 3\Lambda^2 + 12\{\theta_1^2 + \theta_2^2 + \theta_3^2 + \theta_4^2\}|S|^2 + 3d^2|S|^2,\\ &\leq 3\Lambda^2\left(1 + 4\frac{\{\theta_1^2 + \theta_2^2 + \theta_3^2 + \theta_4^2\} + d^2}{\Lambda^2}|S|^2\right).\end{aligned}$$

If

$$\frac{\{\theta_1^2 + \theta_2^2 + \theta_3^2 + \theta_4^2\} + d^2}{\Lambda^2} < 1,$$

then

$$\left|f_1(t,S,...,B)\right|^2 \leq 3\Lambda^2(1+|S|^2),$$
$$\leq K_S(1+|S|^2),$$

where $K_S = 3\Lambda^2$.

$$\left|f_2(t,S,E,...,B)\right|^2 \leq 2|\theta_1 E + \theta_2 I + \theta_3 A + \theta_4 B|^2 \left|\frac{S}{N}\right|^2 + 2(d+\varpi)^2|E|^2,$$
$$\leq 8(\theta_1^2 + \theta_2^2 + \theta_3^2 + \theta_4^2)|S|^2 + 2(d+\varpi)^2|E|^2,$$
$$\leq 8(\theta_1^2 + \theta_2^2 + \theta_3^2 + \theta_4^2)M_S + 2(d+\varpi)^2|E|^2,$$
$$\leq 8(\theta_1^2 + \theta_2^2 + \theta_3^2 + \theta_4^2)M_S$$
$$\times \left(1 + \frac{(d+\varpi)^2}{4(\theta_1^2 + \theta_2^2 + \theta_3^2 + \theta_4^2)}|E|^2\right).$$

If

$$\frac{(d+\varpi)^2}{4(\theta_1^2 + \theta_2^2 + \theta_3^2 + \theta_4^2)} < 1,$$

then

$$\left|f_2(t,S,E,...,B)\right|^2 \leq K_E(1+|E|^2),$$

where $K_E = 8(\theta_1^2 + \theta_2^2 + \theta_3^2 + \theta_4^2)M_S$.

$$\left|f_3(t,S,,...I,...,B)\right|^2 \leq |(1-\xi)\varpi E - (d+d_1+\omega_1)I|^2,$$
$$\leq 2(1-\xi)^2\varpi^2 M_E\left(1 + \frac{(d+d_1+\omega_1)^2}{(1-\xi)^2\varpi^2 M_E}|I|^2\right)$$

If $\frac{(d+d_1+\omega_1)^2}{(1-\xi)^2\varpi^2 M_E} < 1$, therefore,

$$\left|f_3(t,S,,...I,...,B)\right|^2 \leq K_I(1+|I|^2),$$

where $K_I = 2(1-\xi)^2\varpi^2 M_E$.

$$\left|f_4(t,S,,...A,...,B)\right|^2 \leq |\xi\varpi E - (d+\omega_2)A|^2,$$

$$\leq 2\xi^2\varpi^2|E|^2 + 2(d+\omega_2)^2|A|^2,$$
$$\leq 2\xi^2\varpi^2 M_E + 2(d+\omega_2)^2|A|^2,$$
$$\leq 2\xi^2\varpi^2 M_E \left(1 + \frac{(d+\omega_2)^2}{\xi^2\varpi^2 M_E}|A|^2\right).$$

If $\frac{(d+\omega_2)^2}{\xi^2\varpi^2 M_E} < 1$, then

$$\left|f_4(t,S,,...A,...,B)\right|^2 \leq K_A(1+|A|^2),$$

where $K_A = 2\xi^2\omega^2 M_E$.

$$\left|f_5(t,S,...,R,B)\right|^2 \leq |\omega_1 I + \omega_2 A - dA|^2,$$
$$\leq 3\omega_1^2|I|^2 + 3\omega_2^2|A|^2 + 3d^2|R|^2,$$
$$\leq 3\omega_1^2 M_I + 3\omega_2^2 M_A + 3d^2|R|^2,$$
$$\leq (3\omega_1^2 M_I + 3\omega_2^2 M_A)\left(1 + \frac{d^2}{\omega_1^2 M_I + \omega_2^2 M_A}|R|^2\right).$$

If $\frac{d^2}{\omega_1^2 M_I + \omega_2^2 M_A} < 1$, then

$$\left|f_5(t,S,...,R,B)\right|^2 \leq K_R(1+|R|^2),$$

where $K_R = 3\omega_1^2 M_I + 3\omega_2^2 M_A$.

$$\left|f_6(t,S,...,B)\right|^2 \leq 4\kappa_1^2 M_E + 4\kappa_2^2 M_I + 4\kappa_3^2 M_A + 4\delta^2|B|^2,$$
$$\leq (4\kappa_1^2 M_E + 4\kappa_2^2 M_I + 4\kappa_3^2 M_A)$$
$$\times \left(1 + \frac{\delta^2}{(\kappa_1^2 M_E + \kappa_2^2 M_I + \kappa_3^2 M_A)}|B|^2\right).$$

If $\frac{\delta^2}{(\kappa_1^2 M_E + \kappa_2^2 M_I + \kappa_3^2 M_A)} < 1$, then

$$\left|f_6(t,S,...,B)\right|^2 \leq K_B(1+|B|^2),$$

where $K_B = (4\kappa_1^2 M_E + 4\kappa_2^2 M_I + 4\kappa_3^2 M_A)$. Therefore,

$$\max\left\{\frac{\delta^2}{(\kappa_1^2 M_E + \kappa_2^2 M_I + \kappa_3^2 M_A)}, \frac{d^2}{\omega_1^2 M_I + \omega_2^2 M_A}, \frac{(d+\omega_2)^2}{\xi^2\varpi^2 M_E},\right.$$
$$\left.\frac{(d+d_1+\omega_1)^2}{(1-\xi)^2\varpi^2 M_E}, \frac{(d+\varpi)^2}{4(\theta_1^2+\theta_2^2+\theta_3^2+\theta_4^2)}, \frac{4\{\theta_1^2+\theta_2^2+\theta_3^2+\theta_4^2\}+d^2}{\Lambda^2}\right\} < 1.$$

Application of Fractal-Fractional Operators to COVID-19 Infection

Thus, the system satisfies the linear growth property.

We now verify the Lipschitz conditions of the system, by letting $L_1 = (\theta_1 E + \theta_2 I + \theta_3 A + \theta_4 B)$ and $S(t) \leq N(t)$.

$$\begin{aligned}
\left| f_1(t, S_1, ..., B) - f_1(t, S_2, ..., B) \right| &= \left| -L_1 S_1 - dS_1 + L_1 S_2 + dS_2 \right|, \\
&\leq \left| L_1(S_1 - S_2) + d(S_1 - S_2) \right|, \\
&\leq L_1 \left| S_1 - S_2 \right| + d \left| S_1 - S_2 \right|, \\
&\leq (\theta_1 M_E + \theta_2 M_I + \theta_3 M_A + \theta_4 M_B + d) \left| S_1 - S_2 \right|, \\
&\leq K_S \left| S_1 - S_2 \right|,
\end{aligned}$$

where $K_S = (\theta_1 M_E + \theta_2 M_I + \theta_3 M_A + \theta_4 M_B + d)$.

$$\begin{aligned}
\left| f_1(t, S, E_1..., B) - f_1(t, S, E_1..., B) \right| &= \left| \theta_1 E_1 S - (\varpi + d) E_1 - \theta_1 E_2 S + (\varpi + d) E_2 \right|, \\
&\leq \theta_1 M_S |E_1 - E_2| + (\varpi + d)|E_1 - E_2|, \\
&\leq (\theta_1 M_S + (\varpi + d))|E_1 - E_2|, \\
&\leq K_E |E_1 - E_2|,
\end{aligned}$$

where $K_E = (\theta_1 M_S + (\varpi + d))$.

$$\begin{aligned}
\left| f_1(t, S, .., I_1..., B) - f_1(t, S, ..., I_2..., B) \right| &= \left| -(d + d_1 + \omega_1) I_1 + (d + d_1 + \omega_1) I_2 \right|, \\
&\leq (d + d_1 + \omega_1)|I_1 - I_2|, \\
&\leq K_I |I_1 - I_2|,
\end{aligned}$$

where $K_I = (d + d_1 + \omega_1)$.

$$\begin{aligned}
\left| f_1(t, S, .., R_1..., B) - f_1(t, S, ..., R_2..., B) \right| &= \left| d(R_1 - R_2) \right|, \\
&\leq d|R_1 - R_2|, \\
&\leq K_R |I_1 - I_2|,
\end{aligned}$$

where $K_R = d$.

$$\left| f_1(t, S, ..., B_1) - f_1(t, ..., B_2) \right| = \left| \delta(B_1 - B_2) \right|,$$

$$\leq \delta|B_1 - B_2|,$$
$$\leq K_B|B_1 - B_2|,$$

where $K_B = \delta$. So, the Lipschitz condition is satisfied, and hence the system admits a unique solution.

24.4 EQUILIBRIUM POINTS AND THEIR ANALYSIS

The disease-free equilibrium of the system (24.2) is given by

$$U_0 = (S_0, E_0, I_0, A_0, R_0, B_0) = \left(\frac{\Lambda}{d}, 0, 0, 0, 0, 0\right).$$

For the stability analysis of the model (24.2), we need first to compute the basic reproduction number denoted by $\mathscr{R}_0$. We consider the classes, E, I, A, and B, and using the method given in [252], we have

$$F = \begin{pmatrix} \theta_1 & \theta_2 & \theta_3 & \theta_4 \\ 0 & 0 & 0 & 0 \\ 0 & 0 & 0 & 0 \\ 0 & 0 & 0 & 0 \end{pmatrix}$$

and

$$V = \begin{pmatrix} \varpi + d & 0 & 0 & 0 \\ -(1-\xi)\varpi & (d+d_1+\omega_1) & 0 & 0 \\ -\xi\varpi & 0 & (d+\omega_2) & 0 \\ -\kappa_1 & -\kappa_2 & -\kappa_3 & \delta \end{pmatrix}.$$

We have the basic reproduction number, given by

$$\mathscr{R}_0 = \frac{k_2(\varpi\xi(\theta_4\kappa_3 + \theta_3\delta) + k_3(\theta_4\kappa_1 + \theta_1\delta)) + \varpi k_3(1-\xi)(\theta_4\kappa_2 + \theta_2\delta)}{k_1 k_2 k_3 \delta},$$
$$= \mathscr{R}_1 + \mathscr{R}_2 + \mathscr{R}_3,$$

where

$$\mathscr{R}_1 = \frac{\varpi(1-\xi)(\theta_4\kappa_2 + \theta_2\delta)}{k_1 k_2 \delta}, \quad \mathscr{R}_2 = \frac{\varpi\xi(\theta_4\kappa_3 + \theta_3\delta)}{k_1 k_3 \delta}, \quad \mathscr{R}_3 = \frac{\theta_4\kappa_1 + \theta_1\delta}{k_1\delta},$$

where $k_1 = d + \varpi$, $k_2 = \omega_1 + d + d_1$, and $k_3 = \omega_2 + d$. The following theorem is given to show the local asymptomatic stability of the system (24.2) at U_0.

Theorem 7 *The fractal-fractional order model is locally asymptotically stable at U_0 if $\mathscr{R}_0 < 1$.*

Application of Fractal-Fractional Operators to COVID-19 Infection

Proof 9 *We have the Jacobian matrix of the model (24.2) at U_0:*

$$J(U_0) = \begin{pmatrix} -d & -\theta_1 & -\theta_2 & -\theta_3 & 0 & -\theta_4 \\ 0 & -k_1+\theta_1 & \theta_2 & \theta_3 & 0 & \theta_4 \\ 0 & \varpi(1-\xi) & -k_2 & 0 & 0 & 0 \\ 0 & \varpi\xi & 0 & -k_3 & 0 & 0 \\ 0 & 0 & \omega_1 & \omega_2 & -d & 0 \\ 0 & \kappa_1 & \kappa_2 & \kappa_3 & 0 & -\delta \end{pmatrix}.$$

The eigenvalues $-d$ and $-d$ are clearly negative, while the rest of the eigenvalues we will compute from the following equation:

$$\lambda^4 + m_1\lambda^3 + m_2\lambda^2 + m_3\lambda + m_4 = 0,$$

where

$m_1 = -\theta_1 + k_1 + k_2 + k_3 + \delta,$

$m_2 = k_1k_2\left(1 - \dfrac{\theta_2\varpi(1-\xi)}{k_1k_2}\right) + k_1k_3\left(1 - \dfrac{\theta_3\varpi\xi}{k_1k_3}\right) + k_1\delta\left(1 - \dfrac{\theta_4\kappa_1 + \theta_1\delta}{k_1\delta}\right)$
$\quad + (k_2+k_3)(\delta - \theta_1) + k_2k_3,$

$m_3 = k_1k_2(1-\mathscr{R}_1)\delta + k_1k_3(1-\mathscr{R}_2)\delta - \theta_2\varpi k_3(1-\xi) - \theta_3\varpi k_2\xi$
$\quad - (k_2+k_3)(\theta_4\kappa_1 + \theta_1\delta)$
$\quad - \theta_1k_2k_3 + k_2k_3\delta + k_1k_2k_3,$

$m_4 = k_1k_2k_3\delta(1-\mathscr{R}_0).$

We have that $m_j > 0$ for $j = 1, 2, 3,$ and 4 whenever $\mathscr{R}_0 < 1$, and further it needs to satisfy the Routh-Hurwitz criteria $m_j > 0$ for $j = 1, 2, 3,$ and 4 and $m_1m_2m_3 > m_3^2 + m_1^2m_4$. Upon satisfying these mentioned conditions, we conclude that the fractal-fractional order model is locally asymptotically stable at U_0 whenever $\mathscr{R}_0 < 1$.

In order to have the expression for the existence of an endemic equilibria for the system (24.2), we denote it by $U_1 = (S^*, E^*, I^*, A^*, R^*, B^*)$ and obtain

$$S^* = \dfrac{\Lambda}{d+\lambda},$$

$$E^* = \dfrac{\lambda S^*}{d+\varpi},$$

$$I^* = \dfrac{\varpi E^*(1-\xi)}{\omega_1 + d + d_1},$$

$$A^* = \dfrac{\varpi\xi E^*}{\omega_2 + d},$$

$$R^* = \frac{A^*\omega_2 + \omega_1 I^*}{d},$$

$$B^* = \frac{\kappa_3 A^* + \kappa_1 E^* + \kappa_2 I^*}{\delta}.$$

Using the above results into the following:

$$\lambda^* = \frac{A^*\theta_3 + \theta_4 B^* + \theta_1 E^* + \theta_2 I^*}{N^*},$$

we have

$$G(\lambda^*) = r_1 \lambda^* + r_2,$$

where

$$r_1 = \delta(\varpi k_3(1-\xi)(\omega_1+d) + k_2(\varpi\xi(\omega_2+d)+dk_3)),$$

$$r_2 = dk_1 k_2 k_3 \delta(1-\mathcal{R}_0).$$

If $\mathcal{R}_0 > 1$, then obviously there exists a unique endemic equilibrium which implies that the model has no possibilities of the backward bifurcation phenomenon.

24.5 DATA FITTING, NUMERICAL SCHEMES, AND THEIR GRAPHICAL RESULTS

We consider the model [6] where the authors considered the real data to obtain the real parameters as shown in Figure 24.1 and the desired data fitting for the model in Figure 24.1. We refer the reader for more details about the data fitting analysis and their estimations to see [6]. The numerical value of the basic reproduction number obtained using the parameter values shown in Table 24.1 is $\mathcal{R}_0 = 1.2937$.

24.5.1 NUMERICAL SCHEME FOR COVID-19 INFECTION MODEL WITH POWER LAW

In order to study the model numerically with the case of power law, we first give generally a new scheme based on the Euler method. The general representations are shown by

$$\begin{aligned} {}^{FFP}_0 D^{p,q}_t y(t) &= f(t,y(t)), \text{ if } t > 0, \\ y(0) &= y_0, \text{ if } t = 0. \end{aligned} \quad (24.4)$$

$${}^{RL}_0 D^p_t y(t) = qt^{q-1} f(t,y(t)) \text{ if } t > 0$$

Application of Fractal-Fractional Operators to COVID-19 Infection

Table 24.1 Parameters fitted to the real data.

Parameter	Value	Source
Λ	$d \times N(0)$	[6]
d	$\frac{1}{74.87 \times 365}$	[6]
θ_1	0.1233	[6]
θ_2	0.0542	[6]
θ_3	0.0020	[6]
θ_4	0.1101	[6]
ϖ	0.1980	[6]
ξ	0.3085	[6]
d_1	0.0104	[6]
ω_1	0.3680	[6]
ω_2	0.2945	[6]
κ_1	0.2574	[6]
κ_2	0.2798	[6]
κ_3	0.1584	[6]
δ	0.3820	[6]

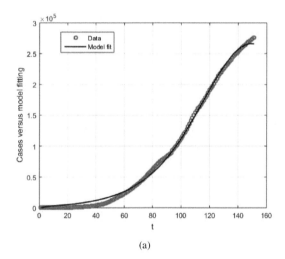

(a)

Figure 24.1 The COVID-19 cases in Kingdom of Saudi Arabia versus model simulation.

$$y(0) = y_0 \text{ if } t = 0. \tag{24.5}$$

We write

$$y(t) = \frac{q}{\Gamma(p)} \int_0^t \tau^{q-1} f(\tau, y(\tau))(t-\tau)^{p-1} d\tau, \; y(0) = y_0; \tag{24.6}$$

at $t = t_{n+1}$, we have

$$y(t_{n+1}) = \frac{q}{\Gamma(p)} \int_0^{t_{n+1}} \tau^{q-1} f(\tau, y(\tau))(t_{n+1} - \tau)^{p-1} d\tau, \quad (24.7)$$

and finally, we write

$$\begin{aligned} y(t_{n+1}) &= \frac{q}{\Gamma(p)} \sum_{j=0}^{n} \int_{t_j}^{t_{j+1}} \tau^{q-1} f(\tau, y(\tau))(t_{n+1} - \tau)^{p-1} d\tau, \\ &\approx \frac{q}{\Gamma(p)} \sum_{j=0}^{n} f(t_j, y_j) \int_{t_j}^{t_{j+1}} \tau^{q-1} (t_{n+1} - \tau)^{p-1} d\tau, \\ &\approx \frac{q}{\Gamma(p)} \sum_{j=0}^{n} f(t_j, y_j) t_{n+1}^{p+q-1} \left[B\left(\frac{t_{j+1}}{t_{n+1}}, q, p\right) - B\left(\frac{t_j}{t_{n+1}}, q, p\right) \right]. \quad (24.8) \end{aligned}$$

The above scheme shown above can be applied to the system (24.2), and the desired scheme will be given as

$$\begin{aligned} S_{n+1} &= \frac{q}{\Gamma(p)} \sum_{j=0}^{n} f_1(t_j, S_j, E_j, I_j, A_j, R_j, B_j) t_{n+1}^{p+q-1} \widehat{B}, \\ E_{n+1} &= \frac{q}{\Gamma(p)} \sum_{j=0}^{n} f_2(t_j, S_j, E_j, I_j, A_j, R_j, B_j) t_{n+1}^{p+q-1} \widehat{B}, \\ I_{n+1} &= \frac{q}{\Gamma(p)} \sum_{j=0}^{n} f_3(t_j, S_j, E_j, I_j, A_j, R_j, B_j) t_{n+1}^{p+q-1} \widehat{B}, \\ A_{n+1} &= \frac{q}{\Gamma(p)} \sum_{j=0}^{n} f_4(t_j, S_j, E_j, I_j, A_j, R_j, B_j) t_{n+1}^{p+q-1} \widehat{B}, \\ R_{n+1} &= \frac{q}{\Gamma(p)} \sum_{j=0}^{n} f_5(t_j, S_j, E_j, I_j, A_j, R_j, B_j) t_{n+1}^{p+q-1} \widehat{B}, \\ B_{n+1} &= \frac{q}{\Gamma(p)} \sum_{j=0}^{n} f_6(t_j, S_j, E_j, I_j, A_j, R_j, B_j) t_{n+1}^{p+q-1} \widehat{B}, \quad (24.9) \end{aligned}$$

where

$$\widehat{B} = \left[B\left(\frac{t_{j+1}}{t_{n+1}}, q, p\right) - B\left(\frac{t_j}{t_{n+1}}, q, p\right) \right]$$

$$\begin{aligned} f_1(t, S, E, I, A, R, B) &= \Lambda - (\theta_1 E + \theta_2 I + \theta_3 A + \theta_4 B)\frac{S}{N} - dS, \\ f_2(t, S, E, I, A, R, B) &= (\theta_1 E + \theta_2 I + \theta_3 A + \theta_4 B)\frac{S}{N} - (\varpi + d)E, \\ f_3(t, S, E, I, A, R, B) &= (1 - \xi)\varpi E - (d + d_1 + \omega_1)I, \end{aligned}$$

Application of Fractal-Fractional Operators to COVID-19 Infection

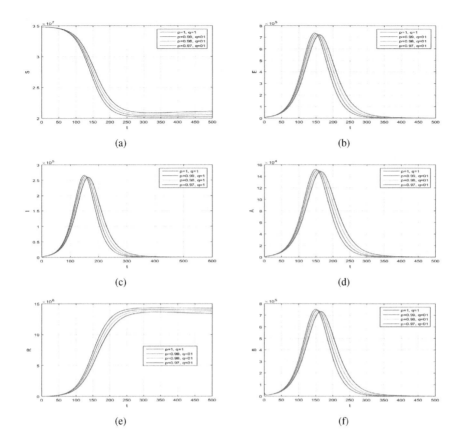

Figure 24.2 Simulation results for fractal-fractional model in Caputo case, when $p = 1, 0.99, 0.98,$ and 0.97 and $q = 1$.

$$f_4(t,S,E,I,A,R,B) = \xi \varpi E - (d+\omega_2)A,$$

$$f_5(t,S,E,I,A,R,B) = \omega_1 I + \omega_2 A - dR,$$

$$f_6(t,S,E,I,A,R,B) = \kappa_1 E + \kappa_2 I + \kappa_3 A - \delta B. \tag{24.10}$$

Using the scheme above for the case of power law kernel, we have the numerical simulations given in Figures 24.2–24.5. The simulation of the model when varying p and fixed q, we have the results in Figure 24.2. Varying both p and q, we give the results in Figure 24.3. The impact of the model parameters is shown in Figure 24.4, while the different phase portraits are shown in Figure 24.5.

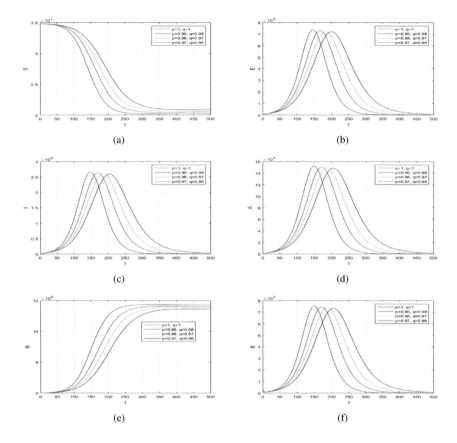

Figure 24.3 Simulation results for fractal-fractional model in Caputo case for various values of p and q.

24.5.2 NUMERICAL SCHEME FOR COVID-19 INFECTION MODEL WITH THE EXPONENTIAL KERNEL

In this subsection, we consider the following Cauchy problem to obtain a numerical scheme for the fractal-fractional exponential kernel. We give the following general scheme:

$$_0^{FFF}D_t^{p,q}y(t) = f(t,y(t)), \text{ if } t > 0,$$

$$y(0) = y_0, \text{ if } t = 0. \quad (24.11)$$

$$_0^{CFR}D_t^p y(t) = qt^{q-1}f(t,y(t)) \text{ if } t > 0$$

$$y(0) = y_0 \text{ if } t = 0 \quad (24.12)$$

Application of Fractal-Fractional Operators to COVID-19 Infection 381

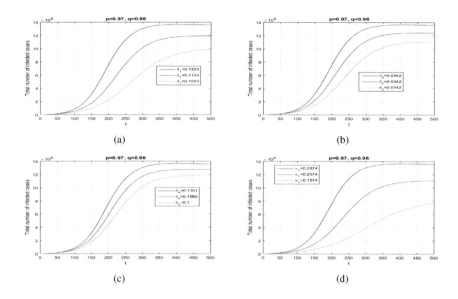

Figure 24.4 Impact of the model parameters on the on the dynamics of total infected population.

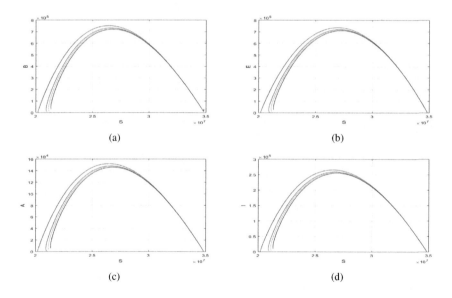

Figure 24.5 Different phase planes of the model variables, when $p = q = 1, 0.97, 0.96,$ and 0.95.

$$y(t) = (1-p)qt^{q-1}f(t,y(t))$$

$$+pq \int_0^t \tau^{q-1} f(\tau, y(\tau)) d\tau, \text{ if } t > 0,$$
$$y(0) = y_0, \text{ if } t = 0; \tag{24.13}$$

at $t = t_{n+1}$, we have

$$y(t_{n+1}) = qt_{n+1}^{q-1} f(t_{n+1}, y(t_{n+1}))$$
$$+ pq \int_0^{t_{n+1}} \tau^{q-1} f(\tau, y(\tau)) d\tau,$$
$$= qt_{n+1}^{q-1} f(t_{n+1}, y^u(t_{n+1}))$$
$$+ pq \sum_{j=0}^n \int_{t_j}^{t_{j+1}} f(\tau, y(\tau)) \tau^{q-1} d\tau \tag{24.14}$$

Further, we obtain

$$y_{n+1} = qt_{n+1}^{q-1} f(t_{n+1}, y^u(t_{n+1}))$$
$$+ p(\Delta t)^q \sum_{j=0}^n f(t_j, y_j) \left\{ (j+1)^q - j^q \right\}, \tag{24.15}$$

where

$$y^u(t_{n+1}) = y_n + p(\Delta t)^q \{(n+1)^q - n^q\} f(t_n, y_n). \tag{24.16}$$

The scheme (24.15) can be used for the numerical solution of the model (24.2) in the case of exponential kernel. We have the final scheme

$$S_{n+1} = qt_{n+1}^{q-1} f_1(t_{n+1}, S^u(t_{n+1}), E^u(t_{n+1}), I^u(t_{n+1}), A^u(t_{n+1}), R^u(t_{n+1}), B^u(t_{n+1}))$$
$$+ p(\Delta t)^q \sum_{j=0}^n f_1(t_j, S_j, E_j, I_j, A_j, R_j, B_j) \left\{ (j+1)^q - j^q \right\},$$
$$E_{n+1} = qt_{n+1}^{q-1} f_2(t_{n+1}, S^u(t_{n+1}), E^u(t_{n+1}), I^u(t_{n+1}), A^u(t_{n+1}), R^u(t_{n+1}), B^u(t_{n+1}))$$
$$+ p(\Delta t)^q \sum_{j=0}^n f_2(t_j, S_j, E_j, I_j, A_j, R_j, B_j) \left\{ (j+1)^q - j^q \right\},$$
$$I_{n+1} = qt_{n+1}^{q-1} f_3(t_{n+1}, S^u(t_{n+1}), E^u(t_{n+1}), I^u(t_{n+1}), A^u(t_{n+1}), R^u(t_{n+1}), B^u(t_{n+1}))$$
$$+ p(\Delta t)^q \sum_{j=0}^n f_3(t_j, S_j, E_j, I_j, A_j, R_j, B_j) \left\{ (j+1)^q - j^q \right\},$$
$$A_{n+1} = qt_{n+1}^{q-1} f_4(t_{n+1}, S^u(t_{n+1}), E^u(t_{n+1}), I^u(t_{n+1}), A^u(t_{n+1}), R^u(t_{n+1}), B^u(t_{n+1}))$$

Application of Fractal-Fractional Operators to COVID-19 Infection

$$+ p(\Delta t)^q \sum_{j=0}^{n} f_4(t_j, S_j, E_j, I_j, A_j, R_j, B_j) \left\{ (j+1)^q - j^q \right\},$$

$$R_{n+1} = q t_{n+1}^{q-1} f_5(t_{n+1}, S^u(t_{n+1}), E^u(t_{n+1}), I^u(t_{n+1}), A^u(t_{n+1}), R^u(t_{n+1}), B^u(t_{n+1}))$$

$$+ p(\Delta t)^q \sum_{j=0}^{n} f_5(t_j, S_j, E_j, I_j, A_j, R_j, B_j) \left\{ (j+1)^q - j^q \right\},$$

$$B_{n+1} = q t_{n+1}^{q-1} f_6(t_{n+1}, S^u(t_{n+1}), E^u(t_{n+1}), I^u(t_{n+1}), A^u(t_{n+1}), R^u(t_{n+1}), B^u(t_{n+1}))$$

$$+ p(\Delta t)^q \sum_{j=0}^{n} f_6(t_j, S_j, E_j, I_j, A_j, R_j, B_j) \left\{ (j+1)^q - j^q \right\}, \qquad (24.17)$$

where

$$S^u(t_{n+1}) = S_n + p(\Delta t)^q \{(n+1)^q - n^q\} f_1(t_n, S_n, E_n, I_n, A_n, R_n, B_n),$$

$$E^u(t_{n+1}) = E_n + p(\Delta t)^q \{(n+1)^q - n^q\} f_2(t_n, S_n, E_n, I_n, A_n, R_n, B_n),$$

$$I^u(t_{n+1}) = I_n + p(\Delta t)^q \{(n+1)^q - n^q\} f_3(t_n, S_n, E_n, I_n, A_n, R_n, B_n),$$

$$A^u(t_{n+1}) = A_n + p(\Delta t)^q \{(n+1)^q - n^q\} f_4(t_n, S_n, E_n, I_n, A_n, R_n, B_n),$$

$$R^u(t_{n+1}) = R_n + p(\Delta t)^q \{(n+1)^q - n^q\} f_5(t_n, S_n, E_n, I_n, A_n, R_n, B_n),$$

$$B^u(t_{n+1}) = B_n + p(\Delta t)^q \{(n+1)^q - n^q\} f_6(t_n, S_n, E_n, I_n, A_n, R_n, B_n). \qquad (24.18)$$

Using the scheme above, we have the numerical simulations for the case of power law in Figures 24.6–24.9. The simulation of the model when varying p and fixed q, we have the results in Figure 24.6. Varying both p and q, we give the results in Figure 24.7. The impact of the model parameters is shown in Figure 24.8, while the different phase portraits are shown in Figure 24.9.

24.5.3 NUMERICAL SCHEME FOR COVID MODEL WITH MITTAG-LEFFLER KERNEL

In this section, a fractal-fractional Cauchy problem with the generalised Mittag-Leffler function is considered,

$$\begin{aligned} {}_0^{FFM} D_t^{p,q} y(t) &= f(t, y(t)), \text{ if } t > 0, \\ y(0) &= y_0, \text{ if } t = 0. \end{aligned} \qquad (24.19)$$

$$\begin{aligned} {}_0^{ABR} D_t^p y(t) &= q t^{q-1} f(t, y(t)) \text{ if } t > 0 \\ y(0) &= y_0 \text{ if } t = 0. \end{aligned} \qquad (24.20)$$

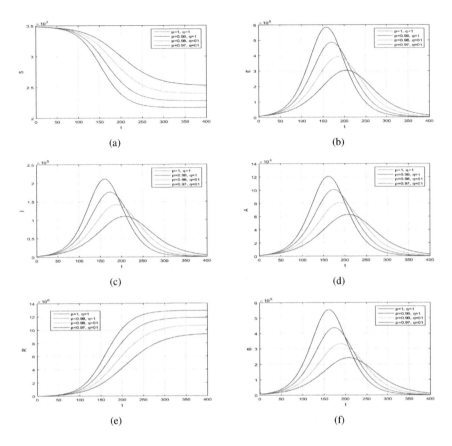

Figure 24.6 Simulation results for fractal-fractional model in Caputo-Fabrizio case, when $p = 1, 0.99, 0.98,$ and 0.97 and $q = 1$.

$$\begin{aligned} y(t) &= qt^{q-1}f(t,y(t))(1-p) + \frac{pq}{\Gamma(p)}\int_0^t \\ &\quad \times \tau^{q-1}(t-\tau)^{p-1}f(\tau,y(\tau))d\tau, \text{ if } t > 0 \\ y(0) &= y_0 \text{ if } t = 0. \end{aligned} \quad (24.21)$$

At $t = t_{n+1}$, we have

$$\begin{aligned} y(t_{n+1}) &= qt_{n+1}^{q-1}f(t_{n+1}, y^p(t_{n+1}))(1-p) \\ &\quad + \frac{pq}{\Gamma(p)}\int_0^{t_{n+1}} \tau^{q-1}(t_{n+1}-\tau)^{p-1}f(\tau,y(\tau))d\tau. \end{aligned} \quad (24.22)$$

Further, we write

$$y(t_{n+1}) = qt_{n+1}^{q-1}f(t_{n+1}, y^p(t_{n+1}))(1-p)$$

Application of Fractal-Fractional Operators to COVID-19 Infection

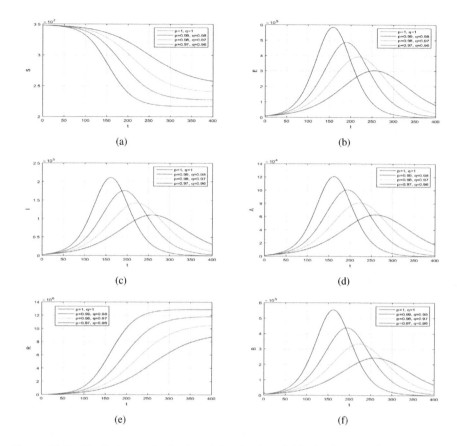

Figure 24.7 Simulation results for fractal-fractional model in the Caputo-Fabrizio case for various values of p and q.

$$+\frac{pq}{\Gamma(p)}\sum_{j=0}^{n}\int_{t_j}^{t_{j+1}}\tau^{q-1}(t_{n+1}-\tau)^{p-1}f(\tau,y(\tau))d\tau. \quad (24.23)$$

Then, we have

$$y(t_{n+1}) \approx y_{n+1} = qt_{n+1}^{q-1}f(t_{n+1},y_{n+1}^p)(1-p)$$

$$+\frac{pq}{\Gamma(p)}\sum_{j=0}^{n}\int_{t_j}^{t_{j+1}}\tau^{q-1}(t_{n+1}-\tau)^{p-1}$$

$$f(\tau,y(\tau))d\tau. \quad (24.24)$$

We write as

$$y_{n+1} = qt_{n+1}^{q-1}f(t_{n+1},y_{n+1}^u)(1-p)$$

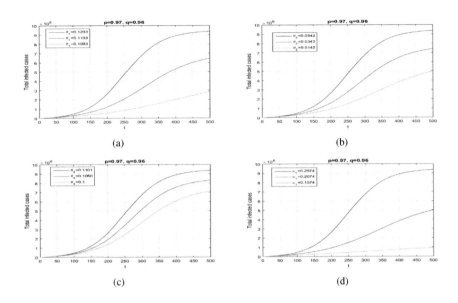

Figure 24.8 Impact of the model parameters on the dynamics of total infected population.

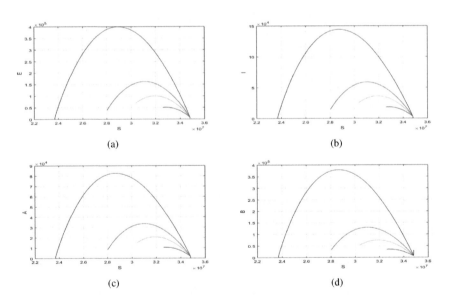

Figure 24.9 Different phase planes of the model variables, when $p = q = 1, 0.97, 0.96$, and 0.95.

$$+\frac{pq}{\Gamma(p)}\sum_{j=0}^{n}f(t_j,y_j)\int_{t_j}^{t_{j+1}}(t_{n+1}-\tau)^{p-1}\tau^{q-1}d\tau. \qquad (24.25)$$

Further, using the concept of generalised Beta-function, we write

$$y_{n+1} = qt_{n+1}^{q-1}f(t_{n+1},y_{n+1}^u)(1-p)+\frac{pq}{\Gamma(p)}\sum_{j=0}^{n}f(t_j,y_j)$$
$$\times t_{n+1}^{p+q-1}\left[B\left(\frac{t_{j+1}}{t_{n+1}},q,p\right)-B\left(\frac{t_j}{t_{n+1}},q,p\right)\right].$$

Further simplifying, we have

$$y_{n+1} = qt_{n+1}^{q-1}f(t_{n+1},y_{n+1}^u)(1-p)+\frac{pq}{\Gamma(p)}(\Delta t)^{p+q-1}(n+1)^{p+q}$$
$$\times \sum_{j=0}^{n}f(t_j,y_j)\left[B\left(\frac{t_{j+1}}{t_{n+1}},q,p\right)-B\left(\frac{t_j}{t_{n+1}},q,p\right)\right], \qquad (24.26)$$

where

$$y_{n+1}^u = \frac{pq}{\Gamma(p)}(\Delta t)^{p+q-1}(n+1)^{p+q-1}\sum_{j=0}^{n}f(t_j,y_j)$$
$$\times \left[B\left(\frac{t_{j+1}}{t_{n+1}},q,p\right)-B\left(\frac{t_j}{t_{n+1}},q,p\right)\right]. \qquad (24.27)$$

Applying the scheme (24.26) to our system (24.2), we have

$$S_{n+1} = qt_{n+1}^{q-1}f_1(t_{n+1},S_{n+1}^u,E_{n+1}^u,I_{n+1}^u,A_{n+1}^u,R_{n+1}^u,B_{n+1}^u)(1-p)$$
$$+\frac{pq}{\Gamma(p)}(\Delta t)^{p+q-1}(n+1)^{p+q}\sum_{j=0}^{n}f_1(t_j,S_j,E_j,I_j,A_j,R_j,B_j)$$
$$\times \left[B\left(\frac{t_{j+1}}{t_{n+1}},q,p\right)-B\left(\frac{t_j}{t_{n+1}},q,p\right)\right],$$
$$E_{n+1} = qt_{n+1}^{q-1}f_2(t_{n+1},S_{n+1}^u,E_{n+1}^u,I_{n+1}^u,A_{n+1}^u,R_{n+1}^u,B_{n+1}^u)(1-p)$$
$$+\frac{pq}{\Gamma(p)}(\Delta t)^{p+q-1}(n+1)^{p+q}\sum_{j=0}^{n}f_2(t_j,S_j,E_j,I_j,A_j,R_j,B_j)$$
$$\times \left[B\left(\frac{t_{j+1}}{t_{n+1}},q,p\right)-B\left(\frac{t_j}{t_{n+1}},q,p\right)\right],$$
$$I_{n+1} = qt_{n+1}^{q-1}f_3(t_{n+1},S_{n+1}^u,E_{n+1}^u,I_{n+1}^u,A_{n+1}^u,R_{n+1}^u,B_{n+1}^u)(1-p)$$
$$+\frac{pq}{\Gamma(p)}(\Delta t)^{p+q-1}(n+1)^{p+q}\sum_{j=0}^{n}f_3(t_j,S_j,E_j,I_j,A_j,R_j,B_j)$$
$$\times \left[B\left(\frac{t_{j+1}}{t_{n+1}},q,p\right)-B\left(\frac{t_j}{t_{n+1}},q,p\right)\right],$$

$$A_{n+1} = qt_{n+1}^{q-1} f_4(t_{n+1}, S_{n+1}^u, E_{n+1}^u, I_{n+1}^u, A_{n+1}^u, R_{n+1}^u, B_{n+1}^u)(1-p)$$
$$+ \frac{pq}{\Gamma(p)} (\Delta t)^{p+q-1} (n+1)^{p+q} \sum_{j=0}^{n} f_4(t_j, S_j, E_j, I_j, A_j, R_j, B_j)$$
$$\times \left[B\left(\frac{t_{j+1}}{t_{n+1}}, q, p\right) - B\left(\frac{t_j}{t_{n+1}}, q, p\right) \right],$$

$$R_{n+1} = qt_{n+1}^{q-1} f_5(t_{n+1}, S_{n+1}^u, E_{n+1}^u, I_{n+1}^u, A_{n+1}^u, R_{n+1}^u, B_{n+1}^u)(1-p)$$
$$+ \frac{pq}{\Gamma(p)} (\Delta t)^{p+q-1} (n+1)^{p+q} \sum_{j=0}^{n} f_5(t_j, S_j, E_j, I_j, A_j, R_j, B_j)$$
$$\times \left[B\left(\frac{t_{j+1}}{t_{n+1}}, q, p\right) - B\left(\frac{t_j}{t_{n+1}}, q, p\right) \right],$$

$$B_{n+1} = qt_{n+1}^{q-1} f_6(t_{n+1}, S_{n+1}^u, E_{n+1}^u, I_{n+1}^u, A_{n+1}^u, R_{n+1}^u, B_{n+1}^u)(1-p)$$
$$+ \frac{pq}{\Gamma(p)} (\Delta t)^{p+q-1} (n+1)^{p+q} \sum_{j=0}^{n} f_6(t_j, S_j, E_j, I_j, A_j, R_j, B_j)$$
$$\times \left[B\left(\frac{t_{j+1}}{t_{n+1}}, q, p\right) - B\left(\frac{t_j}{t_{n+1}}, q, p\right) \right], \tag{24.28}$$

where

$$S_{n+1}^u = \frac{pq}{\Gamma(p)} (\Delta t)^{p+q-1} (n+1)^{p+q-1} \sum_{j=0}^{n} f_1(t_j, S_j, E_j, I_j, A_j, R_j, B_j)$$
$$\times \left[B\left(\frac{t_{j+1}}{t_{n+1}}, q, p\right) - B\left(\frac{t_j}{t_{n+1}}, q, p\right) \right],$$

$$E_{n+1}^u = \frac{pq}{\Gamma(p)} (\Delta t)^{p+q-1} (n+1)^{p+q-1} \sum_{j=0}^{n} f_2(t_j, S_j, E_j, I_j, A_j, R_j, B_j)$$
$$\times \left[B\left(\frac{t_{j+1}}{t_{n+1}}, q, p\right) - B\left(\frac{t_j}{t_{n+1}}, q, p\right) \right],$$

$$I_{n+1}^u = \frac{pq}{\Gamma(p)} (\Delta t)^{p+q-1} (n+1)^{p+q-1} \sum_{j=0}^{n} f_3(t_j, S_j, E_j, I_j, A_j, R_j, B_j)$$
$$\times \left[B\left(\frac{t_{j+1}}{t_{n+1}}, q, p\right) - B\left(\frac{t_j}{t_{n+1}}, q, p\right) \right],$$

$$A_{n+1}^u = \frac{pq}{\Gamma(p)} (\Delta t)^{p+q-1} (n+1)^{p+q-1} \sum_{j=0}^{n} f_4(t_j, S_j, E_j, I_j, A_j, R_j, B_j)$$
$$\times \left[B\left(\frac{t_{j+1}}{t_{n+1}}, q, p\right) - B\left(\frac{t_j}{t_{n+1}}, q, p\right) \right],$$

$$R_{n+1}^u = \frac{pq}{\Gamma(p)} (\Delta t)^{p+q-1} (n+1)^{p+q-1} \sum_{j=0}^{n} f_5(t_j, S_j, E_j, I_j, A_j, R_j, B_j)$$
$$\times \left[B\left(\frac{t_{j+1}}{t_{n+1}}, q, p\right) - B\left(\frac{t_j}{t_{n+1}}, q, p\right) \right],$$

Application of Fractal-Fractional Operators to COVID-19 Infection

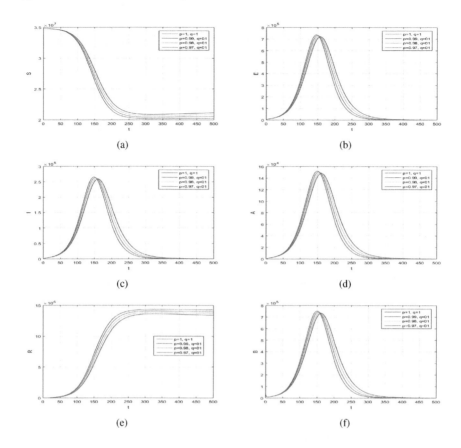

Figure 24.10 Simulation results for fractal-fractional model in AB case, when $p = 1, 0.99, 0.98$, and 0.97 and $q = 1$.

$$B_{n+1}^u = \frac{pq}{\Gamma(p)}(\Delta t)^{p+q-1}(n+1)^{p+q-1}\sum_{j=0}^{n} f_6(t_j, S_j, E_j, I_j, A_j, R_j, B_j)$$
$$\times \left[B\left(\frac{t_{j+1}}{t_{n+1}}, q, p\right) - B\left(\frac{t_j}{t_{n+1}}, q, p\right)\right]. \quad (24.29)$$

Using the scheme above for the case of Mittag-Leffler kernel, we have the numerical simulations given in Figures 24.10–24.13. The simulation of the model when varying p and fixed q, we have the results in Figure 24.10. Varying both p and q, we give the results in Figure 24.11. The impact of the model parameters is shown in Figure 24.12, while the different phase portraits are shown in Figure 24.13.

24.6 CONCLUSION

In this chapter a fractal-fractional model for COVID-19 infection in the sense of various fractal-fractional operators have been analysed. We studied the model

390   Numerical Methods for Fractal-Fractional Differential

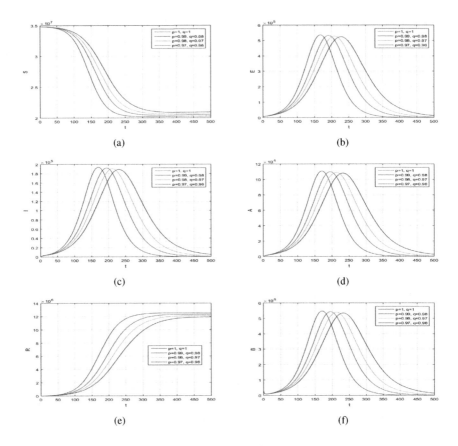

Figure 24.11 Simulation results for fractal-fractional model in AB case for various values of p and q.

equilibrium points and obtained their local asymptotic stability. Further, the realistic parameters were obtained from the published work, and then we computed the numerical value of the basic reproduction number. We considered the new Euler approach and obtained the numerical schemes for the case of power law, exponential law, and the Mittag-Leffler law. The schemes were used further to obtain the graphical results in detail.

Application of Fractal-Fractional Operators to COVID-19 Infection 391

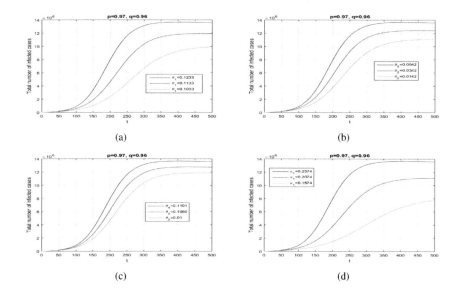

Figure 24.12 Impact of the model parameters on the dynamics of total infected population.

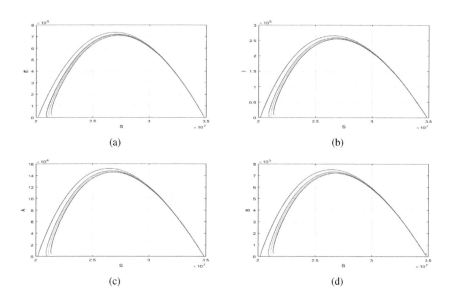

Figure 24.13 Different phase planes of the model variables, when $p = q = 1, 0.97, 0.96,$ and 0.95.

References

1. Saudi Arabia announces the first case of coronavirus, Arab News. Riyadh: Saudi Research and Marketing Group. 3 March 2020, 2020.
2. Saudi Arabia detects second coronavirus case". Arab News. 4 March 2020. Archived from the original on 5 March 2020. Retrieved 7 March 2020.
3. Sabah E Abd-Elraheem, Hayam Hamza Mansour, et al. Salivary changes in type 2 diabetic patients. *Diabetes & Metabolic Syndrome: Clinical Research & Reviews*, 11:S637–S641, 2017.
4. Eugene Ackerman, John W Rosevear, and Warren F McGuckin. A mathematical model of the glucose-tolerance test. *Physics in Medicine & Biology*, 9(2):203, 1964.
5. Akif Akgul, Irene Moroz, Ihsan Pehlivan, and Sundarapandian Vaidyanathan. A new four-scroll chaotic attractor and its engineering applications. *Optik*, 127(13):5491–5499, 2016.
6. Marei Saeed Alqarni, Metib Alghamdi, Taseer Muhammad, Ali Saleh Alshomrani, and Muhammad Altaf Khan. Mathematical modeling for novel coronavirus (covid-19) and control. *Numerical Methods for Partial Differential Equations*, 38(4):760–776, 2022.
7. FT Arecchi, R Meucci, G Puccioni, and J Tredicce. Experimental evidence of subharmonic bifurcations, multistability, and turbulence in a q-switched gas laser. *Physical Review Letters*, 49(17):1217, 1982.
8. Joshua Kiddy K Asamoah. Fractal–fractional model and numerical scheme based on newton polynomial for q fever disease under atangana–baleanu derivative. *Results in Physics*, 34:105189, 2022.
9. Joshua Kiddy K Asamoah, MA Owusu, Zhen Jin, FT Oduro, Afeez Abidemi, and Esther Opoku Gyasi. Global stability and cost-effectiveness analysis of covid-19 considering the impact of the environment: Using data from ghana. *Chaos, Solitons & Fractals*, 110103, 2020.
10. Abdon Atangana. Fractal-fractional differentiation and integration: Connecting fractal calculus and fractional calculus to predict complex system. *Chaos, Solitons & Fractals*, 102:396–406, 2017.
11. Abdon Atangana. Blind in a commutative world: Simple illustrations with functions and chaotic attractors. *Chaos, Solitons & Fractals*, 114:347–363, 2018.
12. Abdon Atangana. Non validity of index law in fractional calculus: A fractional differential operator with markovian and non-markovian properties. *Physica A: Statistical Mechanics and Its Applications*, 505:688–706, 2018.
13. Abdon Atangana. Extension of rate of change concept: From local to nonlocal operators with applications. *Results in Physics*, 19:103515, 2020.
14. Abdon Atangana. Modelling the spread of covid-19 with new fractal-fractional operators: Can the lockdown save mankind before vaccination? *Chaos, Solitons & Fractals*, 136:109860, 2020.
15. Abdon Atangana, Jose Francisco Gomez Aguilar, Matthew Owolabi Kolade, and Jordan Yankov Hristov. Fractional differential and integral operators with non-singular and non-local kernel with application to nonlinear dynamical systems, 2020.
16. Abdon Atangana, Ali Akgül, and Kolade M Owolabi. Analysis of fractal fractional differential equations. *Alexandria Engineering Journal*, 59(3):1117–1134, 2020.

17. Abdon Atangana and Seda İğret Araz. Fractional stochastic modelling illustration with modified chua attractor. *The European Physical Journal Plus*, 134(4):160, 2019.
18. Abdon Atangana and Muhammad Altaf Khan. Validity of fractal derivative to capturing chaotic attractors. *Chaos, Solitons & Fractals*, 126:50–59, 2019.
19. Abdon Atangana, Muhammad Altaf Khan, et al. Modeling and analysis of competition model of bank data with fractal-fractional caputo-fabrizio operator. *Alexandria Engineering Journal*, 2020.
20. Abdon Atangana and Ilknur Koca. Chaos in a simple nonlinear system with atangana–baleanu derivatives with fractional order. *Chaos, Solitons & Fractals*, 89:447–454, 2016.
21. Abdon Atangana and Toufik Mekkaoui. Trinition the complex number with two imaginary parts: Fractal, chaos and fractional calculus. *Chaos, Solitons & Fractals*, 128:366–381, 2019.
22. Abdon Atangana and Aydin Secer. A note on fractional order derivatives and table of fractional derivatives of some special functions. In *Abstract and applied analysis*, volume 2013. Hindawi, 2013.
23. LF Ávalos-Ruiz, JF Gómez-Aguilar, A Atangana, and Kolade M Owolabi. On the dynamics of fractional maps with power-law, exponential decay and mittag–leffler memory. *Chaos, Solitons & Fractals*, 127:364–388, 2019.
24. Agnessa Babloyantz, JM Salazar, and C Nicolis. Evidence of chaotic dynamics of brain activity during the sleep cycle. *Physics Letters A*, 111(3):152–156, 1985.
25. Santo Banerjee, Sanjay K Palit, Sayan Mukherjee, MRK Ariffin, and Lamberto Rondoni. Complexity in congestive heart failure: A time-frequency approach. *Chaos: An Interdisciplinary Journal of Nonlinear Science*, 26(3):033105, 2016.
26. BC Bao, H Bao, N Wang, M Chen, and Q Xu. Hidden extreme multistability in memristive hyperchaotic system. *Chaos, Solitons & Fractals*, 94:102–111, 2017.
27. Bo-Cheng Bao, Quan Xu, Han Bao, and Mo Chen. Extreme multistability in a memristive circuit. *Electronics Letters*, 52(12):1008–1010, 2016.
28. Bocheng Bao, Tao Jiang, Guangyi Wang, Peipei Jin, Han Bao, and Mo Chen. Two-memristor-based chuas hyperchaotic circuit with plane equilibrium and its extreme multistability. *Nonlinear Dynamics*, 89(2):1157–1171, 2017.
29. Bocheng Bao, Tao Jiang, Quan Xu, Mo Chen, Huagan Wu, and Yihua Hu. Coexisting infinitely many attractors in active band-pass filter-based memristive circuit. *Nonlinear Dynamics*, 86(3):1711–1723, 2016.
30. Bocheng Bao, Li Xu, Zhimin Wu, Mo Chen, and Huagan Wu. Coexistence of multiple bifurcation modes in memristive diode-bridge-based canonical chuas circuit. *International Journal of Electronics*, 105(7):1159–1169, 2018.
31. Han Bao, Ning Wang, Huagan Wu, Zhe Song, and Bocheng Bao. Bi-stability in an improved memristor-based third-order wien-bridge oscillator. *IETE Technical Review*, 36(2):109–116, 2019.
32. Kosar Barati, Sajad Jafari, Julien Clinton Sprott, and Viet-Thanh Pham. Simple chaotic flows with a curve of equilibria. *International Journal of Bifurcation and Chaos*, 26(12):1630034, 2016.
33. Roberto Barrio, Fernando Blesa, and Sergio Serrano. Qualitative analysis of the rössler equations: Bifurcations of limit cycles and chaotic attractors. *Physica D: Nonlinear Phenomena*, 238(13):1087–1100, 2009.
34. Roberto Barrio, M Angeles Martínez, Sergio Serrano, and Daniel Wilczak. When chaos meets hyperchaos: 4d rössler model. *Physics Letters A*, 379(38):2300–2305, 2015.

35. Irina Bashkirtseva, Venera Nasyrova, and Lev Ryashko. Noise-induced bursting and chaos in the two-dimensional rulkov model. *Chaos, Solitons & Fractals*, 110:76–81, 2018.
36. Atiyeh Bayani, Karthikeyan Rajagopal, Abdul Jalil M Khalaf, Sajad Jafari, GD Leutcho, and J Kengne. Dynamical analysis of a new multistable chaotic system with hidden attractor: Antimonotonicity, coexisting multiple attractors, and offset boosting. *Physics Letters A*, 383(13):1450–1456, 2019.
37. JR Beddington, CA Free, and JH Lawton. Dynamic complexity in predator-prey models framed in difference equations. *Nature*, 255(5503):58, 1975.
38. Debashis Biswas and Samares Pal. Stability analysis of a delayed hiv/aids epidemic model with saturated incidence. *International Journal of Mathematics Trends and Technology*, 43(3):222–231, 2017.
39. Geoff Boeing. Visual analysis of nonlinear dynamical systems: Chaos, fractals, self-similarity and the limits of prediction. *Systems*, 4(4):37, 2016.
40. Kais Bouallegue. A new class of neural networks and its applications. *Neurocomputing*, 249:28–47, 2017.
41. Michele Caputo. Linear models of dissipation whose q is almost frequency independentii. *Geophysical Journal International*, 13(5):529–539, 1967.
42. Michele Caputo and Mauro Fabrizio. A new definition of fractional derivative without singular kernel. *Progress in Fractional Differentiation and Application*, 1(2):1–13, 2015.
43. Ushnish Chaudhuri and Awadhesh Prasad. Complicated basins and the phenomenon of amplitude death in coupled hidden attractors. *Physics Letters A*, 378(9):713–718, 2014.
44. GR Chen and JH Lü. Dynamics analysis, control and synchronization of generalized lorenz system, 2003.
45. Guanrong Chen and Tetsushi Ueta. Yet another chaotic attractor. *International Journal of Bifurcation and Chaos*, 9(07):1465–1466, 1999.
46. Jerry Chen and Denis Blackmore. On the exponentially self-regulating population model. *Chaos, Solitons & Fractals*, 14(9):1433–1450, 2002.
47. W Chen. Time–space fabric underlying anomalous diffusion. *Chaos, Solitons & Fractals*, 28(4):923–929, 2006.
48. Wen Chen, Hongguang Sun, Xiaodi Zhang, and Dean Korošak. Anomalous diffusion modeling by fractal and fractional derivatives. *Computers & Mathematics with Applications*, 59(5):1754–1758, 2010.
49. Xiangyong Chen, Jianlong Qiu, Jinde Cao, and Haibo He. Hybrid synchronization behavior in an array of coupled chaotic systems with ring connection. *Neurocomputing*, 173:1299–1309, 2016.
50. Xiangyong Chen, Chengyong Wang, and Jianlong Qiu. Synchronization and anti-synchronization of n different coupled chaotic systems with ring connection. *International Journal of Modern Physics C*, 25(05):1440011, 2014.
51. LEONO Chua, Motomasa Komuro, and Takashi Matsumoto. The double scroll family. *IEEE Transactions on Circuits and Systems*, 33(11):1072–1118, 1986.
52. Agnieszka Chudzik, Przemyslaw Perlikowski, Andrzej Stefanski, and Tomasz Kapitaniak. Multistability and rare attractors in van der pol–duffing oscillator. *International Journal of Bifurcation and Chaos*, 21(07):1907–1912, 2011.

53. Meechoke Chuedoung, Warunee Sarika, and Yongwimon Lenbury. Dynamical analysis of a nonlinear model for glucose–insulin system incorporating delays and β-cells compartment. *Nonlinear Analysis: Theory, Methods & Applications*, 71(12):e1048–e1058, 2009.
54. S Çiçek, A Ferikoğlu, and I Pehlivan. A new 3d chaotic system: Dynamical analysis, electronic circuit design, active control synchronization and chaotic masking communication application. *Optik*, 127(8):4024–4030, 2016.
55. B Cuahutenango-Barro, MA Taneco-Hernández, and JF Gómez-Aguilar. On the solutions of fractional-time wave equation with memory effect involving operators with regular kernel. *Chaos, Solitons & Fractals*, 115:283–299, 2018.
56. Sara Dadras and Hamid Reza Momeni. A novel three-dimensional autonomous chaotic system generating two, three and four-scroll attractors. *Physics Letters A*, 373(40):3637–3642, 2009.
57. Yin Dai, Huanzhen Wang, and Haoran Sun. Cyclic-shift chaotic medical image encryption algorithm based on plain text key-stream. *International Journal of Simulation–Systems, Science & Technology*, 17(27), 2016.
58. Christopher M Danforth. Chaos in an atmosphere hanging on a wall. *Mathematics of Planet Earth*, 17, 2013.
59. Lokenath Debnath. A brief historical introduction to fractional calculus. *International Journal of Mathematical Education in Science and Technology*, 35(4):487–501, 2004.
60. Robert Devaney. *An introduction to chaotic dynamical systems*. CRC Press, 2018.
61. Dawid Dudkowski, Sajad Jafari, Tomasz Kapitaniak, Nikolay V Kuznetsov, Gennady A Leonov, and Awadhesh Prasad. Hidden attractors in dynamical systems. *Physics Reports*, 637:1–50, 2016.
62. Dawid Dudkowski, Awadhesh Prasad, and Tomasz Kapitaniak. Perpetual points and hidden attractors in dynamical systems. *Physics Letters A*, 379(40-41):2591–2596, 2015.
63. Leah Edelstein-Keshet. *Mathematical models in biology*. SIAM, 2005.
64. MM El-Dessoky and Muhammad Altaf Khan. Modeling and analysis of an epidemic model with fractal-fractional atangana-baleanu derivative. *Alexandria Engineering Journal*, 61(1):729–746, 2022.
65. S Elaydi. Discrete chaos. Champman & Hall, 2000.
66. Thomas Elbert, William J Ray, Zbigniew J Kowalik, James E Skinner, Karl Eugen Graf, and Niels Birbaumer. Chaos and physiology: Deterministic chaos in excitable cell assemblies. *Physiological Reviews*, 74(1):1–47, 1994.
67. Okan Erkaymaz, Mahmut Ozer, and Matjaž Perc. Performance of small-world feedforward neural networks for the diagnosis of diabetes. *Applied Mathematics and Computation*, 311:22–28, 2017.
68. Duccio Fanelli and Francesco Piazza. Analysis and forecast of covid-19 spreading in china, italy and france. *Chaos, Solitons & Fractals*, 134:109761, 2020.
69. Philippe Faure and Henri Korn. Is there chaos in the brain? i. concepts of nonlinear dynamics and methods of investigation. *Comptes Rendus de l'Académie des Sciences-Series III-Sciences de la Vie*, 324(9):773–793, 2001.
70. Mehmet Onur Fen. Persistence of chaos in coupled lorenz systems. *Chaos, Solitons & Fractals*, 95:200–205, 2017.
71. A Fortin, M Fortin, and JJ Gervais. Complex transition to chaotic flow in a periodic array of cylinders. *Theoretical and Computational Fluid Dynamics*, 3(2):79–93, 1991.

References

72. AC Fowler, JD Gibbon, and MJ McGuinness. The complex lorenz equations. *Physica D: Nonlinear Phenomena*, 4(2):139–163, 1982.
73. James Gleick. *Chaos: Making a new science*. Open Road Media, 2011.
74. JF Gómez-Aguilar. Fundamental solutions to electrical circuits of non-integer order via fractional derivatives with and without singular kernels. *The European Physical Journal Plus*, 133(5):197, 2018.
75. JF Gómez-Aguilar. Chaos and multiple attractors in a fractal-fractional shinrikis oscillator model. *Physica A: Statistical Mechanics and Its Applications*, 122918, 2019.
76. José Francisco Gómez-Aguilar, Abdon Atangana, and Victor Fabian Morales-Delgado. Electrical circuits rc, lc, and rl described by atangana–baleanu fractional derivatives. *International Journal of Circuit Theory and Applications*, 45(11):1514–1533, 2017.
77. John Guckenheimer and Ricardo A Oliva. Chaos in the hodgkin–huxley model. *SIAM Journal on Applied Dynamical Systems*, 1(1):105–114, 2002.
78. Taza Gul, Haris Anwar, Muhammad Altaf Khan, Ilyas Khan, and Poom Kumam. Integer and non-integer order study of the go-w/go-eg nanofluids flow by means of marangoni convection. *Symmetry*, 11(5):640, 2019.
79. Fatemeh Hadaeghi, Mohammad Reza Hashemi Golpayegani, Sajad Jafari, and Greg Murray. Toward a complex system understanding of bipolar disorder: A chaotic model of abnormal circadian activity rhythms in euthymic bipolar disorder. *Australian & New Zealand Journal of Psychiatry*, 50(8):783–792, 2016.
80. Martin Hasler. Engineering chaos for encryption and broadband communication. *Philosophical Transactions of the Royal Society of London. Series A: Physical and Engineering Sciences*, 353(1701):115–126, 1995.
81. Boris Hasselblatt and Anatole Katok. *A first course in dynamics: With a panorama of recent developments*. Cambridge University Press, 2003.
82. Alan Hastings and Thomas Powell. Chaos in a three-species food chain. *Ecology*, 72(3):896–903, 1991.
83. Edward H Hellen and Evgeny Volkov. How to couple identical ring oscillators to get quasiperiodicity, extended chaos, multistability, and the loss of symmetry. *Communications in Nonlinear Science and Numerical Simulation*, 62:462–479, 2018.
84. William H Herman and Paul Zimmet. Type 2 diabetes: An epidemic requiring global attention and urgent action. *Diabetes Care*, 35(5):943–944, 2012.
85. MH Heydari and A Atangana. An optimization method based on the generalized lucas polynomials for variable-order space-time fractional mobile-immobile advection-dispersion equation involving derivatives with non-singular kernels. *Chaos, Solitons & Fractals*, 132:109588, 2020.
86. DY Hsieh. Hydrodynamic instabilty, chaos and phase transition. *Nonlinear Analysis: Theory, Methods & Applications*, 30(8):5327–5334, 1997.
87. Zhixing Hu, Wanbiao Ma, and Shigui Ruan. Analysis of sir epidemic models with nonlinear incidence rate and treatment. *Mathematical Biosciences*, 238(1):12–20, 2012.
88. Changchun Hua and Xinping Guan. Adaptive control for chaotic systems. *Chaos, Solitons & Fractals*, 22(1):55–60, 2004.
89. Vladimir G Ivancevic and Tijana T Ivancevic. *Complex nonlinearity: Chaos, phase transitions, topology change and path integrals*. Springer Science & Business Media, 2008.
90. Mohammad Ali Jafari, Ezzedine Mliki, Akif Akgul, Viet-Thanh Pham, Sifeu Takougang Kingni, Xiong Wang, and Sajad Jafari. Chameleon: The most hidden chaotic flow. *Nonlinear Dynamics*, 88(3):2303–2317, 2017.

91. Sajad Jafari, JC Sprott, and S Mohammad Reza Hashemi Golpayegani. Elementary quadratic chaotic flows with no equilibria. *Physics Letters A*, 377(9):699–702, 2013.
92. Sajad Jafari, JC Sprott, Viet-Thanh Pham, Christos Volos, and Chunbiao Li. Simple chaotic 3d flows with surfaces of equilibria. *Nonlinear Dynamics*, 86(2):1349–1358, 2016.
93. Sajad Jafari, JC Sprott, Viet-Thanh Pham, Christos Volos, and Chunbiao Li. Simple chaotic 3d flows with surfaces of equilibria. *Nonlinear Dynamics*, 86(2):1349–1358, 2016.
94. Rashid Jan, Muhammad Altaf Khan, and JF Gómez-Aguilar. Asymptomatic carriers in transmission dynamics of dengue with control interventions. *Optimal Control Applications and Methods*, 2019.
95. Rashid Jan, Muhammad Altaf Khan, Poom Kumam, and Phatiphat Thounthong. Modeling the transmission of dengue infection through fractional derivatives. *Chaos, Solitons & Fractals*, 127:189–216, 2019.
96. Fahd Jarad and Thabet Abdeljawad. Generalized fractional derivatives and laplace transform. *Discrete & Continuous Dynamical Systems-S*, 13(3):709, 2020.
97. Ryutaro Kanno. Representation of random walk in fractal space-time. *Physica A: Statistical Mechanics and Its Applications*, 248(1–2):165–175, 1998.
98. T Kapitaniak. Stochastic response with bifurcations to non-linear duffing's oscillator. *Journal of Sound and Vibration*, 102(3):440–441, 1985.
99. Tomasz Kapitaniak and Gennadiĭ Leonov. Multistability: Uncovering hidden attractors.
100. Stephen H Kellert. *In the wake of chaos: Unpredictable order in dynamical systems*. University of Chicago Press, 1993.
101. J Kengne, A Nguomkam Negou, and D Tchiotsop. Antimonotonicity, chaos and multiple attractors in a novel autonomous memristor-based jerk circuit. *Nonlinear Dynamics*, 88(4):2589–2608, 2017.
102. J Kengne, ZT Njitacke, and HB Fotsin. Dynamical analysis of a simple autonomous jerk system with multiple attractors. *Nonlinear Dynamics*, 83(1–2):751–765, 2016.
103. MA Khan, Arshad Khan, A Elsonbaty, and AA Elsadany. Modeling and simulation results of a fractional dengue model. *The European Physical Journal Plus*, 134(8):379, 2019.
104. MA Khan, Syed Wasim Shah, Saif Ullah, and JF Gómez-Aguilar. A dynamical model of asymptomatic carrier zika virus with optimal control strategies. *Nonlinear Analysis: Real World Applications*, 50:144–170, 2019.
105. Muhammad Altaf Khan. The dynamics of a new chaotic system through the caputo–fabrizio and atanagan–baleanu fractional operators. *Advances in Mechanical Engineering*, 11(7):1687814019866540, 2019.
106. Muhammad Altaf Khan, Cicik Alfiniyah, Ebraheem Alzahrani, et al. Analysis of dengue model with fractal-fractional caputo–fabrizio operator. *Advances in Difference Equations*, 2020(1):1–23, 2020.
107. Muhammad Altaf Khan and Abdon Atangana. Modeling the dynamics of novel coronavirus (2019-ncov) with fractional derivative. *Alexandria Engineering Journal*, 2020.
108. Muhammad Altaf Khan, Abdon Atangana, Ebraheem Alzahrani, et al. The dynamics of covid-19 with quarantined and isolation. *Advances in Difference Equations*, 2020(1):1–22, 2020.

References

109. Muhammad Altaf Khan and Francisco Gómez-Aguilar. Tuberculosis model with relapse via fractional conformable derivative with power law. *Mathematical Methods in the Applied Sciences*, 42(18):7113–7125, 2019.
110. Muhammad Altaf Khan, Olusola Kolebaje, Ahmet Yildirim, Saif Ullah, P Kumam, and P Thounthong. Fractional investigations of zoonotic visceral leishmaniasis disease with singular and non-singular kernel. *The European Physical Journal Plus*, 134(10):481, 2019.
111. Muhammad Altaf Khan, Saif Ullah, and Muhammad Farooq. A new fractional model for tuberculosis with relapse via atangana–baleanu derivative. *Chaos, Solitons & Fractals*, 116:227–238, 2018.
112. Muhammad Altaf Khan, Saif Ullah, KO Okosun, and Kamil Shah. A fractional order pine wilt disease model with caputo–fabrizio derivative. *Advances in Difference Equations*, 2018(1):410, 2018.
113. Saeed Khorashadizadeh and Mohammad-Hassan Majidi. Chaos synchronization using the fourier series expansion with application to secure communications. *AEU-International Journal of Electronics and Communications*, 82:37–44, 2017.
114. A Anatolii Aleksandrovich Kilbas, Hari Mohan Srivastava, and Juan J Trujillo. *Theory and applications of fractional differential equations*, volume 204. Elsevier Science Limited, 2006.
115. Minseo Kim, Unsoo Ha, Kyuho Jason Lee, Yongsu Lee, and Hoi-Jun Yoo. A 82-nw chaotic map true random number generator based on a sub-ranging sar adc. *IEEE Journal of Solid-State Circuits*, 52(7):1953–1965, 2017.
116. ST Kingni, S Jafari, H Simo, and P Woafo. Three-dimensional chaotic autonomous system with only one stable equilibrium: Analysis, circuit design, parameter estimation, control, synchronization and its fractional-order form. *The European Physical Journal Plus*, 129(5):76, 2014.
117. Henri Korn and Philippe Faure. Is there chaos in the brain? ii. experimental evidence and related models. *Comptes rendus biologies*, 326(9):787–840, 2003.
118. Mustafa RS Kulenovic and Gerasimos Ladas. *Dynamics of second order rational difference equations: With open problems and conjectures*. Chapman and Hall/CRC, 2001.
119. Qiang Lai, Akif Akgul, Chunbiao Li, Guanghui Xu, and Ünal Çavuşoğlu. A new chaotic system with multiple attractors: Dynamic analysis, circuit realization and s-box design. *Entropy*, 20(1):12, 2018.
120. Qiang Lai, Akif Akgul, Xiao-Wen Zhao, and Huiqin Pei. Various types of coexisting attractors in a new 4d autonomous chaotic system. *International Journal of Bifurcation and Chaos*, 27(09):1750142, 2017.
121. Qiang Lai and Shiming Chen. Generating multiple chaotic attractors from sprott b system. *International Journal of Bifurcation and Chaos*, 26(11):1650177, 2016.
122. Yongwimon Lenbury, Sitipong Ruktamatakul, and Somkid Amornsamarnkul. Modeling insulin kinetics: Responses to a single oral glucose administration or ambulatory-fed conditions. *Biosystems*, 59(1):15–25, 2001.
123. GA Leonov, NV Kuznetsov, MA Kiseleva, EP Solovyeva, and AM Zaretskiy. Hidden oscillations in mathematical model of drilling system actuated by induction motor with a wound rotor. *Nonlinear Dynamics*, 77(1-2):277–288, 2014.
124. GA Leonov, NV Kuznetsov, and TN Mokaev. Hidden attractor and homoclinic orbit in lorenz-like system describing convective fluid motion in rotating cavity. *Communications in Nonlinear Science and Numerical Simulation*, 28(1-3):166–174, 2015.

125. GA Leonov, NV Kuznetsov, and TN Mokaev. Homoclinic orbits, and self-excited and hidden attractors in a lorenz-like system describing convective fluid motion. *The European Physical Journal Special Topics*, 224(8):1421–1458, 2015.
126. GA Leonov, NV Kuznetsov, and VI Vagaitsev. Localization of hidden chua's attractors. *Physics Letters A*, 375(23):2230–2233, 2011.
127. GA Leonov, NV Kuznetsov, and VI Vagaitsev. Hidden attractor in smooth chua systems. *Physica D: Nonlinear Phenomena*, 241(18):1482–1486, 2012.
128. Gennady A Leonov and Nikolay V Kuznetsov. Hidden attractors in dynamical systems. from hidden oscillations in hilbert–kolmogorov, aizerman, and kalman problems to hidden chaotic attractor in chua circuits. *International Journal of Bifurcation and Chaos*, 23(01):1330002, 2013.
129. Chunbiao Li, Wen Hu, Julien Clinton Sprott, and Xinhai Wang. Multistability in symmetric chaotic systems. *The European Physical Journal Special Topics*, 224(8):1493–1506, 2015.
130. Chunbiao Li and Julien Clinton Sprott. Multistability in a butterfly flow. *International Journal of Bifurcation and Chaos*, 23(12):1350199, 2013.
131. Chunbiao Li and Julien Clinton Sprott. Coexisting hidden attractors in a 4-d simplified lorenz system. *International Journal of Bifurcation and Chaos*, 24(03):1450034, 2014.
132. Chunbiao Li and Julien Clinton Sprott. Multistability in the lorenz system: A broken butterfly. *International Journal of Bifurcation and Chaos*, 24(10):1450131, 2014.
133. Chunbiao Li, Julien Clinton Sprott, Akif Akgul, Herbert HC Iu, and Yibo Zhao. A new chaotic oscillator with free control. *Chaos: An Interdisciplinary Journal of Nonlinear Science*, 27(8):083101, 2017.
134. Chunbiao Li, Julien Clinton Sprott, Zeshi Yuan, and Hongtao Li. Constructing chaotic systems with total amplitude control. *International Journal of Bifurcation and Chaos*, 25(10):1530025, 2015.
135. Jian-Fen Li, Nong Li, Yu-Ping Liu, and Yi Gan. Linear and nonlinear generalized synchronization of a class of chaotic systems by using a single driving variable. 2009.
136. Mingming Li and Xianning Liu. An sir epidemic model with time delay and general nonlinear incidence rate. In *Abstract and applied Analysis*, volume 2014. Hindawi, 2014.
137. Yanguang C Li. Existence of chaos in evolution equations. *Mathematical and Computer Modelling*, 36(11-13):1211–1219, 2002.
138. Yong-Min Li, Saif Ullah, Muhammad Altaf Khan, Mohammad Y Alshahrani, and Taseer Muhammad. Modeling and analysis of the dynamics of hiv/aids with non-singular fractional and fractal-fractional operators. *Physica Scripta*, 96(11):114008, 2021.
139. Zhongfei Li, Zhuang Liu, and Muhammad Altaf Khan. Fractional investigation of bank data with fractal-fractional caputo derivative. *Chaos, Solitons & Fractals*, 109528, 2019.
140. Zhongfei Li, Zhuang Liu, and Muhammad Altaf Khan. Fractional investigation of bank data with fractal-fractional caputo derivative. *Chaos, Solitons & Fractals*, 131:109528, 2020.
141. Yang Li-Jiang and Chen Tian-Lun. Application of chaos in genetic algorithms. *Communications in Theoretical Physics*, 38(2):168, 2002.
142. Kuang-Yow Lian, Tung-Sheng Chiang, Chian-Song Chiu, and Peter Liu. Synthesis of fuzzy model-based designs to synchronization and secure communications for chaotic systems. *IEEE Transactions on Systems, Man, and Cybernetics, Part B (Cybernetics)*, 31(1):66–83, 2001.

143. CH Lien, S Vaidyanathan, A Sambas, M Mamat, WSM Sanjaya, et al. A new two-scroll chaotic attractor with three quadratic nonlinearities, its adaptive control and circuit design. In *IOP conference series: Materials science and engineering*, volume 332, page 012010. IOP Publishing, 2018.
144. Jan John Liszka-Hackzell. Prediction of blood glucose levels in diabetic patients using a hybrid ai technique. *Computers and Biomedical Research*, 32(2):132–144, 1999.
145. Chongxin Liu, Tao Liu, Ling Liu, and Kai Liu. A new chaotic attractor. *Chaos, Solitons & Fractals*, 22(5):1031–1038, 2004.
146. Shuang Liu, Bin Liu, and Pei-Ming Shi. Nonlinear feedback control of hopf bifurcation in a relative rotation dynamical system. 2009.
147. Wenbo Liu and Guanrong Chen. A new chaotic system and its generation. *International Journal of Bifurcation and Chaos*, 13(01):261–267, 2003.
148. Wenbo Liu and Guanrong Chen. Can a three-dimensional smooth autonomous quadratic chaotic system generate a single four-scroll attractor? *International Journal of Bifurcation and Chaos*, 14(04):1395–1403, 2004.
149. Yan Liu and Ling Lü. Synchronization of n different coupled chaotic systems with ring and chain connections. *Applied Mathematics and Mechanics*, 29(10):1299–1308, 2008.
150. Edward N Lorenz. Deterministic nonperiodic flow. *Journal of the Atmospheric Sciences*, 20(2):130–141, 1963.
151. Jorge Losada and Juan J Nieto. Properties of a new fractional derivative without singular kernel. *Progress in Fractional Differentiation and Applications*, 1(2):87–92, 2015.
152. Jinhu Lü and Guanrong Chen. A new chaotic attractor coined. *International Journal of Bifurcation and Chaos*, 12(03):659–661, 2002.
153. Jinhu Lü, Guanrong Chen, and Daizhan Cheng. A new chaotic system and beyond: The generalized lorenz-like system. *International Journal of Bifurcation and Chaos*, 14(05):1507–1537, 2004.
154. Jinhu Lü, Guanrong Chen, and Suochun Zhang. Dynamical analysis of a new chaotic attractor. *International Journal of Bifurcation and Chaos*, 12(05):1001–1015, 2002.
155. JA Tenreiro Machado, Alexandra MSF Galhano, and Juan J Trujillo. On development of fractional calculus during the last fifty years. *Scientometrics*, 98(1):577–582, 2014.
156. Ravi Kiran Maddali, Divya Ahluwalia, Adwitiya Chaudhuri, and Sk Sarif Hassan. Dynamics of a three dimensional chaotic cancer model. *International Journal of Mathematical Trends and Technology*, 53(5):353–368, 2018.
157. Gamal M Mahmoud, MA Al-Kashif, and Shaban A Aly. Basic properties and chaotic synchronization of complex lorenz system. *International Journal of Modern Physics C*, 18(02):253–265, 2007.
158. Gamal M Mahmoud, Tassos Bountis, and Emad E Mahmoud. Active control and global synchronization of the complex chen and lü systems. *International Journal of Bifurcation and Chaos*, 17(12):4295–4308, 2007.
159. Athena Makroglou, Jiaxu Li, and Yang Kuang. Mathematical models and software tools for the glucose-insulin regulatory system and diabetes: An overview. *Applied numerical mathematics*, 56(3–4):559–573, 2006.
160. Interval Maps. Discrete dynamics and difference equations. In *Proceedings of the twelfth international conference on difference equations and applications*, volume 23, page 27. World Scientific, 2007.
161. VP Maslov. Theory of chaos and its application to the crisis of debts and the origin of inflation. *Russian Journal of Mathematical Physics*, 16(1):103–120, 2009.

162. Elias C Massoud, Jef Huisman, Elisa Benincà, Michael C Dietze, Willem Bouten, and Jasper A Vrugt. Probing the limits of predictability: Data assimilation of chaotic dynamics in complex food webs. *Ecology Letters*, 21(1):93–103, 2018.
163. Robert M May. Chaos and the dynamics of biological populations. *Nuclear Physics B-Proceedings Supplements*, 2:225–245, 1987.
164. Michael D McKenzie. Chaotic behavior in national stock market indices: New evidence from the close returns test. *Global Finance Journal*, 12(1):35–53, 2001.
165. Alfredo Medio and Marji Lines. *Nonlinear dynamics: A primer*. Cambridge University Press, 2001.
166. Christa Meisinger, Barbara Thorand, Andrea Schneider, Jutta Stieber, Angela Döring, and Hannelore Löwel. Sex differences in risk factors for incident type 2 diabetes mellitus: The monica augsburg cohort study. *Archives of Internal Medicine*, 162(1):82–89, 2002.
167. Malihe Molaie, Razieh Falahian, Shahriar Gharibzadeh, Sajad Jafari, and Julien C Sprott. Artificial neural networks: Powerful tools for modeling chaotic behavior in the nervous system. *Frontiers in Computational Neuroscience*, 8:40, 2014.
168. Malihe Molaie, Sajad Jafari, Julien Clinton Sprott, and S Mohammad Reza Hashemi Golpayegani. Simple chaotic flows with one stable equilibrium. *International Journal of Bifurcation and Chaos*, 23(11):1350188, 2013.
169. VF Morales-Delgado, JF Gómez-Aguilar, Khaled M Saad, Muhammad Altaf Khan, and P Agarwal. Analytic solution for oxygen diffusion from capillary to tissues involving external force effects: A fractional calculus approach. *Physica A: Statistical Mechanics and Its Applications*, 523:48–65, 2019.
170. Mark S Mosko and Frederick H Damon. *On the order of chaos: Social anthropology and the science of chaos*. Berghahn Books, 2005.
171. JM Munoz-Pacheco, E Tlelo-Cuautle, I Toxqui-Toxqui, C Sanchez-Lopez, and R Trejo-Guerra. Frequency limitations in generating multi-scroll chaotic attractors using cfoas. *International Journal of Electronics*, 101(11):1559–1569, 2014.
172. Hayder Natiq, MRM Said, MRK Ariffin, Shaobo He, Lamberto Rondoni, and Santo Banerjee. Self-excited and hidden attractors in a novel chaotic system with complicated multistability. *The European Physical Journal Plus*, 133(12):557, 2018.
173. Fahimeh Nazarimehr, Sajad Jafari, Seyed Mohammad Reza Hashemi Golpayegani, and Julien Clinton Sprott. Categorizing chaotic flows from the viewpoint of fixed points and perpetual points. *International Journal of Bifurcation and Chaos*, 27(02):1750023, 2017.
174. Fahimeh Nazarimehr, Batool Saedi, Sajad Jafari, and Julien Clinton Sprott. Are perpetual points sufficient for locating hidden attractors? *International Journal of Bifurcation and Chaos*, 27(03):1750037, 2017.
175. Calistus N Ngonghala, Enahoro Iboi, Steffen Eikenberry, Matthew Scotch, Chandini Raina MacIntyre, Matthew H Bonds, and Abba B Gumel. Mathematical assessment of the impact of non-pharmaceutical interventions on curtailing the 2019 novel coronavirus. *Mathematical Biosciences*, 108364, 2020.
176. Nelson Okafor, Bashar Zahawi, Damian Giaouris, and Soumitro Banerjee. Chaos, co-existing attractors, and fractal basin boundaries in dc drives with full-bridge converter. In *Proceedings of 2010 IEEE international symposium on circuits and systems*, pages 129–132. IEEE, 2010.
177. D Okuonghae and A Omame. Analysis of a mathematical model for covid-19 population dynamics in lagos, nigeria. *Chaos, Solitons & Fractals*, 110032, 2020.

References

178. Alan V Oppenheim, Gregory W Wornell, Steven H Isabelle, and Kevin M Cuomo. Signal processing in the context of chaotic signals. In *[Proceedings] ICASSP-92: 1992 IEEE international conference on acoustics, speech, and signal processing*, volume 4, pages 117–120. IEEE, 1992.
179. World Health Organization. Coronavirus.
180. Kolade M Owolabi and Abdon Atangana. Chaotic behaviour in system of noninteger-order ordinary differential equations. *Chaos, Solitons & Fractals*, 115:362–370, 2018.
181. Kolade M Owolabi and Abdon Atangana. On the formulation of adams-bashforth scheme with atangana-baleanu-caputo fractional derivative to model chaotic problems. *Chaos: An Interdisciplinary Journal of Nonlinear Science*, 29(2):023111, 2019.
182. Lin Pan, Wuneng Zhou, Jianan Fang, and Dequan Li. A new three-scroll unified chaotic system coined. *International Journal of Nonlinear Science*, 10(4):462–474, 2010.
183. Shirin Panahi, Zainab Aram, Sajad Jafari, Jun Ma, and JC Sprott. Modeling of epilepsy based on chaotic artificial neural network. *Chaos, Solitons & Fractals*, 105:150–156, 2017.
184. Ju H Park. Adaptive synchronization of a unified chaotic system with an uncertain parameter. *International Journal of Nonlinear Sciences and Numerical Simulation*, 6(2):201–206, 2005.
185. Louis M Pecora and Thomas L Carroll. Synchronization in chaotic systems. *Physical Review Letters*, 64(8):821, 1990.
186. JH Peng, EJ Ding, M Ding, and W Yang. Synchronizing hyperchaos with a scalar transmitted signal. *Physical Review Letters*, 76(6):904, 1996.
187. Zai-Ping Peng, Chun-Hua Wang, Yuan Lin, and Xiao-Wen Luo. A novel four-dimensional multi-wing hyper-chaotic attractor and its application in image encryption. 2014.
188. Viet-Thanh Pham, Christos Volos, Sajad Jafari, and Tomasz Kapitaniak. Coexistence of hidden chaotic attractors in a novel no-equilibrium system. *Nonlinear Dynamics*, 87(3):2001–2010, 2017.
189. Viet-Thanh Pham, Christos Volos, Sajad Jafari, Sundarapandian Vaidyanathan, Tomasz Kapitaniak, and Xiong Wang. A chaotic system with different families of hidden attractors. *International Journal of Bifurcation and Chaos*, 26(08):1650139, 2016.
190. Viet-Thanh Pham, Christos K Volos, and Sundarapandian Vaidyanathan. Multi-scroll chaotic oscillator based on a first-order delay differential equation. In *Chaos modeling and control systems design*, pages 59–72. Springer, 2015.
191. Alexander N Pisarchik and Ulrike Feudel. Control of multistability. *Physics Reports*, 540(4):167–218, 2014.
192. Awadhesh Prasad. Existence of perpetual points in nonlinear dynamical systems and its applications. *International Journal of Bifurcation and Chaos*, 25(02):1530005, 2015.
193. Guoyuan Qi, Barend Jacobus van Wyk, and Michaël Antonie van Wyk. A four-wing attractor and its analysis. *Chaos, Solitons & Fractals*, 40(4):2016–2030, 2009.
194. Sania Qureshi, Abdon Atangana, and Asif Ali Shaikh. Strange chaotic attractors under fractal-fractional operators using newly proposed numerical methods. *The European Physical Journal Plus*, 134(10):523, 2019.
195. Sania Qureshi, Abdullahi Yusuf, Asif Ali Shaikh, Mustafa Inc, and Dumitru Baleanu. Fractional modeling of blood ethanol concentration system with real data application. *Chaos: An Interdisciplinary Journal of Nonlinear Science*, 29(1):013143, 2019.
196. Karthikeyan Rajagopal, Sajad Jafari, Anitha Karthikeyan, Ashokkumar Srinivasan, and Biniyam Ayele. Hyperchaotic memcapacitor oscillator with infinite equilibria and co-existing attractors. *Circuits, Systems, and Signal Processing*, 37(9):3702–3724, 2018.

197. A Rauh, L Hannibal, and NB Abraham. Global stability properties of the complex lorenz model. *Physica D: Nonlinear Phenomena*, 99(1):45–58, 1996.
198. Flavia Ravelli and Renzo Antolini. Complex dynamics underlying the human electrocardiogram. *Biological Cybernetics*, 67(1):57–65, 1992.
199. Shuili Ren, Shirin Panahi, Karthikeyan Rajagopal, Akif Akgul, Viet-Thanh Pham, and Sajad Jafari. A new chaotic flow with hidden attractor: The first hyperjerk system with no equilibrium. *Zeitschrift für Naturforschung A*, 73(3):239–249, 2018.
200. Gojka Roglic et al. Who global report on diabetes: A summary. *International Journal of Noncommunicable Diseases*, 1(1):3, 2016.
201. Lamberto Rondoni, MRK Ariffin, Renuganth Varatharajoo, Sayan Mukherjee, Sanjay K Palit, and Santo Banerjee. Optical complexity in external cavity semiconductor laser. *Optics Communications*, 387:257–266, 2017.
202. Bertram Ross. A brief history and exposition of the fundamental theory of fractional calculus. In *Fractional calculus and its applications*, pages 1–36. Springer, 1975.
203. Otto E Rössler. Chemical turbulence: Chaos in a simple reaction-diffusion system. *Zeitschrift für Naturforschung A*, 31(10):1168–1172, 1976.
204. Kristina I Rother. Diabetes treatmentbridging the divide. *The New England Journal of Medicine*, 356(15):1499, 2007.
205. Leonid A Safonov, Elad Tomer, Vadim V Strygin, Yosef Ashkenazy, and Shlomo Havlin. Multifractal chaotic attractors in a system of delay-differential equations modeling road traffic. *Chaos: An Interdisciplinary Journal of Nonlinear Science*, 12(4):1006–1014, 2002.
206. Papri Saha, Santo Banerjee, and A Roy Chowdhury. Chaos, signal communication and parameter estimation. *Physics Letters A*, 326(1-2):133–139, 2004.
207. Yasuhisa Saito, Wanbiao Ma, and Tadayuki Hara. A necessary and sufficient condition for permanence of a lotka–volterra discrete system with delays. *Journal of Mathematical Analysis and Applications*, 256(1):162–174, 2001.
208. Aceng Sambas, Mada Sanjaya Ws, Mustafa Mamat, and Rizki Putra Prastio. Mathematical modelling of chaotic jerk circuit and its application in secure communication system. In *Advances in chaos theory and intelligent control*, pages 133–153. Springer, 2016.
209. S Sampath, S Vaidyanathan, Ch K Volos, and VT Pham. An eight-term novel four-scroll chaotic system with cubic nonlinearity and its circuit simulation. *Journal of Engineering Science and Technology Review*, 8(2):1–6, 2015.
210. Sivaperumal Sampath, Sundarapandian Vaidyanathan, Aceng Sambas, Mohamad Afendee, Mustafa Mamat, and Mada Sanjaya. A new four-scroll chaotic system with a self-excited attractor and circuit implementation. *International Journal of Engineering & Technology*, 7(3):1931, August 2018.
211. Syed Azhar Ali Shah, Muhammad Altaf Khan, Muhammad Farooq, Saif Ullah, and Ebraheem O Alzahrani. A fractional order model for hepatitis b virus with treatment via atangana-baleanu derivative. *Physica A: Statistical Mechanics and Its Applications*, 122636, 2019.
212. Syed Azhar Ali Shah, Muhammad Altaf Khan, Muhammad Farooq, Saif Ullah, and Ebraheem O Alzahrani. A fractional order model for hepatitis b virus with treatment via atangana–baleanu derivative. *Physica A: Statistical Mechanics and Its Applications*, 538:122636, 2020.
213. Pooja Rani Sharma, Manish Dev Shrimali, Awadhesh Prasad, Nikolay V Kuznetsov, and Gennady A Leonov. Controlling dynamics of hidden attractors. *International Journal of Bifurcation and Chaos*, 25(04):1550061, 2015.

214. PR Sharma, MD Shrimali, A Prasad, NV Kuznetsov, and GA Leonov. Control of multistability in hidden attractors. *The European Physical Journal Special Topics*, 224(8):1485–1491, 2015.
215. Jonathan E Shaw, Richard A Sicree, and Paul Z Zimmet. Global estimates of the prevalence of diabetes for 2010 and 2030. *Diabetes Research and Clinical Practice*, 87(1):4–14, 2010.
216. Steven W Shaw and Richard H Rand. The transition to chaos in a simple mechanical system. *International Journal of Non-Linear Mechanics*, 24(1):41–56, 1989.
217. Yuankai Shi and Frank B Hu. The global implications of diabetes and cancer. *Lancet (London, England)*, 383(9933):1947–1948, 2014.
218. Leonid P Shil'nikov. *Methods of qualitative theory in nonlinear dynamics*, volume 5. World Scientific, 2001.
219. Alexander Silchenko, Tomasz Kapitaniak, and Vadim Anishchenko. Noise-enhanced phase locking in a stochastic bistable system driven by a chaotic signal. *Physical Review E*, 59(2):1593, 1999.
220. Sudeshna Sinha, Ramakrishna Ramaswamy, and J Subba Rao. Adaptive control in nonlinear dynamics. *Physica D: Nonlinear Phenomena*, 43(1):118–128, 1990.
221. J Clint Sprott. Some simple chaotic flows. *Physical review E*, 50(2):R647, 1994.
222. Julien C Sprott, Sajad Jafari, Abdul Jalil M Khalaf, and Tomasz Kapitaniak. Megastability: Coexistence of a countable infinity of nested attractors in a periodically-forced oscillator with spatially-periodic damping. *The European Physical Journal Special Topics*, 226(9):1979–1985, 2017.
223. Julien Clinton Sprott. Simplest chaotic flows with involutional symmetries. *International Journal of Bifurcation and Chaos*, 24(01):1450009, 2014.
224. Julien Clinton Sprott. Symmetric time-reversible flows with a strange attractor. *International Journal of Bifurcation and Chaos*, 25(05):1550078, 2015.
225. Julien Clinton Sprott, Xiong Wang, and Guanrong Chen. Coexistence of point, periodic and strange attractors. *International Journal of Bifurcation and Chaos*, 23(05):1350093, 2013.
226. Christopher C Strelioff and Alfred W Hübler. Medium-term prediction of chaos. *Physical Review Letters*, 96(4):044101, 2006.
227. HongGuang Sun, Mark M Meerschaert, Yong Zhang, Jianting Zhu, and Wen Chen. A fractal richards equation to capture the non-boltzmann scaling of water transport in unsaturated media. *Advances in Water resources*, 52:292–295, 2013.
228. Junwei Sun, Yi Shen, Guodong Zhang, Chengjie Xu, and Guangzhao Cui. Combination–combination synchronization among four identical or different chaotic systems. *Nonlinear Dynamics*, 73(3):1211–1222, 2013.
229. OI Tacha, Ch K Volos, Ioannis M Kyprianidis, Ioannis N Stouboulos, Sundarapandian Vaidyanathan, and V-T Pham. Analysis, adaptive control and circuit simulation of a novel nonlinear finance system. *Applied Mathematics and Computation*, 276:200–217, 2016.
230. S Tang and JM Liu. Chaos synchronization in semiconductor lasers with optoelectronic feedback. *IEEE Journal of Quantum Electronics*, 39(6):708–715, 2003.
231. Yan-Xia Tang, Abdul Jalil M Khalaf, Karthikeyan Rajagopal, Viet-Thanh Pham, Sajad Jafari, and Ye Tian. A new nonlinear oscillator with infinite number of coexisting hidden and self-excited attractors. *Chinese Physics B*, 27(4):040502, 2018.
232. Yang Tang and Jian-an Fang. Synchronization of n-coupled fractional-order chaotic systems with ring connection. *Communications in Nonlinear Science and Numerical Simulation*, 15(2):401–412, 2010.

233. Ziqi Tao, Aimin Shi, and Jing Zhao. Epidemiological perspectives of diabetes. *Cell Biochemistry and Biophysics*, 73(1):181–185, 2015.
234. Angel A Tateishi, Haroldo V Ribeiro, and Ervin K Lenzi. The role of fractional time-derivative operators on anomalous diffusion. *Frontiers in Physics*, 5:52, 2017.
235. S Jeeva Sathya Theesar, Santo Banerjee, and P Balasubramaniam. Synchronization of chaotic systems under sampled-data control. *Nonlinear Dynamics*, 70(3):1977–1987, 2012.
236. Gheorghe Tigan and Dumitru Opriş. Analysis of a 3d chaotic system. *Chaos, Solitons & Fractals*, 36(5):1315–1319, 2008.
237. E Tlelo-Cuautle, VH Carbajal-Gomez, PJ Obeso-Rodelo, JJ Rangel-Magdaleno, and Jose Cruz Nuñez-Perez. Fpga realization of a chaotic communication system applied to image processing. *Nonlinear Dynamics*, 82(4):1879–1892, 2015.
238. O Torrealba-Rodriguez, RA Conde-Gutiérrez, and AL Hernández-Javier. Modeling and prediction of covid-19 in mexico applying mathematical and computational models. *Chaos, Solitons & Fractals*, 109946, 2020.
239. Murat Tuna and Can Bülent Fidan. Electronic circuit design, implementation and fpga-based realization of a new 3d chaotic system with single equilibrium point. *Optik*, 127(24):11786–11799, 2016.
240. Atta Ullah, Sajjad Shaukat Jamal, and Tariq Shah. A novel scheme for image encryption using substitution box and chaotic system. *Nonlinear Dynamics*, 91(1):359–370, 2018.
241. Saif Ullah and Muhammad Altaf Khan. Modeling the impact of non-pharmaceutical interventions on the dynamics of novel coronavirus with optimal control analysis with a case study. *Chaos, Solitons & Fractals*, 110075, 2020.
242. Saif Ullah, Muhammad Altaf Khan, and Muhammad Farooq. A fractional model for the dynamics of tb virus. *Chaos, Solitons & Fractals*, 116:63–71, 2018.
243. Saif Ullah, Muhammad Altaf Khan, Muhammad Farooq, Zakia Hammouch, and Dumitru Baleanu. A fractional model for the dynamics of tuberculosis infection using caputo-fabrizio derivative. *Discrete & Continuous Dynamical Systems-S*, page 975, 2019.
244. Şahin Utkucan and Şahin Tezcan. Forecasting the cumulative number of confirmed cases of covid-19 in italy, uk and usa using fractional nonlinear grey bernoulli model. *Chaos, Solitons & Fractals*, 109948, 2020.
245. S Vaidyanathan. Mathematical analysis, adaptive control and synchronization of a ten-term novel three-scroll chaotic system with four quadratic nonlinearities. *International Journal of Controlled Theory and Applications*, 9(1):1–20, 2016.
246. Sundarapandian Vaidyanathan. 3-cells cellular neural network (cnn) attractor and its adaptive biological control. *International Journal of PharmTech Research*, 8(4):632–640, 2015.
247. Sundarapandian Vaidyanathan. Adaptive control of the fitzhugh-nagumo chaotic neuron model. *International Journal of PharmTech Research*, 8(6):117–127, 2015.
248. Sundarapandian Vaidyanathan. Lotka-volterra population biology models with negative feedback and their ecological monitoring. *International Journal of PharmTech Research*, 8(5):974–981, 2015.
249. Sundarapandian Vaidyanathan and Suresh Rasappan. Hybrid synchronization of hyperchaotic qi and lü systems by nonlinear control. In *International conference on computer science and information technology*, pages 585–593. Springer, 2011.

250. Sundarapandian Vaidyanathan, Aceng Sambas, and Mustafa Mamat. Analysis, synchronisation and circuit implementation of a novel jerk chaotic system and its application for voice encryption. *International Journal of Modelling, Identification and Control*, 28(2):153–166, 2017.
251. Sundarapandian Vaidyanathan, Aceng Sambas, Mustafa Mamat, and WS Mada Sanjaya. A new three-dimensional chaotic system with a hidden attractor, circuit design and application in wireless mobile robot. *Archives of Control Sciences*, 27(4):541–554, 2017.
252. Pauline Van den Driessche and James Watmough. Reproduction numbers and subthreshold endemic equilibria for compartmental models of disease transmission. *Mathematical Biosciences*, 180(1-2):29–48, 2002.
253. Michel Vellekoop and Raoul Berglund. On intervals, transitivity= chaos. *The American Mathematical Monthly*, 101(4):353–355, 1994.
254. Ch K Volos, Ioannis M Kyprianidis, and Ioannis N Stouboulos. A chaotic path planning generator for autonomous mobile robots. *Robotics and Autonomous Systems*, 60(4):651–656, 2012.
255. Bin Wang, Shihua Zhou, Xuedong Zheng, Changjun Zhou, Jing Dong, and Libo Zhao. Image watermarking using chaotic map and dna coding. *Optik*, 126(24):4846–4851, 2015.
256. Fan-Zhen Wang, Guo-Yuan Qi, Zeng-Qiang Chen, and Zhu-Zhi Yuan. On a four-winged chaotic attractor. 2007.
257. Guangyi Wang, Fang Yuan, Guanrong Chen, and Yu Zhang. Coexisting multiple attractors and riddled basins of a memristive system. *Chaos: An Interdisciplinary Journal of Nonlinear Science*, 28(1):013125, 2018.
258. Jian-Jun Wang, Jin-Zhu Zhang, and Zhen Jin. Analysis of an sir model with bilinear incidence rate. *Nonlinear Analysis: Real World Applications*, 11(4):2390–2402, 2010.
259. L Wang. 3-scroll and 4-scroll chaotic attractors generated from a new 3-d quadratic autonomous system. *Nonlinear Dynamics*, 56(4):453–462, 2009.
260. Lei Wang and Xiao-Song Yang. Global analysis of a generalized nosé–hoover oscillator. *Journal of Mathematical Analysis and Applications*, 464(1):370–379, 2018.
261. Wanting Wang and Muhammad Altaf Khan. Analysis and numerical simulation of fractional model of bank data with fractal–fractional atangana–baleanu derivative. *Journal of Computational and Applied Mathematics*, 369:112646, 2020.
262. Wanting Wang, Muhammad Altaf Khan, P Kumam, P Thounthong, et al. A comparison study of bank data in fractional calculus. *Chaos, Solitons & Fractals*, 126:369–384, 2019.
263. Xiong Wang and Guanrong Chen. A chaotic system with only one stable equilibrium. *Communications in Nonlinear Science and Numerical Simulation*, 17(3):1264–1272, 2012.
264. Xiong Wang, Sundarapandian Vaidyanathan, Christos Volos, Viet-Thanh Pham, and Tomasz Kapitaniak. Dynamics, circuit realization, control and synchronization of a hyperchaotic hyperjerk system with coexisting attractors. *Nonlinear Dynamics*, 89(3):1673–1687, 2017.
265. Zuoxun Wang, Jiaxun Liu, Fangfang Zhang, and Sen Leng. Hidden chaotic attractors and synchronization for a new fractional-order chaotic system. *Journal of Computational and Nonlinear Dynamics*, 14(8), 2019.
266. Zhouchao Wei. Dynamical behaviors of a chaotic system with no equilibria. *Physics Letters A*, 376(2):102–108, 2011.

267. Zhouchao Wei, Irene Moroz, Julien Clinton Sprott, Zhen Wang, and Wei Zhang. Detecting hidden chaotic regions and complex dynamics in the self-exciting homopolar disc dynamo. *International Journal of Bifurcation and Chaos*, 27(02):1730008, 2017.
268. Zhouchao Wei, Viet-Thanh Pham, Tomasz Kapitaniak, and Zhen Wang. Bifurcation analysis and circuit realization for multiple-delayed wang–chen system with hidden chaotic attractors. *Nonlinear Dynamics*, 85(3):1635–1650, 2016.
269. Zhouchao Wei, Pei Yu, Wei Zhang, and Minghui Yao. Study of hidden attractors, multiple limit cycles from hopf bifurcation and boundedness of motion in the generalized hyperchaotic rabinovich system. *Nonlinear Dynamics*, 82(1-2):131–141, 2015.
270. Zhouchao Wei, Wei Zhang, and Minghui Yao. On the periodic orbit bifurcating from one single non-hyperbolic equilibrium in a chaotic jerk system. *Nonlinear Dynamics*, 82(3):1251–1258, 2015.
271. Charlotte Werndl. What are the new implications of chaos for unpredictability? *The British Journal for the Philosophy of Science*, 60(1):195–220, 2009.
272. Jianhong Wu, Biao Tang, Nicola Luigi Bragazzi, Kyeongah Nah, and Zachary McCarthy. Quantifying the role of social distancing, personal protection and case detection in mitigating covid-19 outbreak in ontario, canada. *Journal of Mathematics in Industry*, 10(1):1–12, 2020.
273. Huiling Xi, Yuxia Li, and Xia Huang. Adaptive function projective combination synchronization of three different fractional-order chaotic systems. *Optik*, 126(24):5346–5349, 2015.
274. Wang Xing-Yuan and Wang Ming-Jun. Chaotic control of the coupled logistic map. *Acta Physica Sinica-Chinese Edition*, 57(2):736, 2008.
275. Xiao-Song Yang and Qingdu Li. Generate n-scroll attractor in linear system by scalar output feedback. *Chaos, Solitons & Fractals*, 18(1):25–29, 2003.
276. JW Yau, Sophie L Rogers, Ryo Kawasaki, Ecosse L Lamoureux, Jonathan W Kowalski, T Bek, SJ Chen, JM Dekker, A Fletcher, J Grauslund, et al. Meta-analysis for eye disease (metaeye) study group. global prevalence and major risk factors of diabetic retinopathy. *Diabetes Care*, 35(3):556–564, 2012.
277. Fei Yu, Ping Li, Ke Gu, and Bo Yin. Research progress of multi-scroll chaotic oscillators based on current-mode devices. *Optik*, 127(13):5486–5490, 2016.
278. Simin Yu, Wallace KS Tang, Jinhu Lü, and Guanrong Chen. Design and implementation of multi-wing butterfly chaotic attractors via lorenz-type systems. *International Journal of Bifurcation and Chaos*, 20(01):29–41, 2010.
279. Abdullahi Yusuf, Sania Qureshi, Mustafa Inc, Aliyu Isa Aliyu, Dumitru Baleanu, and Asif Ali Shaikh. Two-strain epidemic model involving fractional derivative with mittag-leffler kernel. *Chaos: An Interdisciplinary Journal of Nonlinear Science*, 28(12):123121, 2018.
280. Yanhui Zhai, Xiaona Ma, and Ying Xiong. Hopf bifurcation analysis for the pest-predator models under insecticide use with time delay. *International Journal of Mathematics Trends and Technology*, 9(2):115–121, 2014.
281. DC Zhang and B Shi. Oscillation and global asymptotic stability in a discrete epidemic model. *Journal of Mathematical Analysis and Applications*, 278(1):194–202, 2003.
282. Zizhen Zhang. A novel covid-19 mathematical model with fractional derivatives: Singular and nonsingular kernels. *Chaos, Solitons & Fractals*, page 110060, 2020.

283. Pengsheng Zheng, Wansheng Tang, and Jianxiong Zhang. Some novel double-scroll chaotic attractors in hopfield networks. *Neurocomputing*, 73(10-12):2280–2285, 2010.
284. Chengqun Zhou, Chunhua Yang, Degang Xu, and Chaoyang Chen. Coexisting attractors, circuit realization and impulsive synchronization of a new four-dimensional chaotic system. *Modern Physics Letters B*, 33(03):1950026, 2019.

Index

A strange attractor, 7

Chaos, 1
Chaotic dynamics, 2
F-F R-L sense with Mittag-Leffler kernel, 26
FF R-L sense with exponential decay, 25

A complex system, 5
A fractal, 7
An attractor, 7
Atangana-Baleanu derivative in RL sense, 15
Atangana-Balenau derivative in Caputo sense, 15

Bifurcation, 8

Caputo derivative, 14
Caputo-Fabrizio, 14
Computer simulation, 4

Dynamical system, 5
Dynamics, 4

Error analysis FF Mittag-Leffler, 47
Error analysis FF-CF, 45
Error analysis FF-RL, 43

FF R-L power law, 25
FF integral Mittag-Leffler, 26
FF integral with exponential decay, 26
FF integral with power law kernel, 26
fractal derivative, 31
fractal derivative with $\alpha, \beta > 0$, 32
fractional derivative, 14
Fundamental theorem of fractional calculus, 16

Laplace of Atangana-Baleanu, 16
Laplace of Caputo, 16
Laplace of Caputo-Fabrizio, 16
Linearisation of a system, 6

Lyapunov exponents, 7

Mathematical system, 4

Non-periodic motion, 8
Non-linear, 5
Numerical scheme Caputo FF model, 34
Numerical scheme FF Atangana-Baleanu, 38
Numerical scheme FF Caputo-Fabrizio, 36

Parameters, 5
Period, 8
Period-doubling, 8
Periodic behavior, 8
Phase space, 4
Physical system, 4

Random, 6
Riemann–Liouville, 14
Riemann-Liouville integral, 15

Sensitivity to initial conditions, 6
Stable, 6
Strange attractors, 3
Summudo transform of Atangana-Baleanu, 16
Summudu transform of Caputo, 16
Summudu transform of Caputo-Fabrizio, 16
System, 4

The Butterfly Effect, 6
The Poincare map, 9

Universality, 8
Unstable, 6